Exploring World Geography Part 1

To Our Grandchildren:
May you live in a better world, and may you help it be so,
until we all live together in the better place God has in store for us.

Exploring World Geography Part 1
Ray Notgrass

ISBN 978-1-60999-154-8

Copyright © 2020 Notgrass History. All rights reserved.
No part of this material may be reproduced without permission from the publisher.

Previous Page: Desert in Namibia by Finding Dan | Dan Grinwis on Unsplash
Front Cover: Reine, Norway, by Francesco Dazzi / Shutterstock.com

All product names, brands, and other trademarks mentioned or pictured
in this book are used for educational purposes only.
No association with or endorsement by the owners of the trademarks is intended.
Each trademark remains the property of its respective owner.

Unless otherwise noted, scripture quotations taken from the New American Standard Bible,
Copyright 1960, 1962, 1963, 1971, 1972, 1973, 1975, 1977, 1995
by the Lockman Foundation. Used by permission.

Cover design by Mary Evelyn McCurdy
Interior design by John Notgrass
Maps by Sean Killen and John Notgrass
Literary introductions by Bethany Poore

Printed in the United States of America

NOTGRASS
HISTORY

975 Roaring River Road
Gainesboro, TN 38562
1-800-211-8793
notgrass.com

Dhow Boats in Doha, Qatar

Part 1

Table of Contents

Why You Should Study Geography vii

How to Use This Curriculum ix

Advice on Writing xiii

Assigned Literature xx

1 Welcome to Your World 1

1 - Seeing What You Have Never Seen Before 3
2 - Who Says Geography Doesn't Make Any Difference? 10
3 - The World According to Strabo 16
4 - Geography Is Not Set in Stone 20
5 - It's a Matter of Worldview 27

2 It Begins with a Map 33

6 - What's a Map Worth to You? 35
7 - You Can Learn a Lot from a Map 41
8 - What a Map Is and What It Is Not 48
9 - The Business of Geography 53
10 - Recipe for a Worldview 57

3 The Middle East Part 1 61

11 - The Physical Geography of the Middle East 63
12 - Drawing Lines: The Making of the Modern Middle East 70
13 - The Toughest Geographic Issue 75
14 - "We Have No Jam": The Saga of the Kurds 83
15 - Faith System: Judaism 88

4 The Middle East Part 2 95

16 - Armenia and Its People 97
17 - Winston Churchill and the Gallipoli Campaign 102
18 - To Be a Woman in Saudi Arabia 108
19 - Sacred Geography: The Meaning of Pilgrimages 112
20 - Faith System: Christianity and the Christian Worldview 117

iii

5   North Africa   123

- 21 - Of Jasmine and Spring: The Story of Tunisia   125
- 22 - A Man and a Canal   131
- 23 - The Berbers   138
- 24 - The North Africa Campaign in World War II   144
- 25 - Faith System: Islam   149

6   West Africa   157

- 26 - Cocoa Growing in Côte d'Ivoire   159
- 27 - Riches and Poverty in Mali   163
- 28 - The Trees of Lzake Volta   168
- 29 - The Music of Nigeria   172
- 30 - Faith System: Folk Religion   178

7   Central Africa   183

- 31 - Life in a Refugee Camp   185
- 32 - The Mbuti: A Threatened Way of Life   189
- 33 - Cameroon: Legacy of Colonialism   194
- 34 - Beautiful and Deadly: Ebola in the DRC   199
- 35 - Kent Brantly, Physician and Missionary   203

8   East Africa   207

- 36 - Hope Instead of Hate in Rwanda   209
- 37 - William and His Windmill   214
- 38 - Give Water, Give Hope, Give Life in Kenya   219
- 39 - Long Distance Runners from Ethiopia   224
- 40 - Where Did You Get That Worldview of Yours?   230

Drummers in Accra, Ghana

Fortress of Guaita on Monte Titano in San Marino

9   Southern Africa   235

41 - They Say a Diamond Is Forever   237
42 - This Is Our Land: The Story of the Zulus   242
43 - Roll the Gospel Chariot Along   247
44 - Can People Once Enemies Get Along?   251
45 - Truth   257

10   Southern Europe   263

46 - The Basques: One People in Two Countries   265
47 - What's In a Name? The Saga of North Macedonia   270
48 - You Go Here and You Go There   274
49 - Microstates: Vestiges of Earlier Times   281
50 - Paul's Sermon on Human Geography   289

11   Western Europe   295

51 - Not a Fairy Tale   297
52 - The Flower That Made the Netherlands Famous   301
53 - When Weather Helped Make History   307
54 - Helping a Continent Be Strong and Free   312
55 - Faith   319

12   Northern Europe   323

56 - Living on the Edge: The Sami of Finland   325
57 - Is This What Tomorrow Looks Like?   330
58 - They Do Things Their Way: The Faroe Islands   334
59 - Surveying the Matter: The Struve Geodetic Arc   338
60 - The Existence of God   342

13   Eastern Europe   347

61 - The Jews of Eastern Europe   349
62 - Defiant Hungarians   357
63 - The Vltava (The Moldau)   364
64 - A War They Didn't Want: Ukraine   368
65 - The Nature of God   374

14   Russia   379

66 - The Bear: Russia   381
67 - Snapshots from the Urals   387
68 - The Deepest Lake in the World   391
69 - The Toughest Project   395
70 - Creation vs. Evolution   400

15   The Arctic and The Antarctic   405

71 - To the North Pole and UNDER   407
72 - Sitting on Top of the World   413
73 - A Homeland for the Inuit   417
74 - *Endurance*   421
75 - "Have You Entered the Storehouses of the Snow?"   427

Sources   S-1

Credits   C-1

Ivolginsky Datsan, Buddhist Temple in Buryatia, Russia

Oculus Station House, New York

Why You Should Study Geography

Imagine going into a large room that is filled with people who are engaged in several different conversations.

Over here, two people really seem to be connecting deeply with each other. In that corner, two other people are in a heated discussion and seem ready to come to blows. Four people in the middle are working together to set up some sort of display. Over there, a small group is looking at pictures on someone's phone; and the people involved really seem to be enjoying themselves.

But you've just arrived, and you have no idea what people are talking about or what they are doing. You think, "If I only knew what is going on here, I'd feel more comfortable. Maybe I could even help or contribute something to a conversation."

So you move around the room, listen to people, and ask some questions. Slowly you get an idea of the different interactions that are taking place. As you get to know people better, you can offer something to some of those conversations. Now that you know what the group setting something up is doing, you offer to help with that project.

This description of entering a large room is something like what you are going to be doing in a few years. You are going to enter a big room called the adult world. Lots of things are going on there.

Some folks get along well, while others have come to blows. Some are developing a project together, but others don't know each other and perhaps don't even speak to each other. If you can understand why things are the way they are in our world, you will be better able to make a positive difference in it.

Helping you better understand our world is the purpose of *Exploring World Geography*. Traditionally, geography deals with such topics as what is a volcano, how much of the world is covered with water, and the definitions of a desert and plate tectonics. Those subjects are elements of physical geography, and you need to understand those subjects to understand the physical world that God made. However, the modern field of geography has expanded to include many elements of human life on this planet, especially as it relates to geography or geographic place. For instance, you can study population geography, cultural geography, political geography, economic geography, rural and urban geography, and, in this course, human geography.

This course deals with two main questions: (1) How has the physical world made a difference in what people have done, how they have lived, and how they live today? (2) How have people made a difference in the physical world, and how are they making a difference in it today?

vii

Journalist and geography writer Robert Kaplan says that geography is the backdrop to human history. Geography is where we live, and it impacts how we live. Geography affects all of us. Some people deny this. They say that geography is only incidental to human interaction. Moreover, with the invention of faster travel and instant worldwide communication, some people say that geography doesn't really matter anymore.

Try telling a soldier who has fought on the desert mountains of Afghanistan that geography doesn't matter. Explain how cotton became king in the pre-Civil War American South without referring to geography. Describe the role of New York City as an international trade and immigration center without referring to its geographic location. Understand the modern Middle East without referring to oil or the existence of Israel. Discuss the immigration issue without referring to a geographic setting. When you look into these and many other issues regarding life on this planet, you will find that geography *does* matter.

Every person can make a profound difference in the lives of others by how they respond to their setting and to the opportunities that God places in their path—a path that runs through geography. Our world—especially the part of it where you live—is where God has placed you. You might wish that you lived in a place with greater natural beauty or with more opportunities for economic advancement, but He has placed you in the location where you are. If you move to another place or region, in that new setting you will find other opportunities and limitations that geography will influence.

The task of living effectively as an adult is before you; serve God by fulfilling that task where you are. We hope that we encourage you to take geography seriously. Be assured of this: if we as Americans do not take geography seriously, other people—including some who have decided to be our enemies—will take it seriously; and that will affect us negatively.

A few key factors help to explain why history and current events have happened the way they have. The most important is God. The second most important are the decisions and actions that people make. We might also cite family, cultural influences, and religious beliefs. In this mix of factors, one key factor is geography.

We hope that this curriculum will help you understand better this fascinating world in which we live. We hope you will think, "Oh, that's why this country has done this or that" or "It's amazing how that mountain range affects that region." Knowing something about the geographic settings in which people live will give you empathy. Knowing inspiring stories will encourage you in your own life. Learning what God teaches us about our world helps us live in it. Learning the power of the gospel that offers hope and truth to every tribe and tongue and people and nation will help you redeem the time that you have.

Geography has relevance to current affairs. In the time that we worked on this curriculum, the ruler of Swaziland changed that country's name to Eswatini. The country once known as Macedonia changed its name to the Republic of North Macedonia. Wildfires devastated the geography of Australia. China's Belt and Road Initiative, which many countries accepted eagerly its early stages, has changed shape and encountered opposition from some countries. Huge demonstrations in Hong Kong protested the way China was governing that city. As we were finishing the development of the curriculum, the COVID-19 pandemic swept the world. As you move into adult life, you will need to remain informed and discerning about the issues in which geography plays a part.

Our goal is to inspire you to think broadly and to act boldly, to see your own setting in time and place as an opportunity for growth and adventure or as an obstacle to overcome. We hope that you will envision what God might do through you as He has done through countless others who have lived in their own times and places. Preparing you to live successfully in that big room called the adult world is what we have tried to do in the lessons, the readings, the literature, and the assignments.

Lake Wanaka, New Zealand

How to Use This Curriculum

As you both, parent and student, plan your study using this curriculum, here are some ideas to help you get the most out of it.

This curriculum provides one year's credit in three subjects: geography, English (literature and composition), and worldview. The 150 lessons are divided into thirty units of five lessons each. Each unit has four lessons on geography, and a final lesson on worldview.

Since a typical school year has thirty-six weeks, you have some flexibility in completing the course. The student can take two weeks to complete a unit if they find a topic particularly interesting or when your schedule is especially busy. Families are free to choose how they want to schedule the course, but many families choose to begin a unit on Monday and finish it on Friday.

On the first day of a unit, read the unit introduction. Here you will find a brief overview of the unit; a list of lessons for that unit; a Bible passage to memorize; a list of books used with that unit; choices for a project for that unit; and, when a literature title is begun, an introduction to that book.

After reading the introduction, choose a project to complete by the end of the unit and make a schedule for how to complete it. Find the memory work for the week in the Bible translation of your choice.

Complete the following each day:

- Read the lesson for the day.

- Complete all of the *Gazetteer*, geography, worldview, and literature assignments for the lesson.

- If you are using the optional *Student Review*, complete the assignment(s) for that lesson.

- Work on your Bible memorization and on your chosen project.

On the last day of each unit, the student will recite or write the memory work and complete the project for the unit.

The curriculum includes the *Exploring World Geography Gazetteer*. This volume contains maps of the continents and regions we discuss in the text, a section on each country of the world, and original source material that we assign once per unit. The Assignments section at the end of each day's lesson includes the *Gazetteer* assignment when appropriate.

You will need to plan carefully what the student does each day. For instance, every fifth day includes

ix

reading the worldview lesson and answering the review questions, finishing the project for the unit, writing or reciting the memory verse for that unit, and taking the geography quiz for that unit.

In twelve of the units the student will also need to finish the literature title they have been reading, answer the review questions on it, and read the literary analysis for that book. In six units, the student will also need to take the geography, English, and worldview exams over the previous five units. Plus, the student will need to study for all of these exams.

Instead of waiting until the last day of a unit to complete all of these assignments, students can spread out the work load and make it easier to complete. For example, students can complete their unit project on Day 4 of the unit. She might also want to plan her reading so that she finishes the literature title on Day 4 of the unit and completes the review questions and literary analysis that day. We have provided the tools for your study of these subjects. How you complete the curriculum is ultimately up to your family determining what is the best approach for you to take.

An assignment checklist is available as a free download on our website (notgrass.com/ewglinks).

Worldview Lessons

In the assignments for several of the lessons in each unit are thought questions regarding worldview. We recommend that the student have a Bible notebook (wire-bound or 3-ring binder) in which she copies each question and writes a response to the question. Alternatively, the parent may choose to have the student read the question aloud and give an answer orally. However, writing down the questions and answers will probably help the student remember the questions and answers better.

As part of our worldview survey, we look at several religious systems that people practice in the world besides Christianity. We look at these other faith systems from the perspective of outsiders. We have never been part of these groups, and the information we share is the result of our research on these subjects. Those who adhere to these faith systems might see inaccuracies or misplaced emphases in our treatment of them. It is not our purpose to misrepresent these faiths or to create straw men that we can easily knock down in an attempt to show the superiority of Christianity.

We have attempted to refrain from using demeaning language or from saying anything like, "We can't believe that intelligent people believe these ridiculous ideas, but apparently they do." We want to show respect for the people who hold these beliefs, even as we express our disagreement with these beliefs and why we believe that Christianity is true. We are not ashamed of the gospel, and we want to keep the door open for civil discussions with those of other faiths in the hope that we can encourage everyone in the pursuit of truth.

Map Skills Assignments

A map skills assignment comes at the end of one lesson in most units, usually on the fourth day of the unit. Their purpose is to help the student better understand and utilize maps. The lessons in Unit 2 have a good deal of information about maps, so the map skills assignments begin in Unit 3.

We recommend that the student create a map skills notebook or folder for these activities.

Tips on Bible Memorization

Each unit of *Exploring World Geography* gives a Bible passage to memorize. Here are some tips on memorization for the student. Pay attention and internalize what the verses mean. It will be much easier to memorize thoughts that you understand than to see them as a string of words that have no meaning to you. Write the verses on index cards. Keep these handy to use when you have a spare moment. Copying out the verses is a good exercise, especially if you learn visually.

How to Use This Curriculum

Draw pictures illustrating the verses. Ask another person to read the verses to you. Ask another person to listen to you and correct your recitation. Working on memorization consistently in small chunks of time over several days works much better than last-minute cramming.

Unit Projects

Each unit (except Unit 3) has three choices for a project, always including a writing assignment. Parents can decide how many writing assignments the student must complete to fulfill the English credit of *Exploring World Geography*. We recommend that you choose the writing assignment as the project a minimum of six times throughout the course. The other project choices include a wide variety of activities: building models, cooking, field trips, volunteer opportunities, and more, all of which will enhance and expand what the student is learning in the course.

The projects relate to the material in the unit. Where applicable, we note the lesson from which the project is drawn. The student should choose a project at the beginning of the unit and work on it throughout the unit. The student may need to look ahead at the relevant section of the lesson to get started on the project.

As you choose projects unit by unit, take the opportunity for the student to try new things and expand her skills. If she has never made a model out of STYROFOAM™, or seldom done any cooking, or doesn't know how to make a video, this is a great opportunity!

The student should complete each project at a high school level. Some of these assignments could be given to an elementary school student and the results would be on an elementary school level. The student should complete the work with care and research and attention to accuracy, creativity, and excellence. Throwing something together in a haphazard fashion is not appropriate. Whether the student spends his time writing an essay or building a model, he should use his mind and hands to create something he can be proud of.

Student Review Pack

The Student Review Pack includes three books to help the parent and student measure the student's progress through the course and understanding of the material: the *Student Review Book*, the *Quiz and Exam Book*, and the *Guide for Parents and Answer Key*. Using these books is optional, but you will likely find them useful.

The *Student Review Book* contains review questions on each lesson, review questions on some of the source documents in the *Gazetteer*, review questions and analysis of the literature, and a map skills assignment for most units.

The *Quiz and Exam Book* has:

- a geography quiz for each unit that covers the first four lessons of that unit and is drawn from the lesson review questions for those four lessons,

- a geography exam covering every five units that is drawn from the quizzes for those units,

- an English exam covering the literary analysis and questions for the books read every five units, and

- a worldview exam every five units covering the review questions for the five worldview lessons in those units.

How We Present Scripture

The most important material in this course are the studies from God's Word. Understanding world geography and literature is important, but how we live before God is the most important issue before each one of us. We want to help you as you do that.

We believe in the inspiration and authority of the Bible, and our desire is to present the Bible in all of its truth, wisdom, and power. We strive in all we do simply to be Christians. We are on a quest to understand the truth that God has provided in His Word.

If you read something in this curriculum that differs from what your family believes, take the opportunity to discuss the issue and search the Scriptures together. We welcome your feedback. If you believe that we have written something in error, please email us so that we can learn together the truth that will set us free.

Notes on the Literature

We chose works of literature that illustrate geography in various places around the world. As the student reads the books, she should take special note of geographic features such as lakes, rivers, mountains, the region of the world, the culture of the people, and how these features fit into the story. The setting of a work of literature is a place in geography, so the study of geography and the study of literature will enhance each other.

Worldview also plays a part in the study of literature. As the student reads each work, he should notice the worldview of the characters; clues to the worldview of the author; and how the book supports, informs, or challenges his worldview. Thus the study of literature and the study of worldview will enhance and support each other.

Appreciation

I am indebted to all those who have helped with this project. My wife, Charlene, wrote the lesson about her ancestor, Pierre Boucher of Boucherville, Quebec, Canada. She proofread the curriculum with me and provided invaluable input in many other ways. Our son John collected illustrations and laid out the pages, lessons, and units. Our daughter Bethany helped to develop the original plan for the curriculum, selected the literature to include, and wrote most of the literary analysis and the literature review questions. Our daughter Mary Evelyn designed the covers and proofread the curriculum. I also want to thank Dena Russell and David Shelton for their vital assistance in developing this curriculum and Sean Killen for producing the beautiful maps in the *Gazetteer*.

Exploring World Geography completes the cycle of Notgrass high school social studies curriculum that includes *Exploring World History, Exploring America, Exploring Government,* and *Exploring Economics*. This series began with the publication of *Exploring America* in 2002. What a joyful ride it has been.

I will forever be grateful for the thousands of students who have used these materials and for the countless words of appreciation we have received. May the Lord receive all the praise. Thank you and thank Him.

Ray Notgrass
Gainesboro, Tennessee
December 2020
ray@notgrass.com

Chinese Typewriter (1970s)

Advice on Writing

Composition is part of most high school English courses. It usually involves learning how to express ideas, write themes, and do research papers. Practicing writing helps you to develop your style and skill, just as practicing any activity will help you to be better at it. I make my living by writing, so I appreciate the importance of this skill.

One goal of high school composition is to prepare you for college composition. I have taught college students who never learned to construct a good sentence, let alone a good paragraph. However, learning to write just for high school and college composition assignments is a limited goal. Life does exist beyond school.

You will probably have many occasions to engage in research and to prepare your thoughts on a vital subject such as abortion or capital punishment. You will have numerous opportunities to write: letters to friends and family, journals, letters to the editor, social media posts, advertisements for your business, and reviews and articles for periodicals, to mention just a few. The Internet has created new possibilities for sharing your ideas in written form. Desktop publishing has made getting a book published within the reach of many people who might not get a contract from a big-name publisher.

Writing helps you express what you understand about a subject. If you can't explain something to another person, you probably don't understand it well yourself. The writing assignments in this course will help you learn to pull your thoughts together.

Good writing style is important in getting your ideas across to other people. Writing skills will be helpful in your job or in conducting your own business. You will bless your spouse and children if you write thoughtful letters to them often. You can help others by expressing yourself well in writing.

Three ways to improve your writing are to read good writing, to write often yourself, and to receive criticism of your writing with humility and a desire to do better. Reading and applying the guidance in good books on writing will also help you refine your technique. I recommend *The Elements of Style* by William Strunk Jr. and E. B. White.

Writing Assignments in This Course

Each week you do a writing assignment (instead of one of the other suggested projects), you will have two or three possible topics from which to choose. Some of the essay prompts refer to topics that one of the lessons in the unit discusses.

A basic way to compose an essay is to write five paragraphs: an opening paragraph that states your purpose, three paragraphs that develop three different points or arguments, and a closing paragraph that summarizes your position or topic. If you are floundering on a particular assignment, using this outline can get you started.

The usual target length of your writing projects for this course is 250 to 300 words, which is about two typed, double-spaced pages.

Writing Tips to Implement

Here are some tips I have learned that have helped my writing.

Write with passion. Believe in what you are saying. People have plenty to read, so give them something that will grip them. If you don't believe deeply in what you are saying, you give others no reason to do so either. This raises an issue that relates to many writing assignments. Assigned writing is like assigned reading: we often approach it as a chore. Deep emotion and a passion for convincing others may be difficult to express in a theme on "The American Interstate System" or "The Internal Hierachy of International Organizations."

Writing with passion means that you should not soft-pedal what you say. Phrases such as "It seems to me," "I think that it would be good if," or "My personal opinion, for what it is worth," take the fire out of your message. It is your piece, so we know it is your opinion. Just state it. Related to this is the common use of quotation marks to highlight a word. Save quotation marks for when you are actually quoting something.

Develop your paper in an orderly and logical way. Using an outline helps me to structure what I am writing. Identify the major points you want to make, the order in which you need to make them, and what secondary points you want to include to support your major points. Be sure that each paragraph has one main point, expressed in a topic sentence, with the other sentences supporting that point. In a narrative, tell what happened first before you tell what happened later. In an essay, make your points in the order of their importance to your overall theme.

Don't try to put everything you believe into one piece. Trust that you will have the opportunity to write again, and stay focused on your topic. Your challenge is to narrow your topic sufficiently to be able to cover it completely.

Use short, simple sentences. Longer sentences do not necessarily show greater intelligence or convey ideas more effectively. You are trying to teach or convince a reader who perhaps has not been thinking about the topic the way you have. He or she will need to see your ideas expressed simply and clearly. Shorter sentences generally stay with people longer: "These are the times that try men's souls." "The only thing we have to fear is fear itself."

Writing Habits to Avoid

Do not begin sentences with "There is" or "There are." Find a more forceful way to cast the sentence. Compare "Four score and seven years ago our fathers brought forth upon this continent a new nation" to "There was a country begun by our ancestors 87 years ago."

Do not habitually begin sentences with "and" or "but." This practice has become a trendy habit in informal writing, but the grammar books tell you never to do this.

Avoid the word "would." Such usage is an attempt to soft-pedal, to indicate customary behavior, or to describe something that is not a reality. "That would be a good idea" is less powerful than "That is a good idea." "Americans would often violate the terms of treaties made with native nations" is not as sharp as "Americans often violated the terms of the treaties."

Avoid using passive voice. "The cow jumped over the moon" is more forceful than "The moon was jumped over by the cow."

Don't imitate someone else's style. That person didn't become a good writer by copying someone

Advice on Writing

else's style; he or she developed his or her own style. You might become enamored with the writing of a favorite author and want to write the way he or she does. Learn from that author, but be yourself.

Additional Suggestions

C. S. Lewis, a prominent 20th-century British author, had good suggestions about writing (*Letters of C. S. Lewis*, edited by W. H. Lewis, first published in 1966; this edition New York: Harcourt Brace, revised edition 1988; pp. 468-9, 485):

- Write with the ear. Each sentence should read well aloud.
- Don't say something is exciting or important. Prove that it is by how you describe it.
- Turn off the radio (in our day, he might say the smartphone and television).
- Read good books and avoid nearly all magazines.

A key to good writing is rewriting. Writing is hard work, and you shouldn't let anyone tell you otherwise. You will not get every word and phrase just right the first time you put them down on paper or type them on the computer. Great, famous, well-paid writers have to rewrite their work and often have editors who revise and critique what they write. Don't be impatient, and don't wait until the last minute. Write something; then go back and rewrite it; then go back a day or two later to consider it again. This is where another pair of loving and honest eyes is helpful. People who have read my writing and who were willing to point out the faults in it have often helped me (although I admit that I have winced inside when I heard their criticism).

Find someone who is willing to take a red pen to your work; a favorite uncle or grandparent might not be that person. You might know exactly what you mean by a particular statement, but someone else might not understand what you said at all. I have often found that when someone doesn't understand a statement I have written, it is because I have tried to say something without really saying it. In other words, I have muddied what should have been a clear statement; and that fuzzy lack of commitment showed through.

Your writing will improve with practice, experience, and exposure to good writing. I hope that in ten years you will not write the same way you do now. The only way you can get to that point is to keep writing, keep learning, and keep reading. I hope that this course helps you on your journey.

Writing a Research Paper

We recommend that you write a research or term paper of eight to ten typed double-spaced pages (about 2,000-2,500 words) during several weeks in the second semester of *Exploring World Geography*. Waiting until the second semester gives you time to prepare and to practice writing shorter papers for your weekly projects.

This section guides you step-by-step through the process of writing a research paper. You and your parents should discuss whether you think a research paper assignment is appropriate for you. Also discuss with your parents whether you should skip the project for each unit during the time you are working on your research paper.

When you are ready to begin, refer to this section. If you feel a need for more detailed guidance, we recommend the section on research papers in *Writer's Inc.* by Great Source. You can also find sample research papers online. The Purdue University Online Writing Lab (OWL) has a sample. (Visit notgrass.com/ewglinks for more details.)

Choosing a Topic

A research paper combines the work of investigation with the task of writing. Choosing your topic is the first step. When you write a research paper, you must define your topic as clearly as possible.

You could expand on an essay you have already written. You might want to concentrate on this topic instead of doing unit essays for a few weeks, with your parents' permission. You might have to narrow a topic for the purposes of your paper. For example, instead of writing on "Art in Armenia," you might choose the narrower topic of "Rugmaking in Armenia."

You can choose to write about a place your ancestors came from, a country you want to visit, an individual who inspires you, or an ethnic group that interests you. Here are some other possible topics that might spark your imagination:

1. China and Geography (focusing on the South China Sea, the Uighurs, Hong Kong, or how China is impacting the world)
2. How I Would Solve the Middle East Dilemma
3. This Is How I See the World (your worldview statement)
4. They Brought My People from There to Here: The Meaning of Slavery in America
5. Is Globalism Good, Bad, or Both?
6. A Unified Korea: Can It Be Done? Should It Be Done?
7. The Distinctive Music of a Region (such as jazz in New Orleans or polka music from Eastern Europe)
8. The Geographic Impact of COVID-19 (or the economic, educational, religious, social, or other impact)
9. We Have to Fix This (addressing an environmental issue)
10. Unraveling Babel: Language in Our Modern World
11. Why I Like (or Dislike) International Organizations
12. What You Would See on the Pan-American Highway
13. The Building of the Panama Canal
14. How We Get from Here to There (Land, Sea, Air, and/or Space Navigation)
15. The Geography of War

If you have another topic you would like to write about, go for it! Focus on something you are passionate about; why take time to do all this work for something you don't really care about and may never look at again? Think about what you might do with your paper once it is finished: send a copy to your congressman or senator, contact your local newspaper to see if they would publish it (newspapers are always looking for material to print), present it orally to a local club, or put it in circulation some other way. Here's your chance to make a difference!

Doing the Research

Research involves finding legitimate, authoritative sources on the subject and gathering information from those sources. The modern researcher has a wealth of material available to him, some good and some worthless.

Sources include books, magazines, newspapers, encyclopedias, scholarly articles, and original sources. Original or primary sources are materials written or developed by someone involved at the time of history you are investigating. A diary written by a sailor on a trading vessel during the Victorian Era is an example of an original source. You probably will not be able to hold the actual document in your hands, but many transcriptions of original source materials can be found in print and online. Secondary sources are materials written later about the subject in question.

Use caution with online sources, as many are not authoritative. A comment by a reader on a blog about the Roman Empire is not necessarily based on fact, and you cannot use information gathered from such a source in a research paper. It might give you an idea about something to research yourself, but just because someone posted it online doesn't make it accurate or relevant.

Wikipedia is the classic example of a non-authoritative source for research. A great deal of the material found on Wikipedia is accurate; but because of the way people create and edit the articles, you cannot use Wikipedia as as an authoritative source. Websites maintained by universities, government entities, and reputable publishers of reference materials are good sources for online research. Google Books and Project Gutenberg have many historic books available in their entirety online.

Do not neglect looking in print resources, such as encyclopedias, for information. A good old-fashioned one-hour visit to the library might provide much more valuable material than hours of sifting through material online. However, you need to be sure that your print sources are reliable also.

The researcher must give proper credit to her sources. Plagiarism is using someone else's words or ideas without giving proper credit to that source. The Internet contains information that you could simply copy and paste into your paper. Though this might be tempting, it is absolutely wrong. Plagiarism is at once lying, stealing, and cheating.

You do not have to cite a source for basic information, such as the fact that Ankara is the capital of Turkey. However, you do need to cite sources for detailed information and for unique perspectives about a topic. As you take notes while doing research, indicate clearly what is a direct quote and what is your paraphrase of another person's writing. Do not copy another person's exact words into your paper without showing that you are quoting and giving credit to the source.

A research paper is a big project that can seem overwhelming. Divide the project into manageable steps. We have provided a schedule that will help you do this. You might need extra time on some steps while you breeze quickly through others. You must stay on track to meet your deadline. Look ahead to the finished product and take it step-by-step.

Your paper should be based on historical fact and should not primarily be an opinion piece. Sometimes differentiating between the two is difficult. A simple list of facts that can be found elsewhere is not interesting. Your paper should have a point, and you should bring your own thoughts to bear on the facts you gather in your research. Your paper will be dull if you do not draw interesting conclusions. Noting how Roman architecture expressed Roman ideals and impacted the concept of beauty and form centuries later is excellent; on the other hand, listing reasons why you like Roman architecture is irrelevant to this paper. Your task for your research paper is to provide information, make observations, and draw conclusions on the topic in an interesting, readable format that is worth someone's time to read.

Day 1: Read the previous two pages and the daily plan on the opposite page. Make a list of at least seven ideas for topics. Discuss ideas for topics with a parent. Select topics that you would like to spend the next few weeks studying and writing about. The index of this curriculum is a source for possible topics.

Day 2: Investigate possible sources for your top three topic ideas to make sure you will be able to find enough material. Choose your topic and write a one-sentence summary of your purpose for the paper. Don't say, "This paper is about what you would see on the Pan-American Highway." Instead, state the substance of your paper: "A journey on the Pan-American Highway gives the traveler a window into the rich cultures of North, Central, and South America."

Day 3: Gather possible sources for research. Make a list of places to look. You can bookmark websites, visit the library, and look through relevant periodicals. Develop a preliminary outline for your paper.

Day 4: Learn how to cite your sources properly. Your research paper should follow MLA (Modern Language Association) guidelines for source citations. Your paper needs to have footnotes or in-text citations for your sources of information and a separate bibliography or works cited page at the end of your paper. Look online for the most up-to-date MLA guidelines. We recommend Purdue University's Online Writing Lab (OWL).

Practice some example citations. Whether you use note cards, copy and paste to a computer document, or a combination of these approaches, be consistent and accurate in your in-text and bibliography citations. Look over the guidelines and your examples with a parent to make sure you are on the right track.

Day 5: Make a general outline for your paper to help guide your research. Make some notes about what you want to say in your paper, questions you hope to answer in your research, and ideas for the main point of your paper. This plan will enable you to make the most of your research time. You want to immerse yourself in the topic you will be writing about. Your final paper will not include every bit of information you read, but you want to write from a position of overflow instead of scraping together just enough facts to fill up your paper.

Day 6: Begin your research. Develop a system to stay organized, keeping track of the source for every quote or fact. For example, if you are using the book, *Tea for the Queen*, note which facts and quotations come from that specific work and the relevant page numbers. You need to know clearly where every item of information came from: book, website, article, etc. Use a minimum of six different sources for your paper.

Day 7: Continue your research.

Day 8: Continue your research.

Day 9: Finish your research. Where do you want this paper to go? What do you want to say? Decide what information you gathered in your research is relevant and what isn't. Highlight key findings in your research. Set aside (but don't throw away) information that does not seem relevant to what you want to say. Talk about your general ideas for your paper with a parent.

Day 10: Work on the final outline for your paper. Jot down the points you want to make in the introduction, the main sections of your paper, what you want to include in each section, and what you want to emphasize in the conclusion. Organize these into an outline. Your research might have shown you that you need to emphasize a point that you had not previously realized was important, or you might not be able to find much information about what you thought was a main idea.

Look through the information you gathered in your research to make sure you didn't leave anything important out of your outline. Finalize your outline and talk about it with a parent. A good, detailed outline will ease your writing process significantly.

Day 11: Re-read "Advice on Writing" on pages xiii-xv of this book. Begin writing your paper, starting with your introduction and conclusion. Your introduction should give a general idea of what your paper is about and the main points you will make. Your conclusion will re-emphasize your main points. Include proper citations as you go, both in-text and on your Works Cited page.

Day 12: Continue work on your first draft.

Day 13: Continue work on your first draft.

Day 14: Continue work on your first draft.

Day 15: Finish the first draft of your paper. Check your in-text source citations and Works Cited page against your research notes, and make sure your formatting is correct. Proofread your paper and make corrections. Give your paper a title. Ask a parent to read and correct your paper and make suggestions for improvement.

Day 16: Discuss the paper with your parent. Think about improvements that you can make. Begin working on the final draft of your paper. Fix mistakes and polish your style.

Day 17: Continue working on your final draft.

Day 18: Continue working on your final draft.

Day 19: Finish writing your final draft. Read your paper carefully for spelling and grammatical errors.

Day 20: Read your paper aloud. Make any final corrections. Save it, print it off, and turn it in. Good work!

Daily Plan				
Day 1	**Day 2**	**Day 3**	**Day 4**	**Day 5**
Investigate possible topics.	Choose a topic and write a purpose sentence.	Research sources, make preliminary outline.	Learn how to give credit.	Make a research plan.
Day 6	**Day 7**	**Day 8**	**Day 9**	**Day 10**
Begin research.	Continue research.	Continue research.	Finish research.	Finalize outline.
Day 11	**Day 12**	**Day 13**	**Day 14**	**Day 15**
Begin writing.	Work on first draft.	Work on first draft.	Work on first draft.	Finish first draft.
Day 16	**Day 17**	**Day 18**	**Day 19**	**Day 20**
Work on final draft.	Work on final draft.	Work on final draft.	Finish final draft.	Polish and turn it in!

Manitoba, Canada

Assigned Literature

Units 1-2	*Know Why You Believe*	Paul Little
Units 3-4	*Blood Brothers*	Elias Chacour with David Hazard
Units 5-7	*Patricia St. John Tells Her Own Story*	Patricia St. John
Unit 8	*A Long Walk to Water*	Linda Sue Park
Units 10-11	*The Day the World Stopped Turning*	Michael Morpurgo
Units 12-13	*Kidnapped*	Robert Louis Stevenson
Units 14-15	*Lost in the Barrens*	Farley Mowat
Units 16-17	*Boys Without Names*	Kashmira Sheth
Units 18-19	*Revolution Is Not a Dinner Party*	Ying Chang Compestine
Units 20-21	*Ann Judson: A Missionary Life for Burma*	Sharon James
Units 24-25	*The Country of the Pointed Firs and Other Stories*	Sarah Orne Jewett
Units 27-28	*Tales from Silver Lands*	Charles Finger

Malé, Maldives

1 Welcome to Your World

In this course we help you see the world in a way that you might never have seen it before. Geography makes a difference in how we live, and it always has. Geographers such as Strabo have studied the earth for millennia. You might think that geography is a subject that doesn't change, but actually geography undergoes significant changes all the time. The worldview lesson in this unit introduces the concept of worldview and tells what difference a person's worldview makes in his or her life.

Lesson 1 - Seeing What You Have Never Seen Before
Lesson 2 - Who Says Geography Doesn't Make a Difference?
Lesson 3 - The World According to Strabo
Lesson 4 - Geography Is Not Set in Stone
Lesson 5 - It's a Matter of Worldview

Memory Verse Memorize Psalm 98:7-9 by the end of the unit.

Books Used The Bible
Exploring World Geography Gazetteer
Know Why You Believe

Project (Choose One)

1) Write a 250-300 word essay on one of the following topics:
   - Tell what you hope to gain from a study of geography. Perhaps include questions you have about the earth and the people who live on it.
   - Tell how geography contributes to your community, whether in economic activity, tourism, or living patterns. Also explore ways people could use geography to make a greater difference, as in developing an industry or tourism activity. See Lesson 2.
2) Collect pictures of geographic features you find beautiful or unique. This could include rivers, mountains, valleys, waterfalls, deserts, and much more. You might find some pictures on postcards, in brochures, online, or in photographs you or your family have taken. Make a scrapbook of your pictures. You might want to write appropriate verses of Scripture next to some of the pictures.
3) Write and deliver a five-minute newscast describing some change in geography: volcano, earthquake, political change in a country, or some other change as described in Lesson 4.

Literature

Paul E. Little wrote *Know Why You Believe* to answer the most common questions he encountered about the Christian faith as he ministered to college students. He does not shy away from the honest doubts, troubling questions, and common confusions many people grapple with in their search for truth. He responds to each query with confidence, knowing that, "After 2,000 years, no question is going to bring Christianity crashing." *Know Why You Believe* helps Christians define what they believe and why. Seekers will find frank and intelligent answers to aid their quest for truth. This book has helped tens of thousands of people understand who God is and define their worldview.

Paul Little (1928-1975) spent his career working with InterVarsity Christian Fellowship, a ministry on college campuses. Little was a popular speaker and writer and also ministered through one-on-one interaction. Paul Little met his wife Marie as they served together in InterVarsity. After Little died in a car accident at age 46, his wife Marie continued in ministry and ensured his helpful books on evangelism and apologetics remained available.

Plan to finish *Know Why You Believe* by the end of Unit 2.

Himalayas, Nepal

1 Seeing What You Have Never Seen Before

Have you ever heard the mountains singing for joy? Have you been listening?
The psalmist wrote:

Let the sea roar and all it contains,
The world and those who dwell in it.
Let the rivers clap their hands,
Let the mountains sing together for joy
Before the Lord, for He is coming to judge the earth;
He will judge the world with righteousness
And the peoples with equity.

—Psalm 98:7-9

The Goal of This Curriculum

To hear the rivers clapping their hands and the mountains singing for joy requires thinking about geography and perceiving the earth in a way you might never have before. Through this course, we want to help you do just that.

Merriam-Webster defines geography as "a science that deals with the description, distribution, and interaction of the diverse physical, biological, and cultural features of the earth's surface." But we want to take the study of geography beyond the description of the physical features of the earth and consider what geographers call human geography.

Human geography examines the activities of human beings, especially how physical geography impacts human activity and how human activity impacts physical geography.

In other words, we'll be talking about YOU and the nearly eight billion other people who share the earth with you, and the impact that geography has on all of us. Where people live and the geographic realities of those places influence who those people are, what they do, and how they live. Geographic places and features influence culture, language, economics, rural and urban life, politics, and international relations. Our goal is to help you better understand the world in which we live, and human geography is a big part of that picture.

In addition, we want to help you see how the earth and all Creation praise God and how they reveal to us the power, love, and wisdom of the Creator. You might not have thought about the world in these terms, but this is the worldview we want to help you adopt. In other words, we want to help you hear the rivers clapping their hands and the mountains singing for joy.

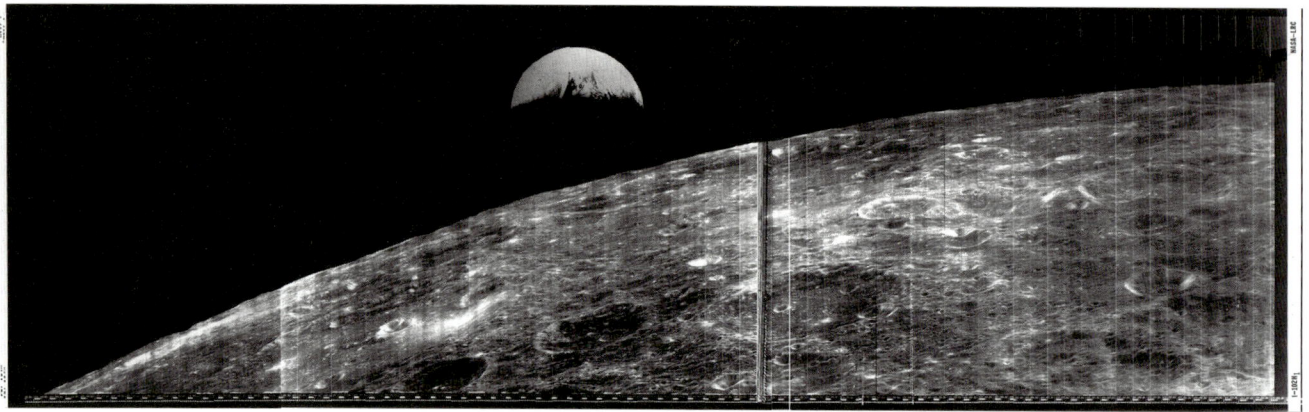

Seeing the Earth in a New Way

For many years, people have been able to see large parts of the earth from mountaintops, balloons, and airplanes. One such person was Katharine Lee Bates. In 1893 she pondered the inspiring view from the top of Pikes Peak in Colorado. As a result of that experience, she wrote a poem that became the song, "America the Beautiful." However, throughout the history of mankind, no one was able to see the entire earth until the 1960s.

In that decade, the American and Russian space programs were in full swing. The American goal, to fulfill President John F. Kennedy's challenge that he issued in 1961, was to land men on the moon and return them to the earth within the decade. Part of the careful process that the American space program NASA followed to achieve this goal was sending unmanned spacecraft to orbit the moon. In 1966, Lunar Orbiter 1 took the above picture of the earth as the orbiter came around the moon's surface.

Two years later, NASA launched a manned spacecraft, Apollo 8, again to orbit the moon but not to land on it. The spacecraft entered lunar orbit on Christmas Eve. That evening, the crew transmitted a live television broadcast back to the earth that included the view of our planet shown below. The broadcast ended with the three astronauts taking turns reading from Genesis 1. It was a thrilling, dramatic moment in American history and in the history of space travel.

Lesson 1 - Seeing What You Have Never Seen Before

In July of 1969, Neil Armstrong and Buzz Aldrin met Kennedy's challenge and became the first men to walk on the moon. People all over the world watched Armstrong take his historic "small step." However, despite all of these amazing accomplishments, people had still not seen the entire planet in sunlight. Because of the angle of the sun, all photos had shown only a portion of the earth's surface.

In 1972 the crew of Apollo 17 was able to capture this image soon after they left the earth's orbit and headed for the moon. The picture has come to be called the "Blue Marble." It is one of the most widely published pictures in history.

Apollo 17 was the last manned mission to the moon. No human has been far enough away from the earth since then to take a picture of the whole planet, although unmanned spacecraft have done so.

The Overview Effect

Every astronaut who has traveled into space, whether around the earth or to the moon, has reported being deeply affected by looking back at the earth beneath them. This has been called the overview effect. As a result of this effect, astronauts have realized many different truths.

Some have perceived the fragility of the earth as it hangs in the darkness of space. Others have pondered how the entire human population—which is divided by nations, languages, and cultures which are often at war with each other—in reality shares a unity of life on this single planet that we inhabit together. Another astronaut recalled his thoughts about the millions of people who live each day without enough food or water, of the injustice and conflicts that humans inflict on each other, and how this larger perspective on the place we all live might help the people of our planet accomplish more good for each other.

Now that you have begun seeing the earth in a way that you might not have ever seen it before, let's get the big picture of this planet we share.

Earth's Setting in the Universe

The sun is a star of moderate size in the Milky Way galaxy. It is nearer the outer edge of the galaxy than the middle of it.

Around the sun orbit eight planets, the dwarf planet Pluto, a band of asteroids, several comets, and other objects. Around the eight planets revolve a total of 173 moons, some of which are very small. The earth is one of those planets. It has one moon, while Jupiter has eight round moons and about 55 smaller, irregular-shaped moons.

The earth is the third planet from the sun. It orbits the sun at an average distance of 93 million miles. Mercury and Venus orbit closer to the sun, while Mars and the other planets are farther away.

The Goldilocks Planet: Just Right

We can see God's wise and loving design of the earth as the dwelling place for humans in many ways. The earth has the attributes required to sustain life. Some scientists call it the Goldilocks planet.

This illustration shows the relative size of the planets but not their relative distance from the sun. Moving from the sun on the left, you can see Mercury, Venus, Earth, Mars, Jupiter, Saturn, Uranus, Neptune, and the dwarf planet Pluto. The sun's diameter is about ten times that of Jupiter.

An astronaut aboard the International Space Station took this photo of the earth's atmosphere with the moon in the background.

It is not too big, not too small, not too cold, and not too hot, but just right. For instance, its distance from the Sun is just right. If it were much closer, it would be too hot; if it were much farther away, it would be too cold.

Most of the earth's atmosphere is within ten miles of the surface. This is another aspect of the overview effect: the realization that the earth's entire population lives in this narrow band of atmosphere that covers the planet. The earth is large enough for its gravity to hold the atmosphere, but it is not so large that its gravity would make the atmosphere too dense for life. This breathable blanket has just the right elements, primarily carbon, hydrogen, oxygen, and nitrogen, to sustain life. In sum, the atmosphere has the content and density necessary for humans and animals to be able to breathe it.

On the other hand, the atmosphere on Venus is too dense and does not have the right combination of elements to sustain human life. Mars is barren and also cannot sustain life. The larger planets farther away from the Sun do not have the atmosphere or the surface makeup that would sustain life.

The moon orbits the earth around 240,000 miles away, which is just the right distance. It influences the tides on the earth and helps to stabilize the earth's rotation. If the moon were a different size or orbited at a different distance, it would not have the same effect on the earth.

The earth's polar axis is tilted about 23 degrees from vertical. This tilt is just right. It enables the earth to have seasons as it orbits the sun. The tilt also gives us differing lengths of days and enables temperatures on the planet to be moderate enough for human life. If it were not tilted, water vapor would flow to the poles and create ice mountains, and the rest of the planet would be too hot to sustain life. If the earth rotated much more slowly than it does, it could not sustain life because of the temperature extremes that would result.

About 71 percent of the earth's surface is covered with water. This amount of water and the fact that it is liquid make our planet's water content just right to sustain life. Life as we know it could not exist with much more or much less of the earth's surface covered with water. The earth's rocky mantle and hot molten core contribute to the right surface temperature for life.

The weather patterns on the earth are just right to sustain life. The climate in different parts of the earth—moderate, tropical, desert, and polar—combine to make large areas of the earth's surface inhabitable.

The earth's magnetic field protects the planet from harmful radiation from the sun. In addition, the fact that the earth's crust and outer mantle are made up of plates that move (a feature called plate tectonics) helps to maintain a moderate surface temperature and helps to regulate the carbon level in the atmosphere. The huge planet Jupiter protects the earth from objects flying through space because its strong gravitational pull attracts many objects that could otherwise strike the earth.

Not Just Simple Life

The makeup of the earth, then, is just right to sustain life, but not just simple forms of life. The requirements for simple forms of life, such as microbes, are not the same as the requirements for intelligent life. The earth is an environment that supports intelligent, complex life. The earth has minerals such as coal and oil and other natural resources that people have learned to use. It has dirt that sustains the growing of food.

The complex interconnected relationship between the various forms of life on the earth make intelligent, complex life possible. For instance, animals breathe in oxygen and breathe out carbon dioxide, while plants take in carbon dioxide and give off oxygen. In this way, plants and animals sustain each other.

This intelligent life is on a far different level from mere existence. Microbes exist, but they do not know that they exist. Beavers build dams, but they do not stand back and admire their work when they finish. Human beings know that they exist, they can admire their handiwork, and they have even used their intelligence to travel to the moon and to send spacecraft to other planets in the solar system.

Benefits from Disasters

The "just right" nature of the earth does not mean that it has no aspects that threaten human life. Some conditions on the earth make life difficult. Earthquakes, tsunamis, volcanoes, hurricanes, tornadoes, strong rain and snow storms, droughts, and floods threaten human life. Many people have lost their lives in such disasters, and these calamities have also destroyed much property.

Even so, some scientists believe that disasters have a positive impact on the earth's ecology, such as by redistributing resources and helping regulate the earth's temperatures. In terms of human response, disasters give people the opportunity to help others in a time of need. This does not mean that disasters are a good thing, but it does mean that good can come even from tragedy.

Goats in Morocco have a interdependent relationship with argan trees. The goats climb up into the trees to eat the trees' fruit. They then spit out the seeds.

Lesson 1 - Seeing What You Have Never Seen Before

What the Overview Means

This awe-inspiring overview of the planet on which we live gives rise to at least two important central truths. One, the earth has been home for every human that has ever lived. Since we share this planet and have no other choice about where to live, we need to respect and understand one another and the planet on which we live.

The second and even more important central truth is that the one God made the one earth for all of mankind, and He sent one Savior to redeem us all. As Paul said, in God we live, move, and have our being. It is obvious that God designed the earth to be a hospitable home for human beings. When we consider this single inhabited planet, surrounded by a thin atmosphere, in the vast reaches of space, it does seem fragile. On the other hand, the earth is strong and enduring. It gives us what we need, and it has survived all of the natural and man-made disasters that have occurred. Our responsibility is to live on the earth in such a way that we honor its Maker.

The Scriptures portray God as Creator of heaven and earth. The psalmist heard the voice of God claiming ownership and control of the earth:

The Mighty One, God, the Lord, has spoken, and summoned the earth from the rising of the sun to its setting. . . . "For every beast of the forest is Mine, The cattle on a thousand hills. I know every bird of the mountains, and everything that moves in the field is Mine."
Psalm 50:1, 10-11

Assignments for Lesson 1

Worldview Copy this question in your Bible notebook and write your answer: When have you seen two people have conflict because they saw a matter differently?

Project Choose your project for this unit and start working on it. Plan to finish it by the end of this unit.

Literature Begin reading *Know Why You Believe*. Plan to finish by the end of Unit 2. The book has 12 chapters, so you might want to read one chapter most days and two chapters on three days so you can finish the book by the fourth day of the next unit and answer questions and read the literary analysis on that day. Today, read the introductory material and Chapter 1. As you read each chapter in the book, write down a two-or-three sentence summary of the case Little makes in that chapter and how it helps support a Christian worldview.

Student Review Answer the questions for Lesson 1 in the *Student Review Book*. Read "What Do You Think About What He Thinks? A Primer for Analysis of Non-Fiction" on pages 1-3.

Egypt from Space

2 Who Says Geography Doesn't Make Any Difference?

How big is the earth? How would you find out? How would you have figured it out over 2,000 years ago, before people had satellites and sophisticated measuring equipment?

Eratosthenes of Alexandria

When Alexander the Great was conquering a large part of the known world in the 300s BC, he founded a city of learning in Egypt called (of all things) Alexandria. Alexander had, after all, been tutored by Aristotle, so he understood the value of learning. The center of learning in Alexandria was the library, which probably held the largest collection of books and other resources in the world at the time.

The library long outlasted Alexander's relatively brief time as conqueror of the world. The head librarian in the latter half of the 200s BC was Eratosthenes. He was from Cyrene in nearby Libya (Remember Simon of Cyrene, who carried Jesus' cross as recorded in Mark 15:21? Same place).

Eratosthenes noticed that at noon on the summer solstice (June 21), the sun shone directly down a well at Syene (now Aswan), Egypt. Thus the sun was at a 90-degree angle or perpendicular to the earth's surface. At Alexandria at the same time, the sun was at an angle of 82.8 degrees. The difference of 7.2 degrees is about 1/50th of the 360 degrees of the earth's circumference. Eratosthenes calculated the distance between the two cities, multiplied it by 50, and got about 25,000 miles.

This calculation by Eratosthenes was amazingly accurate. Modern sophisticated calculations tell us that the earth's circumference is about 24,859 miles through the poles. It's a little more, 24,901 miles, around the equator. In other words, the earth is a little pudgy around the middle.

By the way, Eratosthenes invented a word that we will use many times in this course. It means writing about, or the study of, the earth: geography.

Eratosthenes Teaching in Alexandria
by Bernardo Strozzi (Italian, 1635)

10

Lesson 2 - Who Says Geography Doesn't Make Any Difference?

You might be thinking that geography is one of those requirements that you have to endure to be able to get the credits you need to graduate. We hope to show you that human geography has had and continues to have a profound impact on the real lives of people all over the world.

Ready to start this exciting adventure? Let's go!

What to Call This Place

The year was 1507. It had been fifteen years since Christopher Columbus sailed west and landed among some islands that he thought were off the west coast of Asia. Another Italian explorer, Amerigo Vespucci, had made two voyages west between 1499 and 1502. He landed on and explored the coast of a large continent, which he also thought was part of Asia.

However, many scholars and probably Vespucci himself became convinced that he had landed on a continent that Europeans had not known about. In 1507, the mapmaker Martin Waldseemuller, a German who was living in France, published Vespucci's account of his voyages as well as a four-by-eight-foot map of his discovery.

Waldseemuller confronted an important question. What should the new land be called? He could have called it Columbia in honor of Columbus; but instead he suggested that it be named in honor of Amerigo Vespucci. As a result, a geographic term we know well came into being: America. Waldseemuller only applied it to what we know as South America; its additional use for the continent to the north came later.

Martin Waldseemuller's 1507 world map was printed using woodblocks on twelve large sheets of paper, each about 16" x 24". Out of some one thousand copies of the map that were originally printed, only one is known to survive today. It was discovered in the private collection of a German family. The Library of Congress purchased it in 2003 for $10 million.

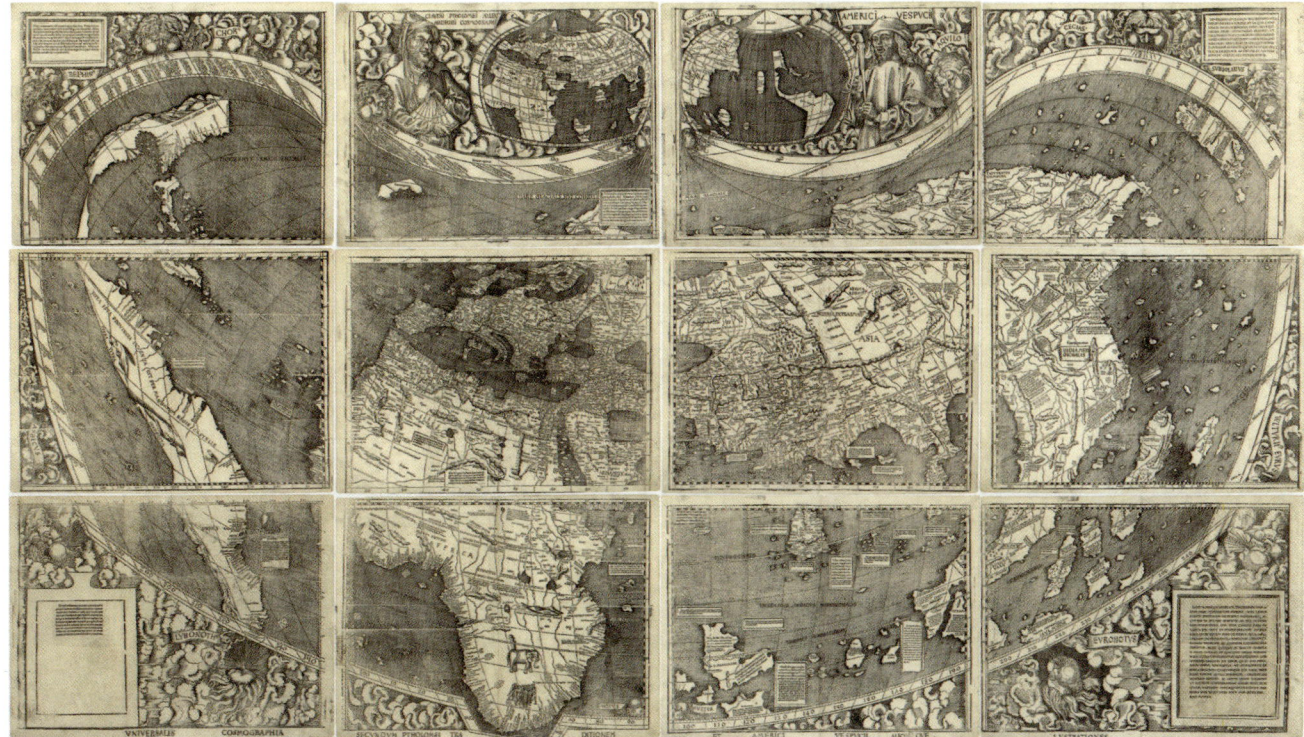

How Big Is the Nation?

The newborn nation could have been confined to a limited scope and size or even have been broken up altogether, but one man saw what it could become and resolutely insisted on making his vision a reality. The crucial question involved geography.

The fighting in the War for American Independence had ended in 1781, but two years later the two opposing sides, an upstart nation and one of the most powerful empires in the world, still had not finalized a peace treaty. A key issue was how big the new United States was going to be.

The thirteen former colonies along the Atlantic coast were a given, but Great Britain had claimed all the land south of the Great Lakes, north of Florida, west of the Appalachians, and east of the Mississippi River. The United States insisted that all of this land now belonged to it.

John Jay *by Gilbert Stuart and John Trumbull (American, c. 1818)*

Britain was not used to negotiating from the position of a vanquished nation, and its government still had an interest in the land to the west of the Appalachians. British troops regularly ventured south of the Great Lakes and tried to stir up native nations there against the new United States, warning them that the Americans wanted to take their land.

Spain, which controlled Florida, had a long-standing interest in lands to the west of the Appalachians. Spain had allied itself with France against Great Britain during the recent War for American Independence. Despite that alliance, both Great Britain and Spain would have liked nothing more than to limit and weaken the new United States, which was now their rival in exploring, settling, and developing the continent.

American and British representatives met in Paris, France, to negotiate a treaty to end the war. One man on the American team understood what was at stake: John Jay. Jay had served in the Continental Congress during the war and was from New York, one of the states that claimed land extending to the Mississippi River. Jay believed that the hand of Providence had led America's rise to nationhood. He understood the rich resources that the North American continent offered. During the treaty negotiations, Jay insisted that the treaty declare that the United States owned all the land south of the Great Lakes, north of Florida and the Gulf of Mexico, and east of the Mississippi. He also insisted that Spain have no say in finalizing the terms of the treaty. Jay's perspectives on these issues became incorporated in the Treaty of Paris that officially ended the Revolutionary War.

The geography of the new nation and of the continent created a new day in world history. The area that the United States eventually controlled encompassed roughly a thousand miles by a thousand miles. The area claimed by the U.S. was larger than England, France, and Spain combined. The agreement meant that the United States could profit from the settlement of the western territory through land sales by the federal government (once the states relinquished their individual claims to it) and from the natural resources that would be found there. The American government created the way for new states to be formed on an equal basis with the original thirteen. The small coastal nation of three million people had immense possibilities before it. The most visionary idea—and to some the most ridiculous and impossible idea—was that one day the nation might spread all the way to the Pacific.

Because of Jay's insight and his role in negotiating the treaty, and because of the way the new nation settled this question of geography, the United States embarked on the course that helped it become the nation it is today.

The Difference One Woman Can Make

Until recent decades, humans had little accurate knowledge of the geography of the ocean floor. Most believed the ocean floor was smooth like a bathtub. In the last half of the 1900s, Marie Tharp overcame social discrimination to do pioneering work in mapping the ocean floor.

Tharp received a master's degree in geology at the University of Michigan in 1944. The field of geology was generally limited to men, but Tharp was able to enroll because so many men were serving in the military during World War II. She began working for an oil company in Oklahoma; but because she was a woman, she was not allowed to do geological field work. As a result she produced maps in her office with data that men gathered.

Later, Tharp analyzed data on the ocean floor that scientists aboard research ships had collected. Again, Tharp was not permitted to take part in shipboard data collection herself, so she utilized the data that male researchers collected.

In 1977 Tharp and colleague Bruce Heezen published the World Ocean Floor Panorama, a map which revealed the complex and varying nature of the ocean floor, including its mountains, ridges, and canyons. They discovered a chain of mountains that runs continuously throughout the world's oceans. Tharp later operated a map-distribution business. She died in 2006.

Marie Tharp and Bruce Heezen (shown at left) also built this handmade globe revealing the ocean floor.

Let's Map the Planet!

All they wanted to do was map the entire planet. For the most part, they have succeeded.

We live in what has been called the Information Age. Sources of information such as the Internet, television channels, and print materials are everywhere. We also live in what has been called the Technological Age, which means that gadgets such as smartphones, tablets, and computers that deliver information are also everywhere. Using technology to gather and distribute information is at the core of how we live in our world today, and Google is one of the leading companies in this field.

Google Earth is a project that uses satellite images to create a 3-D globe of our planet. Their goal was to create a map of the entire earth that people could browse on their computers in the same way that they browse websites for information or for items to purchase. Google Earth is available to download free online.

Google Maps is a navigation program that helps travelers reach their destinations. It has a local focus. At one point, Google had 1,100 employees and 6,000 hired contractors working on mapping projects. Google has utilized satellite images, aerial photography, and data gathered by people driving around making notes and taking pictures to build its database. This latter source provides what has been called "ground truth," which reports one-way streets (something a satellite image won't tell you), construction sites, and other information travelers want. Google has even experimented with mapping malls and departments within a single store to give people the information they are looking for.

These projects, Google's investment in them, and the popularity of these programs all demonstrate just how important geographic information is in our world today. Google wasn't the first and isn't the only source of map and navigation information. Other sources such as MapQuest, Yahoo! Maps, and Apple Maps have been available for years. Global Positioning System (GPS) devices such as TomTom and Garmin offer drivers detailed route information. The mapping field has fierce competition because people want the information that computer mapping programs provide.

Depending on a mapping program such as Google Maps can have its downside. For instance, what shows up as the shortest route between two points can include rough, curvy country roads for the unsuspecting traveler. In addition, a small map on a smartphone screen doesn't offer the larger perspective and tourist information that a printed travel atlas provides, such as where travelers will be in three hours and what they might see just off their chosen path. But overall, taxi drivers, delivery people, salesmen, repairmen, tourists, and many others have

Collecting data for Google Maps in Japan (left) and Kenya

Lesson 2 - Who Says Geography Doesn't Make Any Difference?

come to depend on navigation smartphone apps, which provide information delivered by modern technology.

You interact with geography every time you admire the landscape around you, buy something that local workers made or that traders brought from somewhere else, speak the language of the people around you, use a navigation app to get from here to there, or take part in the culture in which you live.

We've come a long way from Eratosthenes staring down a well and Martin Waldseemuller drawing a somewhat-accurate map from limited information, but people still make maps and multitudes still depend on them. Geography still affects our lives every day, and people interact with geography with more immediacy and on a broader scale than ever before.

We hope you enjoy our trip around the rich, wonderful, amazing, varied world God made and come to appreciate the people who live in this world in ways you might never have before. For all of this, we give God the glory and honor.

The earth is the Lord's, and all it contains,
The world, and those who dwell in it.
Psalm 24:1

Assignments for Lesson 2

Worldview Copy this question in your Bible notebook and write your answer: What are five statements about life and the world that you believe are rock solid truth?

Project Continue working on your project.
Read "Advice on Writing" on pages xiii-xv if you haven't already.

Literature Continue reading *Know Why You Believe*.

Student Review Answer the questions for Lesson 2.

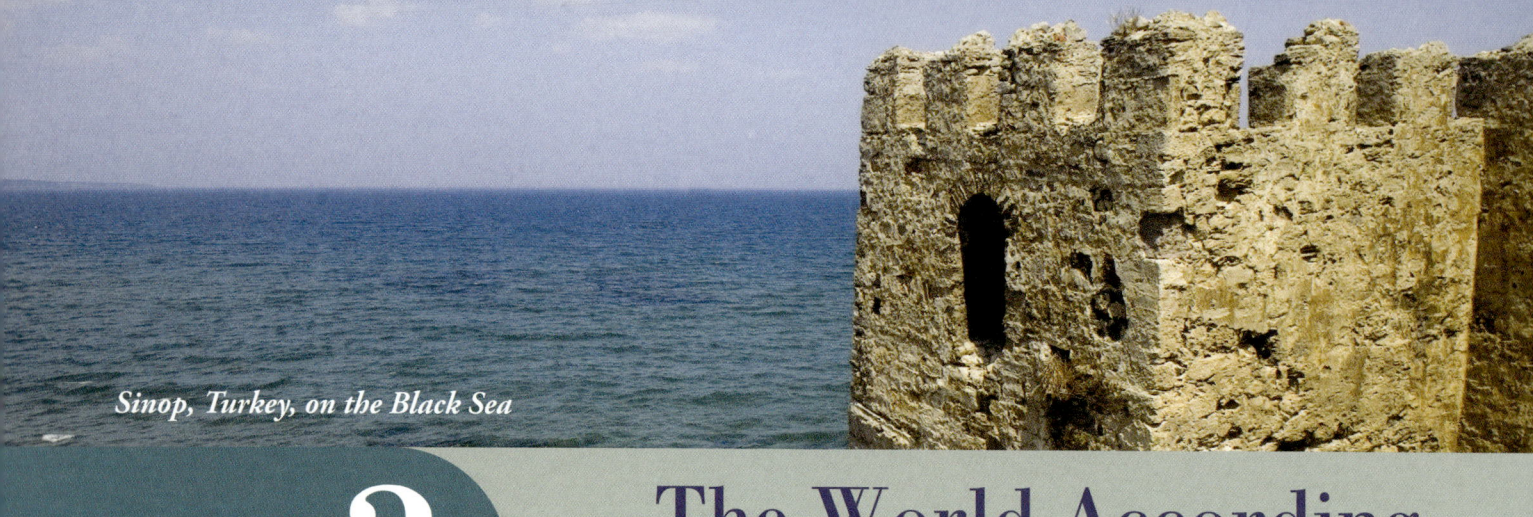

Sinop, Turkey, on the Black Sea

3 The World According to Strabo

Eratosthenes measured the world. Strabo described it.

The political and military leaders of nations have been dealing with each other at least since the five kings went to war in Genesis 14. In those dealings, the leaders want to know the lay of the land and what kind of people and leaders inhabit other places. Today we have such resources as the Central Intelligence Agency's World Factbook (cia.gov). In the ancient world, that information was not so easily obtainable. This is where Strabo comes in.

The historian and geographer Strabo was born in 64 BC in Amasya, the capital of the region of Pontus in what is now northeastern Turkey. A statue of Strabo stands in Amasya today. The statue, seen on the opposite page, portrays Strabo holding a book and standing next to a world globe.

Pontus

The region of Pontus is a lesson in human geography all by itself. It sits beside the Black Sea, and because of its strategic location the armies of many nations have invaded it and the people of many nations have taken up residence there. When Persian settlers came, they brought the Persian lifestyle and the worship of the Persian god Mithra (in fact, a dynasty of kings of Pontus shared the name Mithridates). After the conquests of Alexander the Great, Greeks settled there and brought their culture. Then as Rome expanded its reach in the century before Christ, Roman armies moved in to take control of Pontus. Military exploits led by the general Pompey in the years before the birth of Christ established Rome's authority in Pontus beyond question.

The boundaries of Pontus varied over time. This map shows an outline of an area inhabited by Greek citizens who sought independence in the early 1900s as the Republic of Pontus.

Lesson 3 - The World According to Strabo

A little later than the time of Strabo, Pontus figured in the story of the New Testament. Some of those present on the Day of Pentecost described in Acts 2 were from Pontus (Acts 2:9). It could be that those Jews were converted to Christ that day and took the gospel back to their homeland. We know that Christians were living in Pontus when Peter addressed them in his first epistle (1 Peter 1:2). In addition, Paul's fellow worker Aquila was from Pontus (Acts 18:2). These references illustrate the importance of Pontus geographically and culturally in the first century.

Life of Strabo

Strabo was born into a wealthy and politically connected family of Greek background. They were part of the royal court. This enabled Strabo to receive the education that private tutors could provide. He was also able to travel widely. His first tutor was Aristodemus, who lived in the city of Nysa in southwestern Asia Minor near Ephesus and who had taught the sons of Pompey, a Roman statesman and general. Nysa was situated on the winding Meander River. The river's name became a word that means to wander aimlessly: meander. This was a fitting start for someone who would travel widely later in his life.

Strabo then moved to Rome to study under Tyrannion. Tyrannion had tutored Roman senator Cicero and his sons. The year was 44 BC, the year in which Julius Caesar was assassinated—an ominous time to take up residence in Rome! Tyrannion was a geographer himself, so he probably exerted a strong influence on Strabo, who was 19 or 20 years old at the time. Strabo became an adherent of the philosophy of Stoicism. Stoics held that a single principle called reason controlled all the actions of the universe. Strabo also became a strong supporter of Roman authority.

Around 20 BC, Strabo published *Historical Sketches*, a collection of various stories that dated from 145 BC, when Rome conquered Greece, to

Statue of Strabo in his hometown of Amasya, Turkey

the beginning of the reign of Augustus in 27 BC. The work reportedly consisted of 47 volumes, but we only know a few scattered quotations in other sources. (This, by the way, is a strong argument in favor of making a backup copy of your work.)

During his life, Strabo traveled extensively. We know that he visited Corinth and went up the Nile to the frontier of Ethiopia. He spent time studying in Alexandria and made several additional trips to Rome. He also visited several cities around the Mediterranean coast. In fact, Strabo claimed to have traveled more than any other person.

Strabo's *Geography*

Later in life, Strabo compiled the information he had gathered from his own travels as well as information from the writings of others into a seventeen-volume work he called *Geography*, published around 21 AD. Unlike *Historical Sketches*, we still have some relatively ancient manuscripts of this work, imperfect as they are. We do not know for sure how long Strabo lived after its publication.

Geography is an invaluable resource. It is the best and most complete description that we have of the Mediterranean and European world in the time of Augustus and, as a result, the time of Christ and the early church. The work has some inaccuracies and some limitations based on Strabo's dependence on other sources, but it is a tremendously helpful and informative work. The Jewish writer Josephus of the first century AD referred to Strabo in his writings.

Strabo's travels were not just the meanderings of someone with wanderlust. Instead, he traveled as he did to gather information for political and military leaders, especially those of Rome. According to Strabo, geography impacts warfare in both planning and engagements and is "essential to all the transactions of the statesman." He says that his work is of value for any educated person; for, as he put it, how could such a person "be satisfied with anything less than the whole world?"

The regions that Strabo discussed include Spain, Gaul (France), and Britain; Italy and Sicily; northern, eastern, and central Europe; Greece, Russia, northwest Iran, central Asia, Anatolia (Asia Minor or Turkey), the area of the Aegean Sea, Persia, India, what we call the Middle East, and North Africa. For Strabo, the region stretching from Spain to India and from Ireland to Ethiopia were the world, although he did have some concept of the Arctic Circle.

Strabo discussed physical features of populated and unpopulated regions of the earth; but he also described plants and animals, people groups, and human activities such as trade, mining, warfare, clothing and personal appearance, housing, and diet. No doubt influenced by his extensive travels by sea, Strabo said that it is "the sea more than anything else that defines the contours of the land and gives it its shape, by forming gulfs, deep seas, straits and likewise isthmuses, peninsulas, and promontories; but both the rivers and the mountains assist the seas herein. It is through such natural features that we gain a clear conception of continents, nations, favourable positions of cities and all the other diversified details with which our geographic map is filled."

A reconstructed version of Strabo's map of the world, from the 1903 reference work Encyclopaedia Biblica.

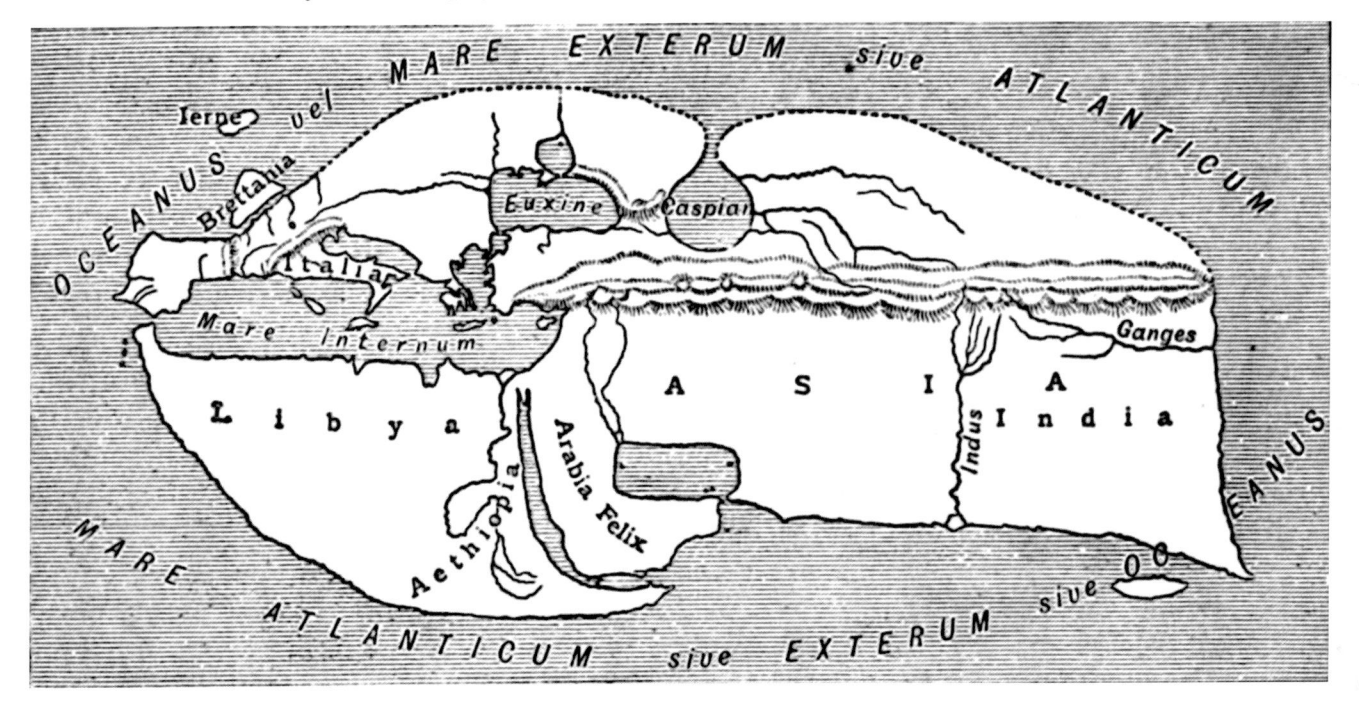

Lesson 3 - The World According to Strabo

Strabo relied heavily on the works of his fellow Greek writer, Homer, whom Strabo considered to be the father of geography. In his poetry, Homer described the geographic features and the people of many lands. As impressive as these descriptions are, however, they are poetry; and even Strabo admits they are inaccurate in some small details. Strabo didn't get everything right himself; for instance, he repeated the belief that the earth (meaning the dry land of the earth) is "entirely encompassed by the ocean, as in truth it is."

Geography isn't perfect, but Strabo gave it his best effort and the result was a landmark study. Physical and human geography still influence all human activity, and educated young men and women such as yourself still need to know the physical and human geography of the whole earth, even though we know that it is larger than Strabo thought. After all, how could such a person as yourself be satisfied with anything less than an understanding of the whole world?

The praise of God extends throughout the whole world.
Oh give thanks to the Lord, for He is good,
For His lovingkindness is everlasting.
Let the redeemed of the Lord say so,
Whom He has redeemed from the hand of the adversary
And gathered from the lands,
From the east and from the west,
From the north and from the south.
Psalm 107:1-3

Assignments for Lesson 3

Gazetteer Read the excerpt from Strabo's *Geography*, pages 243-244, and answer the questions about the excerpt in the *Student Review*.

Worldview Copy these statements in your notebook and complete them:
1. I treat my parents the way I do because . . .
2. I obey the laws of my country because . . .
3. When I am with other people, I try to do what is right because . . .

Project Continue working on your project.

Literature Continue reading *Know Why You Believe*.

Student Review Answer the questions for Lesson 3.

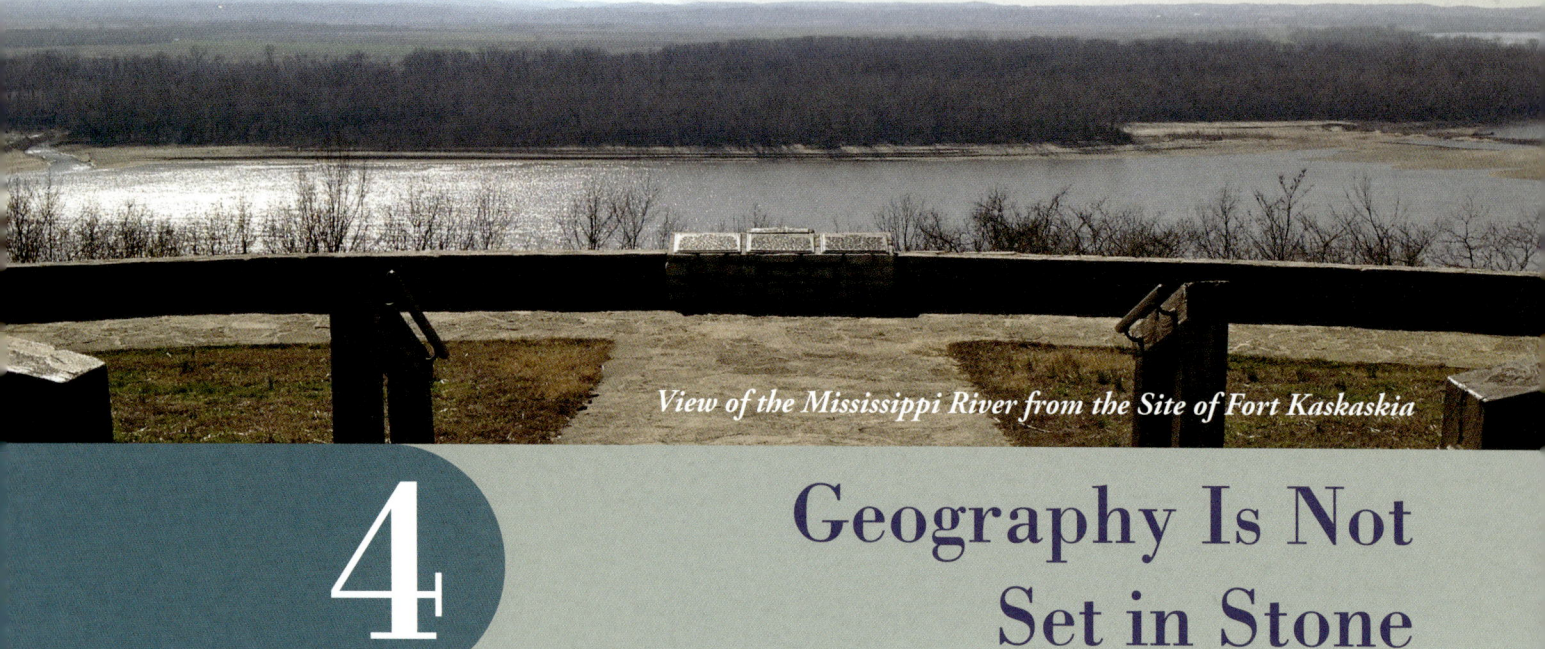

View of the Mississippi River from the Site of Fort Kaskaskia

4 Geography Is Not Set in Stone

Illinois lies along the eastern bank of the Mississippi River. Long before the area became a state, French trappers, traders, and priests entered the region. French settlers established a settlement on the Mississippi and named it Kaskaskia for a native nation that lived there.

Kaskasia grew into a busy river port and became the home of flour mills and other thriving businesses. King Louis XV of France honored the town with the gift of a 650-pound bronze bell in 1741. Kaskaskia grew to be home to over 7,000 people.

After the United States gained control of the area, Kaskaskia became the capital of Illinois Territory in 1809. It was still the capital when Illinois became a state in 1818; but the state government soon moved the capital to Vandalia, a city that is northeast of Kaskaskia and closer to the center of the state.

As time went on, the Mississippi River that helped build the town also caused its near destruction by frequent flooding. In the aftermath of an 1881 flood, the river itself divided. This left Kaskaskia on an island. The eastern course of the river flowed into the channel of the Kaskaskia River, which had formerly joined the Mississippi a few miles downstream. Eventually the western flow became a bayou, and the Mississippi followed the eastern course. This left Kaskaskia on the western, Missouri side of the river. The site of Fort Kaskaskia is still on the eastern side, on a bluff overlooking the river and town.

The population of Kaskaskia declined as the town lost political and economic importance and as flooding made residency there uncertain. By 1990 the population was 32. A major flood in 1993 covered the town with about 10-12 feet of water. The 2010 census of the one-time state capital showed a population of 14 people.

Kaskaskia is the only incorporated Illinois town on the western side of the Mississippi. It has an Illinois telephone area code and a Missouri postal zip code. Illinois provides law enforcement; Missouri provides fire protection. A bridge over the bayou enables access to and from the outside world, while another bridge, one that crosses the Mississippi at nearby Chester, Illinois, keeps it connected to its home state.

Geography changes. But the little town of Kaskaskia still has its bell.

Lesson 4 - Geography Is Not Set in Stone

You might think of human geography as a snapshot of life on the earth, including physical features, political boundaries, and world cultures. Actually, human geography is more like a video of a moving object. Change is a major part of life, so how the earth and life on it are changing is a major part of the study of human geography. Geography is not just about characteristics; it is about processes as well.

Actually, this should not be surprising. After all, you are changing. When was the last time you looked at a photograph of yourself from five years ago, or even three years ago? You've changed since then, and you will keep on changing. So it is with the earth and its human population. Let's think about how this is true.

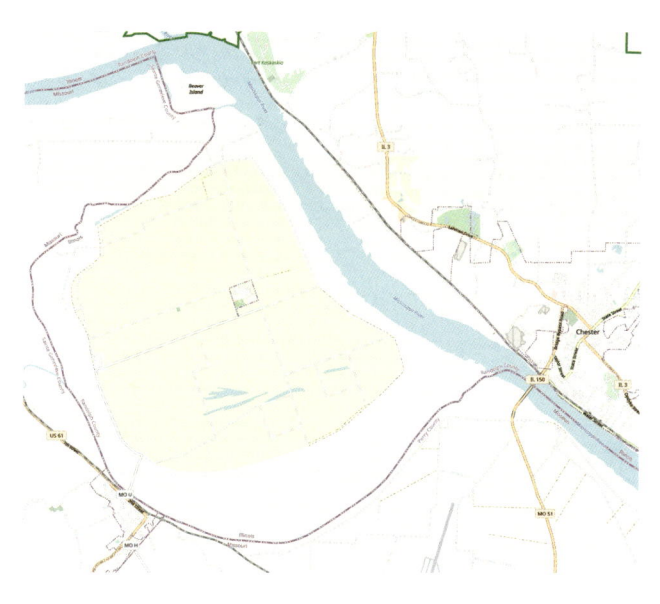

Above is a map of Kaskaskia. You can see the Missouri/Illinois border diverting from the middle of the Mississippi River to go around Kaskaskia. Below is an image from the 1993 flooding.

Changes in Physical Geography

The physical geography of the earth experiences slow, incremental changes in many ways. The motions of the seas change the shape of shorelines. The seas rise and fall as a result of the melting and freezing of the polar ice packs. The slow movement of huge glaciers changes the contours of the land over which they move. When a glacier "calves" and a portion of it breaks off and falls into the sea, that chunk of ice becomes an iceberg. Slow movements in the tectonic plates of the earth result in changes in the earth's surface.

Besides such incremental changes, the earth occasionally experiences dramatic changes as a result of earthquakes and the eruption of volcanoes. A major volcanic eruption can result in the spread of ash and lava that changes the nature and appearance of the surrounding land. In 1943 a volcano deep in the earth erupted in a level cornfield near the town of Paricutin, Mexico, about 200 miles west of Mexico City. The volcano spewed rock and ash off and on for several years. Before the eruption ceased, the mountain had destroyed two towns and the peak stood about 1,353 feet above the valley floor.

In 1963 the eruption of an undersea volcano off the coast of Iceland resulted in a new island being formed that now covers about a half square mile and reaches about five hundred feet above the sea. Seeds washed up on its shore, and the first plants appeared on the island in 1965. The island is called Surtsey and has been dedicated to scientific study. In another example of change, a new island of mud appeared off the coast of Pakistan in 2013 as the result of an earthquake.

Change happens in downward movement also. In 2018 a large crack appeared in the Rift Valley of southwest Kenya. Speculation abounded as to what the giant opening meant for the continent.

Certainly the surface of the earth is stable for the most part and remains the same year after year. Yet the earth does undergo physical changes, some of which you can expect to see in your lifetime.

At left is a 1943 photo of the Paricutin volcano. Below is a 2013 photo showing vegetation growing on the lava flows around the bell tower of the old church of San Juan Parangaricutiro.

Lesson 4 - Geography Is Not Set in Stone

Changes in Political Geography

Nations rise and fall; they grow in power and sometimes disappear. (Have you ever heard of the Roman Empire or the Ottoman Empire?) Part of the identity of ethnic groups comes from the geography of the land on which they live. People who live along a seacoast often become fishermen and traders. Those who live in mountainous regions may become isolated and inwardly focused.

The political boundaries of nations change. If you compared a world map from one hundred or even sixty years ago to a map of the world today, you would see many changes. New nations have emerged. Old nations have changed their names. Areas that were once colonies have become independent countries.

After the fall of Communism in Eastern Europe, the ethnic identities of groups in the Balkan region in southeastern Europe—along with their historic conflicts—re-emerged. This led to significant changes in the drawing of national borders from the time when Communist rulers controlled those peoples with authoritarian rule and with little regard for ethnic identity.

Thirteen British colonies in America declared their independence as a new nation, the United States of America, in 1776. Venezuela declared its independence from Spain in 1811. The disparate German kingdoms became the unified nation of Germany in 1871. Timor-Leste, or East Timor, gained its independence from Indonesia in 2002. The Republic of South Sudan separated from Sudan in 2011.

Alliances among nations change. Countries in Europe that were enemies during World War II formed the European Common Market in 1957. This alliance of nations grew and became the European Union (EU) in 1993. However, a majority of the voters in the United Kingdom voted in 2016 to leave the EU. One factor in this "Brexit" (British exit) from the EU was the geographic and cultural separation that the United Kingdom has long felt between it and continental Europe.

UK newspapers reported the results of the "Brexit" referendum on June 24, 2016.

Governments change. The makeup of the American government changes every two to four years. In some countries, the very nature of government can change almost overnight. The twentieth century saw dramatic changes in the form of government in several countries, including Germany, Italy, the Soviet Union, and South Africa. For years Hong Kong was part of the British Empire. In 1997 the city came under the control of China through an agreement between China and the United Kingdom.

Consider the Middle East. At the center of the conflict between Israel and the Muslim nations is the issue of geography: who controls (or who should control) the area known as Palestine? God gave the land to the nation of Israel, conditional on their faithfulness. Israel was unfaithful to God, and conquering nations took the people away from their homeland. Centuries later Muslim conquerors gained control of it. After the Holocaust of World War II, many Jews and others wanted to see the creation of a Jewish homeland. The United Nations declared its support for the division of Palestine into Jewish and Palestinian Arab homelands. Jews responded by taking some of the land and declaring the state of Israel in 1948. Palestinian Muslims

Individuals can choose to work together in peace, even when governments cannot find agreement. In 1970 a Roman Catholic monastery provided land for a new community in Israel, as seen above. It is called Neve Shalom in Hebrew and Wahat al-Salam in Arabic. Both names mean "Oasis of Peace." Jewish and Arab settlers have chosen to live together and learn together, seeking to create lasting change throughout the region with the School for Peace.

---

and the rest of the Muslim world rejected the UN resolution because it took away what Arabs saw as their land. The history of the last several decades in the Middle East have largely been the result of Israel defending their claim to the land and the Palestinians and their Muslim allies attempting to regain control of the land and push the Israelis out.

At the start of World War I, Belgium was a major player in international politics and economics. It had been an important factor in economic activity for centuries, and it had colonies in other parts of the world. At the same time, most people saw the region we call the Middle East as a backwater area that had little impact on world affairs.

Then major changes occurred. Europe was devastated by two world wars. The United States became the dominant economic and military power in the world. Other countries, such as the Soviet Union and China, grew in power and world significance. The increase in the world demand for oil and the continuing conflict between Muslims and Jews in the Middle East placed that region at the center of world politics. One hundred years after World War I, which has been more prominent in world politics and economics, Belgium or the Middle East? The answer is obvious.

Cultural Changes

Cultural geography examines the human cultures of various geographic places and regions. We can describe the cultures of various places as they exist now, but cultures are continually changing. The pace and nature of change depend on the number of people from other cultures who come to a place and their influence on the people in that place through the media, product availability in stores, and other factors.

The culture and language of England began changing in 1066 when Norman invaders from France conquered the Anglo-Saxon defenders. French became the language of the English royal court. With European colonization of North America in the 1600s and 1700s, the dominant culture on the continent was no longer that of native nations. Instead, the dominant cultures were British and (in parts of Canada) French. Later in the United States, Catholic Irish and Italian immigration influenced the dominant Protestant British culture. Today, American cultural influences in clothing, music, and restaurants are strong in many other countries.

Lesson 4 - Geography Is Not Set in Stone

Changes in Economics

Countries and regions become known for the economic activity in those places, such as potato growing in Ireland, automobile manufacturing in Michigan, chinaware in the cities called "The Potteries" in northern England, and electronics in Japan. Cultural and geographic factors play a part in how this economic activity develops.

The growth of textile mills in the early years of the Industrial Revolution occurred in places where rivers were large enough to power waterwheels that aided textile production. Cities grew where developers could build good ports. New York City has good harbors and was well positioned in the new United States. It received goods from and sent them to inland locations via the Erie Canal-Hudson River route. New York City didn't just happen. The geography of New York City led to its role as a major economic center of the world.

Seagoing ships use the oceans to carry products in millions of sea containers to port cities. Geography that enables rail and truck lines to carry those goods inland is an essential feature for a major port city.

Women working in a Chinese factory

Industrial production does not always remain in the place where it began. Many corporations and manufacturers which were once located in the northeastern United States, from New England through Ohio and Michigan, moved to the South and Southwest where the climate is more favorable and where the culture is more accepting of nonunion labor arrangements. This allowed companies to save on labor costs.

Much of the production of American consumer goods has moved to other, less wealthy and less industrialized countries, such as China and Mexico, where production and labor costs are lower than they are in the U.S. This process is sometimes called the deindustrialization of a region or nation. The increasing interconnectedness of production and trade among the nations and the increasing dependence of many nations on economic activity outside of their own borders is called globalization.

The Human Factor

Mankind's use of the earth's natural resources has undoubtedly improved the living standards of much of the world's population. We can live better, healthier, safer, and more productive lives by the use of coal, oil, trees, clay (for dishes), water (for hydroelectric production), and other resources in our geography. Manmade dams have aided in flood control and the production of electricity. Dams have also created lakes for recreation and to serve as reservoirs.

However, mankind's interaction with our environment has not been completely beneficial. People have polluted the land, water, and air. For many years, lumber companies cut trees and did not plant new ones. This deforestation still occurs in some countries. The clearing of jungles in places such as Brazil negatively affects the environmental dynamics in those localities. Strip mining and open-pit mining have led to considerable damage to the land. We know that human irresponsibility is the reason why several animal species have become extinct.

One human activity for which the consequences are thus far unclear is China's building of artificial islands in the South China Sea and their constructing military installations on these islands. China has done this by dumping huge amounts of sand on coral reefs. Their purpose is to exert a greater dominance throughout the South China Sea. But the long-term consequences for the environment and for international relations are not yet clear.

Geography in Motion

Human geography is not a static, once-for-all subject. Instead, it is a dynamic subject that influences how we live and that is influenced by how we live. As an adult, you will live in a world that will probably be different in significant ways from the world in which your grandparents lived.

May we live well to honor the God of Creation, who always cares for the world He made. After the flood, the Lord told Noah that, even as the earth experiences changing seasons, it will endure.

While the earth remains,
Seedtime and harvest,
And cold and heat,
And summer and winter,
And day and night
Shall not cease.
Genesis 8:22

Assignments for Lesson 4

Worldview Copy this question in your notebook and write your answer: What is a religious belief that you think everyone should accept?

Project Continue working on your project.

Literature Continue reading *Know Why You Believe*.

Student Review Answer the questions for Lesson 4.

5 It's a Matter of Worldview

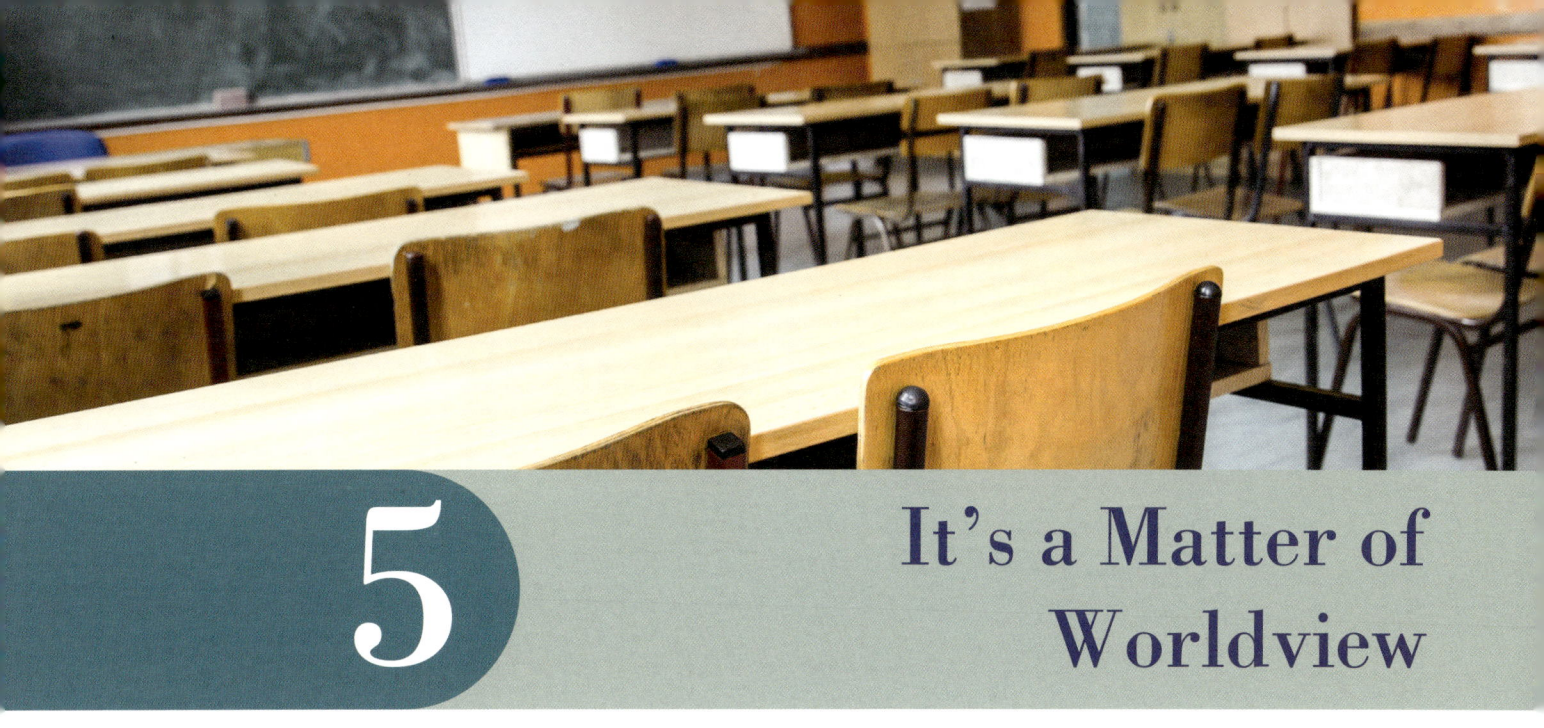

The scene is the office of Jeremy Hankins, assistant principal in an American public elementary school. In his office are Ryan and Sarah Thompson, parents of Brett Thompson, a third grade student at the school. The Thompsons have decided to withdraw Brett from the school and teach him at home. Mr. Hankins has asked to meet with them.

Mr. Hankins speaks:

"Thank you very much for coming in to talk with me. Mr. and Mrs. Thompson, I know you have the legal right to homeschool Brett, but I'm concerned about what this will mean for him. Children today grow up in a world where they hear many ideas. I don't believe that it's the role for adults, even for a child's parents, to tell children what to think. I believe parents and teachers need to help children learn how to think.

"Many people believe that the main responsibility for teaching children lies with the parents, but in our day schools have the trained staff and the professionally produced materials to teach children what they need to learn and how they need to learn. In our school we teach the facts of academic subjects, but we also teach values like diversity, acceptance, equality, tolerance, and social justice. We help children understand history in a way that teaches them to consider people who have long been overlooked, the people who are the victims of history.

"And we certainly can't favor one religious belief over another. We have children from all sorts of backgrounds and family situations, and we want to affirm their worth as people. Our goal is to prepare children to live in the world as it is, to be ready to hold down a good job, to get along with others, and to help them develop decision-making skills. We want them to learn how to weigh all the factors in every situation and make the best decision for that situation.

"The world isn't like a family's living room, and parents won't always be there to protect their children. We're helping them leave the nest and fly on their own. We don't want to shelter children from the real world, but to help them live successfully in it. I have serious doubts that homeschooling will accomplish these goals for Brett."

Mr. Thompson replies:

"Mr. Hankins, Brett's mother and I have decided to homeschool Brett because we want to provide the environment and the teaching that will support and encourage him in his life, primarily in his Christian faith. That is the most important part of our lives, and we hope it will be for Brett also.

"Brett has received a good education in the academic subjects here and is probably ahead of his grade level in many areas. But we have two main concerns. First, some of what he has been taught we find objectionable. In science, the teachers and the textbooks assume that evolution is true, that the world began from purely natural causes, and that it is billions of years old. We disagree with that.

"We've also seen homework sheets that assume climate change is the result of human activity and that only worldwide government action can change it. We certainly see how people can harm the environment. We know people have done so, but we don't agree with the assumptions and models that many people use to blame what they call climate change on human activity.

"We have concerns about the general social and—I guess the word is philosophical—atmosphere of the school. As Christians, we can't just go along with lessons about accepting alternative lifestyles that we believe are sinful. We've noticed lessons that seem to say white American men are the cause of most of the world's problems. Brett loves God and is learning about how God works in the world. What he hears at school tears down his faith, which is not yet mature. He's talked to us about the confusion he feels."

What is the conflict here? It is not that one person wants what he believes to be best for the child and the other does not. It's a matter of worldview: how each person sees the world, what he values, and the best way to train children.

The next scene is the end of a debate between two candidates for president. They each make their closing statement.

Candidate A:

"I believe in freedom. Every American should have the freedom to pursue his or her dreams. Government often gets in the way of that pursuit with regulations and burdensome taxes. Government does not create economic growth. Central planners cannot know everything they need to know in order to run the economy so that it can grow. We need government to protect us and to provide basic services for the common good and maybe to help people temporarily who have fallen on hard times, but government has gotten too big and too invasive.

"I want to roll government back, make government spend less, make it spend responsibly, and cut the deficit as I would encourage any family to do. I want to let Americans keep more of what they earn and spend it how they think best and not have some bureaucrat make those decisions. Individuals and companies create jobs, and they shouldn't be taxed to death. People should be able to make their own health care decisions in a free market."

Candidate B:

"I believe in security. I believe that our society, through government action, ought to provide security for every American. Every American deserves economic security, and the security of having excellent health care free of charge. Every American should have the security to live their chosen lifestyle with the support of government and society. Americans deserve personal freedom, and that includes freedom from gun violence.

"Millions of Americans can't make it on their own, and they need help from the rest of us, a safety net. We can do this if the wealthy will pay their fair share of taxes. It's great to talk about balancing the budget, but we can't do that on the backs of the poor and the people who depend on government services just to get by. We have way too much waste and unnecessary spending on the military, and too many tax breaks for corporations and wealthy individuals.

Lesson 5 - It's a Matter of Worldview

We're all in this together, and each of us needs to give what we can to help everyone."

These candidates seem to be talking past each other. They do not even seem to be dealing with the same issues. What is the difference between them? It's not that one candidate loves his country and the other doesn't. It's not that one wants what is best for American citizens and the other doesn't. It's a matter of worldview: how they understand the way the world works and the way they think government should work.

The third scene finds two acquaintances talking in a coffee shop. The first man speaks:

"I just don't see how you can say that you know God exists. Did He speak to you from heaven and tell you? Science says that the world evolved over a long period of time. We know that changes take place in nature, so why couldn't the world have changed a lot over a long period? Who knows how it all started? No one was there to watch.

"And I'm sorry, but I just can't accept a 2,000-year-old book, written by people promoting one particular religious viewpoint, as my absolute final authority. Seems to me we have to figure out life as best we can. There are so many cultures and belief systems around the world that we can't insist that there is only one right way for everyone. If Christianity works for you, fine. I see how people who call themselves Christians really treat each other and other people, and I'll find some other approach, thank you. My goal in life is to be as happy and comfortable as possible, and I'll do whatever it takes to have that."

The other man replies:

"I don't see how you can say that God doesn't exist. Our world started somehow. Nothing in this world starts from nothing, so how could the world itself just start in some irrational process? We have too much order in the world for it to be the result of chance and mere accidents. And by the way, someone was there at Creation: God, and He has told us about it.

"My goal is to honor God with the life He gave me. That will give me the greatest satisfaction I could ever find, but if honoring God brought me pain and death I could still do nothing else because of what He did for me.

"My highest good is not self but love because that is the nature of the God who created me. The great thing about Christianity is that it answers the same critical needs of everyone, everywhere, in every culture: Where did I come from? Why am I here? How can I best relate to other human beings? How do I deal with sin? Where am I headed?"

What is the difference between these two perspectives? It is not that one person is intelligent and the other isn't. It's not that one has thought through the big issues of life and the other hasn't. It's a matter of worldview.

It is sometimes amazing to see how different people have such widely varying opinions about the same person, event, or idea. This happens because people have widely divergent worldviews. When you hear someone make a comment, state his opinion, or express a belief, and you ask yourself, "Where is he coming from on that?" you are wondering about his worldview.

What Is a Worldview?

A worldview is just what the word implies. Your worldview is how you see the world. It is the set of presuppositions you have that leads you to see the world as you do. Your worldview is the glasses through which you evaluate what happens and how you assign meaning to it. Your worldview is how you think. It is the filter through which you judge whether a statement of information or belief is right or wrong or merely an opinion. It is how you decide whether an action is important or inconsequential, and whether a situation demands your response and what that response should be.

Another word for worldview is paradigm. Your paradigm is the pattern you use to organize your world. Your paradigm or worldview is your inner compass that helps you determine the right way to go. A paradigm is not necessarily reality, but it is how you view reality. Your goal should be to conform your paradigm to reality.

"I Didn't Realize I Have a Worldview. Do I?"

Yes! Everyone has a worldview. You might not have thought about it, but you have a worldview. You might call it your point of view, your belief system, your outlook, or your philosophy.

You don't even have to make a conscious effort to develop a worldview in order to have one. You develop one subconsciously as a result of your experiences and influences. The wise and godly parent begins at a child's birth to lead the child's heart and mind into the path of godliness, just as Timothy's mother and grandmother did for him (2 Timothy 1:5).

What Difference Does a Worldview Make?

A worldview is more than just deciding how you would answer a question on an opinion survey. Worldviews make a difference because worldviews have consequences. Thoughts, beliefs, and perceptions lead to actions. In other words, you develop your worldview as a consequence of what happens to you and around you, and you apply your worldview in your response to what happens to you and around you.

Your worldview is the basis for how you live: how you think, how you act, how you interact with others, how you respond to events and ideas, how you make choices, how you live in a family, how you do your work, how you save and spend money—in short, your worldview is the basis for how you understand the world in which you live and how you do everything in your life.

Here is an illustration of the importance of worldview. If you were to be convinced of a need to change your actions in some way, just trying to change your actions without changing your worldview probably won't work. You will probably default to your old actions because you will tend to act on the basis of your worldview and not on the basis of a stated belief that does not come from your actual worldview.

Examples of How Worldviews Matter

1) Consider a scenario that has happened many times. Imagine that Fred and Dorothy Smith and their children were the first black family to move into the neighborhood. Some of their neighbors were openly cold and hostile toward them. Most didn't like it but generally kept their opinions to themselves—most of the time, anyway. Bob and Betty Johnson thought, talked, and prayed about what they should do. The Johnsons were Christians who took their faith seriously. They knew that Christ accepted them because of their worth in God's eyes. Bob and Betty had been newcomers to a neighborhood once, when Bob had been stationed at an Army base a long way from their hometown. They knew how much they appreciated it when another couple had welcomed them and had become fast friends. Bob and Betty knew that the Smiths were made in the image of God and that God loved them. Betty baked some cookies, and she and Bob walked up to the Smiths' house and knocked on the door. The Johnsons' worldview influenced their actions.

2) In lamenting what he saw as the decline of traditional values, author C. S. Lewis noted that we laugh at honor but are shocked to find traitors in our midst. In other words, many people mock traditional moral standards and then are surprised to see so much immorality in our society. People should not express the belief over and over that there is no absolute truth, that what is right for one person is not necessarily what is right for others, and that

everything is relative, and then be surprised when people begin acting as though there is no absolute truth and that everything is relative. Worldview has practical, real-life, everyday, sometimes life-and-death consequences.

3) The Nazi worldview was based on the belief that they were a superior race. As a consequence, they believed that they were correct in eliminating people of other ethnic groups, such as Jews and gypsies. Some Muslims believe that non-Muslims are infidels and should either become Muslims or be put to death. This belief lies behind the acts of terrorism that some Muslims have committed. Terrorists throughout history have acted out of their worldview, which held that others didn't deserve to live before the righteous wrath of their belief system.

These illustrations demonstrate how worldview makes a difference in how people live.

A Worldview Survey

The fifth lessons in the units of *Exploring World Geography*, taken together, comprise a worldview survey. We give particular attention to the elements of a worldview that is based on the teachings of the Bible. Developing a solid, consistent, Biblical worldview is an essential part of studying the earth and man's interaction with it (in other words, geography) in a wise and godly manner. Some of the worldview lessons specifically relate to geography. Thus you could say that this is a worldview course that helps you develop a view of the world.

We will also survey some of the worldviews that people hold other than the Christian worldview. We believe that it will be helpful for you to have some grasp of these worldviews for two reasons: (1) you will be at least somewhat familiar with these worldviews when you run into them in your life, and (2) sometimes seeing another viewpoint helps you clarify what you believe yourself. Throughout this survey, we hope that you will be able to see the impact of holding particular worldviews, especially a Christian worldview.

In a letter to the Corinthians, Paul described the obedience of Christ as the filter he used as his worldview.

We are destroying speculations and every lofty thing raised up against the knowledge of God, and we are taking every thought captive to the obedience of Christ. . . .
2 Corinthians 10:5

Assignments for Lesson 5

Worldview Write or recite the memory verse for this unit from memory.

Project Finish your project for this unit.

Literature Continue reading *Know Why You Believe*.

Student Review Answer the questions for Lesson 5.
Take the geography quiz for Unit 1 in the *Quiz and Exam Book*.

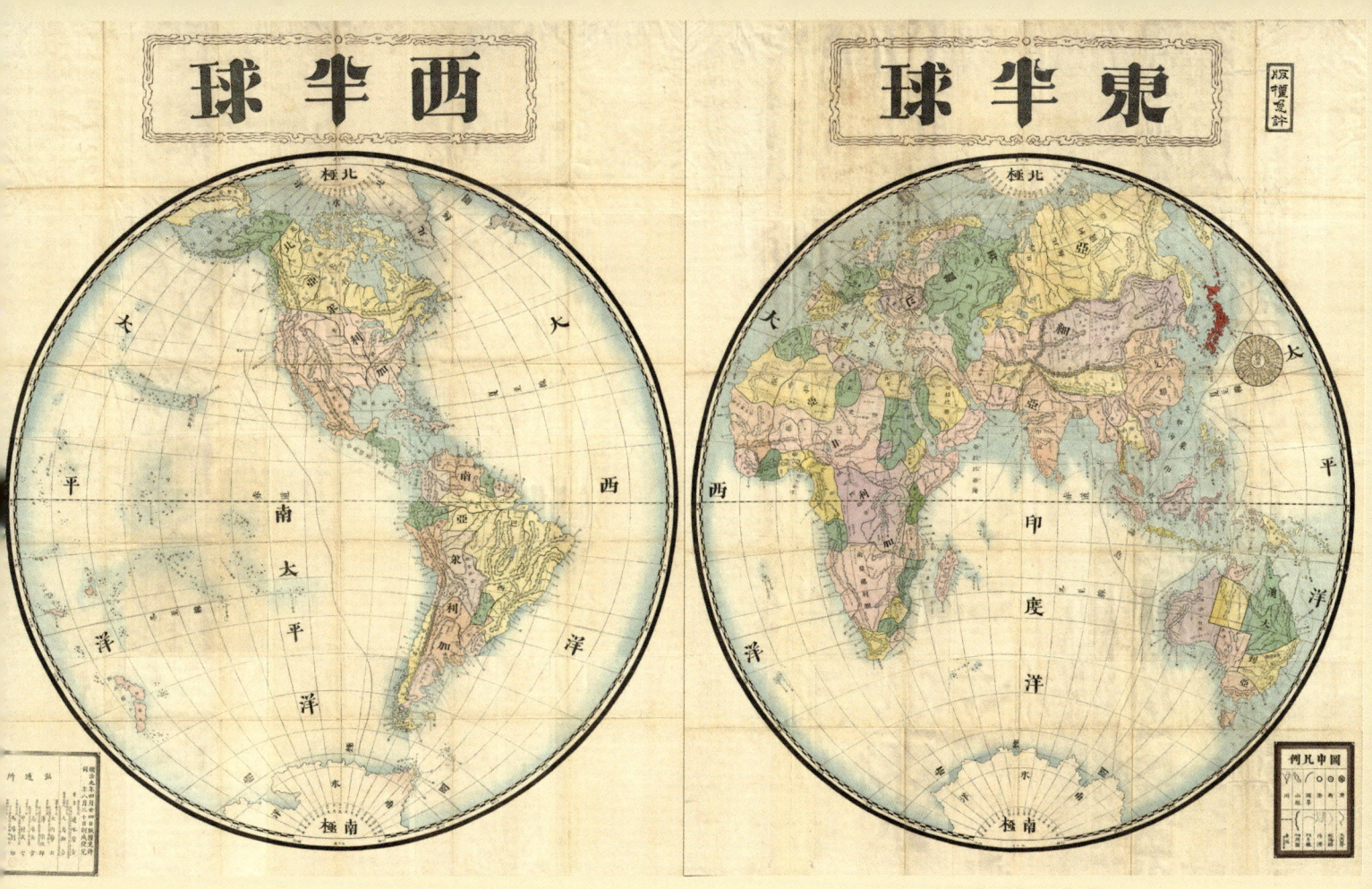

Japanese Map of the World (c. 1879)

2 It Begins with a Map

Maps are an essential element in studying geography. People have been making and using maps for many centuries, but the format of these maps has changed considerably. They reflect our ever more accurate understanding of the earth. People make maps for many different purposes, and they produce them in many different ways. Mapmakers include certain agreed-upon elements in most maps. While a map provides much useful information, they can't include every kind of information we might want. William Rand and Andrew McNally started a map company in the 1850s that still makes maps today. This unit's worldview lesson gives the basic elements of a person's worldview.

Lesson 6 - What's a Map Worth to You?
Lesson 7 - You Can Learn a Lot from a Map
Lesson 8 - What a Map Is and What It Is Not
Lesson 9 - The Business of Geography
Lesson 10 - Recipe for a Worldview

Memory Verse   Memorize Psalm 8 by the end of the unit.

Books Used   The Bible
Exploring World Geography Gazetteer
Know Why You Believe

Project (Choose One)

1) Write a 250-300 word essay on one of the following topics:
   - Tell ways that you, your family, or people in general use maps, such as in determining distance and directions, seeing what lies in a certain geographic area, and side trips you might take when traveling. Tell what you find helpful and what you find frustrating about different forms of maps and things you think a map should show.
   - Describe Americans' love affair with automobiles, our dependence on them, and the role maps play in that relationship.
2) Write a short story in which a map is a key element of the plot.
3) Draw a map illustrating something you are familiar with that someone else might not be familiar with. For example, you could draw a street map of your town or neighborhood, a map showing the route between your home and a place where you often go, or a map that highlights museums and historic sites in your area.

Soviet Military Officers, c. 1950s

6 What's a Map Worth to You?

Men died making them. An army officer could go to prison for losing one.

A foreign buyer could purchase thousands of them—but for cash only.

What was this precious item? A map.

The Soviet Map Project

As the Soviet Union was unraveling in 1989, its military men were selling off what they had control over for what they could get. American map dealer Russell Guy flew to Estonia with a quarter million dollars in cash. He bought thousands of Soviet-made maps, each marked "SECRET." Over the succeeding years Guy, retired British software developer John Davies, and a handful of other people uncovered an amazing fact. During the Cold War (1945-1989) the Soviet military undertook one of the biggest mapmaking projects in history. Their goal was to map the entire world, in some places in amazing detail.

The Soviets had several motivations. First, they wanted detailed, accurate information about the huge country they ruled. The Communist leaders wanted to know the placement and condition of roads, the size of lakes and rivers, and the layout of cities. They wanted to determine where their resources were and how they could develop them. They wanted to know how they could best move men and material where they might be needed. The Communist leaders utilized tens of thousands of surveyors and cartographers on the project. They didn't all survive. One Russian surveyor reported finding a note left by a fellow surveyor on a tree trunk in Siberia in 1948 that indicated the man and his co-worker were near death. They likely did not make it home.

Second, the Soviets' main enemy was the United States. As important as information about Russia was, they believed that information about the country they saw as their primary threat was even more vital. The Russians created detailed maps of the United States, including street maps of the major cities that identified buildings along the streets and noted the construction materials and load limits of bridges. The Soviets considered the maps so valuable that a Russian officer could face prison or worse if he misplaced one. Where did the Soviets get such detailed information about the United States? Some was information published by the United States Geological Survey and other sources. Other information no doubt came from spies in our country. The Russians probably also utilized high-flying reconnaissance planes and spy satellites.

But Soviet interest in maps didn't end there. The Russians dreamed of world domination, so they wanted accurate maps of the entire world: cities, military installations, factories, roads, ports—any information that could help if the opportunity came for them to move their tanks and troops into other countries. Again, some information was freely available, while they obtained other information by espionage.

One expert on the subject estimated that the Soviet military produced about 1.1 million maps. Not copies, but distinct, different maps, of which they made untold copies. Many of the maps show astounding details, such as the width of roads and the depth of lakes. It's a little unnerving to see a detailed Soviet-era map of the Pentagon with the roads around it and the Potomac River lettered in Russian. Or maps of New York, Chicago, San Francisco, and San Diego that identify buildings and military installations—in Russian.

Sometimes the differences between Soviet military maps and maps from other sources are intriguing. For instance, a 1984 Soviet map of Chatham, England, shows a dockyard where the Royal Navy built submarines. A British map from the same period shows a blank space there because the British did not want information about that location to become known—but the Soviets knew it anyway. On the other hand, maps published in the Soviet Union and available to the general Russian public were deliberately inaccurate in terms of distances, directions, and other details. The Communist government did not want such information available to its own people or to others who might obtain those maps.

Today a few map dealers around the world offer the Soviet maps for sale. Customers include governments, university libraries, and telecommunication companies that need to know the topography of an area where they plan to set up cell phone towers. The United States intelligence community used some of these Soviet-made maps to aid our operations in Afghanistan in the early 2000s.

Of course, the United States had detailed maps of the Soviet Union as well. We gathered information with technology such as reconnaissance planes and spy satellites, and we had our share of spies on the ground in Russia, too. But the information our government sought was harder to come by because the Soviet Union was a closed society.

This 1972 image was taken by a U.S. spy satellite of an area near Moscow.

Lesson 6 - What's a Map Worth to You?

A Long Tradition

The Soviet map project is just one example of the ways that maps provide vital information. The tradition of cartography has a long history. One of the oldest maps we know about is on a clay tablet from what is now northern Iraq. It dates from about 2300 B.C. The map shows waterways, mountains, and communities where people lived. A little later in history, Babylonians in what is now southern Iraq began producing maps. They developed a technique that we still use today. They divided a circle into 360 degrees. This is the basis for the lines of latitude and longitude we use today.

Elsewhere in the world, Egyptians were making maps before 1000 B.C. Egypt also developed techniques that enabled them to survey property boundaries accurately.

Scholars in the Greek civilization thought a great deal about the size and shape of the earth. Around the time of Christ, Greeks developed systems of map projections, ways to represent on flat sheets the reality of a round earth. One of the most influential geographers in history was Ptolemy (90-168 AD), a Greek who studied in Alexandria, Egypt, for many years. About 150 AD, he published the eight-volume work *Geography*, which gave extensive instructions about drawing maps. On one map he drew, Europe, Asia, and the Mediterranean Sea were larger than they really are, and his estimate of the size of the earth was smaller than it really is. Despite these errors, Ptolemy's work was influential for centuries. About the same time, Chinese astronomer Zhang Heng developed a grid system for maps, which made drawing maps and finding locations on maps easier.

In the thousand years from about 400 to 1400 AD, most advances in mapmaking took place among the Arabs and Chinese. The work of Ptolemy was translated into Arabic in the 800s. The Arab scholar al-Idrisi (c. 1100-c. 1165) was born in Morocco but moved to Sicily in 1145 and served in the court of the Sicilian king. He drew a rectangular map of the world made up of seventy sheets pieced together. His map wasn't very accurate, but it reflected the

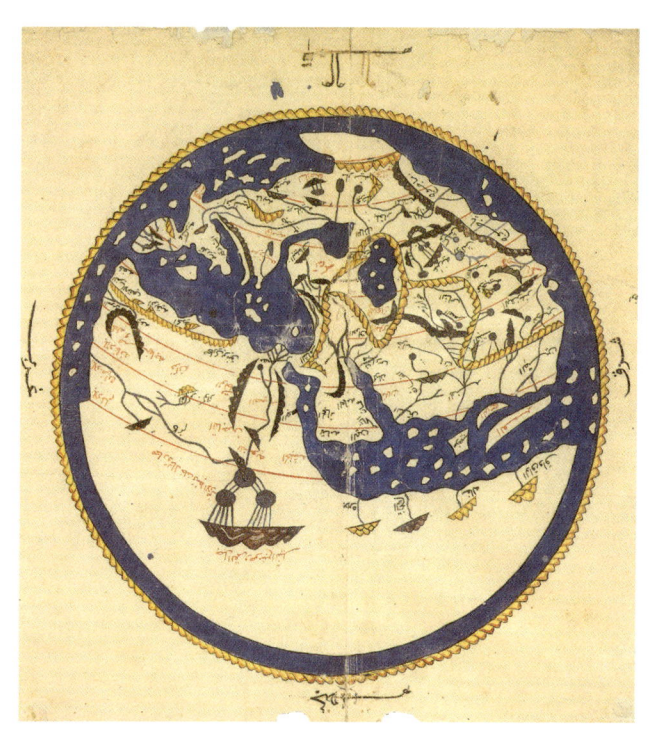

This world map is from a 1456 copy of al-Idrisi's work. This image is rotated 180 degrees so you can more easily identify Africa, Europe, and Asia.

interest people had in representing the world in map form. Al-Idrisi understandably placed the Arabian Peninsula at the center of his map, and he drew Africa on the upper half, which meant that north was pointing down. Around the same time, someone in China drew a map on a stone that showed coastlines, rivers, and settlements. China produced the first printed map about 1155, some three hundred years before the first printed maps appeared in Europe.

Medieval Maps

In the Middle Ages in Europe, a common map form was the "T and O" (see the example on the next page). This map reflected worldview more than it did geographic reality. The T and O map was bounded by a circle (the "O") that represented the ocean that many thought made up the outer boundary of the world. The three known continents, Asia, Europe, and Africa, were separated by a "T". One bar of the "T" represented the Mediterranean Sea separating Europe and Africa. The other bar represented the

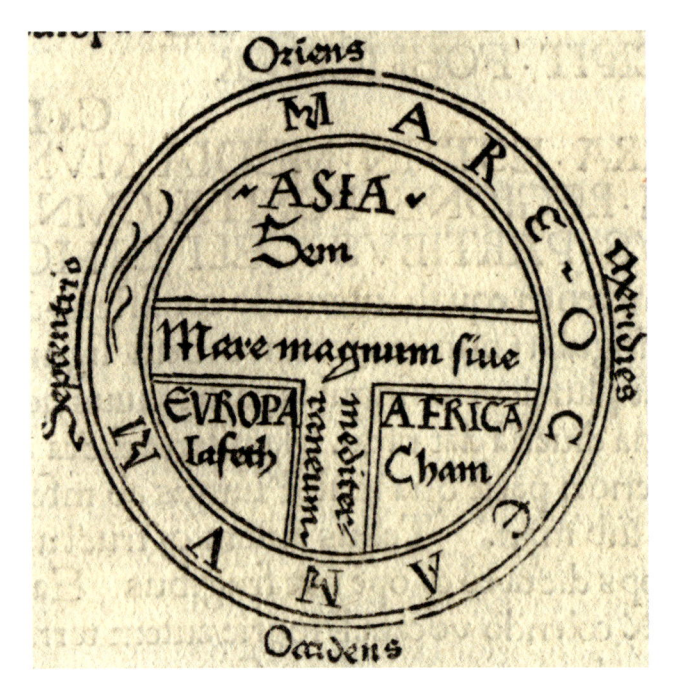

This T and O map from 1472 is the earliest known map printed in Europe.

Don River in Europe and the Nile River in Egypt, separating Asia from the other two continents. The map made no effort to portray coastlines, rivers, and other geographic features. The T and O showed Asia, which was called the Orient, at the top. This practice is the origin of the word that we still use to describe how a map is positioned: *orientation*.

As Europeans ventured further and further into the Atlantic Ocean in the 1200s and 1300s, mapmakers used the growing body of knowledge to produce *portolan* charts to help sailors with navigation, especially into harbors. These charts used grid lines and portrayed coastlines accurately. The word portolan comes from *porto*, the Italian word for harbor.

Several developments in the 1400s encouraged mapmaking. First, Ptolemy's work was translated into Latin, the language of scholarship. This enabled many European scholars to study it. Second, the printing press encouraged the wide distribution of maps and the sharing of knowledge. Third, the age of exploration reflected the growing interest in the world and furthered mankind's shared knowledge about the world. People were learning the nature of their world, but their knowledge was incomplete. Some maps showed sea monsters in areas that had not yet been explored, images that were the product of speculation and inaccurate reports.

Developments in mapmaking during this period had profound impact. In 1492, slightly before Columbus sailed west in the fall of that year, the German merchant and navigator Martin Behaim produced the first known surviving representation of the earth as a globe. The globe represented the earth as slightly smaller than it really is, and of course it did not show the American continents; but it did demonstrate that most educated people understood that the earth was a sphere.

In 1570 Abraham Ortelius, a mapmaker from Flanders, published a compilation of maps he called *Theatrum Orbis Terrarum*, or *Theatre of the World*. This was the first published atlas.

Behaim's globe, restored in the 19th century, is now housed at the German National Museum in Nuremberg.

From the New World

The exploration of the New World produced more maps. In 1607 French explorer Samuel de Champlain produced a map showing the area from

Lesson 6 - What's a Map Worth to You?

Hubbard / Foster Map of New England (1677)

what is now Maryland and the northeastern United States through what is now Eastern Canada. John Smith published a map of the Chesapeake Bay in 1612 and later rendered a map of New England.

The first map to be engraved and published in America was of New England. John Hubbard drew it, and John Foster of Boston published it in 1677. The map shows the ocean at the bottom with New England at the top. North is to the right. This was the perspective English settlers had as they approached the New World.

In the 1700s and 1800s, French mapmakers sought to make their profession more scientific. Around 1847 they began using the word *cartographie* for the science of making maps. The word comes from *carte*, meaning chart or map, and *graphie*, meaning to write or to represent. Associated with this term is the word for a person who makes maps: *cartographer*.

The 1800s were another period when interest in the geography of the world increased rapidly. As the United States expanded west, surveying was important because surveyors determined the precise location of property and political boundaries and developed intimate knowledge of geographic features. Scientific study of human activity also increased, and thematic maps showed patterns of population density, agricultural production, and other information. In 1879 Congress established the United States Geological Survey, which produced maps of the United States. The USGS is still an agency of the U.S. government today.

What We Know

Geography continues to be an important subject. One evidence of this is the fact that Prince William of the United Kingdom, second in line of succession to the British throne behind his father, Prince Charles, earned a master's degree in geography at St. Andrews University in 2005. It makes sense that the future head of state of a major country would learn all he could about the subject of geography, which plays such an important role in today's world.

Maps represent the accumulation of data compiled by countless hours of work. The form of the information has changed to some degree; numerous digital maps now accompany our trove of analog ones.

What do we know? What do our enemies know? Maps are still important to us.

And to them.

When the Lord told Moses to send spies into the land of Canaan, their mission was to find out all they could about its natural and man-made geographic features.

When Moses sent them to spy out the land of Canaan, he said to them, "Go up there into the Negev; then go up into the hill country. See what the land is like, and whether the people who live in it are strong or weak, whether they are few or many. How is the land in which they live, is it good or bad? And how are the cities in which they live, are they like open camps or with fortifications? How is the land, is it fat or lean? Are there trees in it or not?"
Numbers 13:17-20

Assignments for Lesson 6

Gazetteer — Review the collection of historic maps and read the captions with them (pages 245-248).

Worldview — Copy this question in your notebook and write your answer: What is the most basic, central, and important reality in the world?

Project — Choose your project for this unit and start working on it. Plan to finish it by the end of this unit.

Literature — Continue reading *Know Why You Believe*. Plan to finish by the end of this unit.

Student Review — Answer the questions for Lesson 6.

7 You Can Learn a Lot from a Map

Look at the map of the world at the front of the *Gazetteer*. What are some things that you notice?

You will see Europe and Africa at the center. Not everybody sees the world this way; in fact, most people in the world do not. For centuries, the Chinese referred to their country as the Middle Kingdom because they saw their civilization as the center of the world.

You will probably see that Africa is huge (twenty percent of the earth's land mass). It has larger countries in the northern part and relatively smaller ones to the south.

The southern tip of Africa to the eastern tip of Siberia in Asia used to be one huge landmass, but the digging of the Suez Canal changed that. A map doesn't tell you how the world has changed, but only what it looks like at the time of its publication.

The part of the world that gave birth to what we know as Western Civilization, from the eastern Mediterranean through Europe, is a relatively small geographic area considering the worldwide influence it has exerted.

Chile and Argentina in southern South America and Australia and New Zealand in the South Pacific are far away from the more populated areas of the world.

Brazil takes up over half of the continent of South America.

The Indian Ocean coastline is a giant arc that extends from southern East Africa around to Indonesia.

Some borders between countries are straight lines, while other borders are very irregular.

What other information do you learn from looking at this map?

People draw informal maps for many reasons. You might draw a map to show a new friend how to get to your house. Then there are maps that show the location of buried treasure on a desert island . . .

A map can tell you many things about our world. This is why having good map-reading skills is important.

The Map

Geographers have used many tools to practice their profession. These tools have included a compass, surveying equipment, a ruler, and a sextant. Geographers also conduct research and use the research of others, such as that reflected in censuses, surveys, and other sources of data. Modern geographers have many advanced tools available to them, such as computers and aerial and satellite

Members of the U.S. Air Force 2nd Space Operations Squadron work with the third-generation of GPS satellites in 2020 at Schriever Air Force Base in Colorado.

photographs. The Global Positioning System (GPS) uses information transmitted by radio emissions from a web of satellites above the earth to pinpoint a receiver's location. A Geographic Information System (GIS) gathers huge amounts of data from many sources and makes that data available for many purposes. All of these tools contribute to the geographer's most important tool, which is the map.

A map is a representation of our world or part of it. Maps help us understand our world. They

Unit 2: It Begins with a Map

show where things are, both in absolute terms (by themselves) and in relation to other things.

People have created many different kinds of maps that show different areas, use different scales of size, and have different purposes. One map you commonly see is a world map, which shows the continents and oceans and, if it is large enough, many islands. Another typical map is a political or geopolitical map, which shows borders that people have created, such as the borders of the states that make up the United States or the borders of the countries in Europe and Africa.

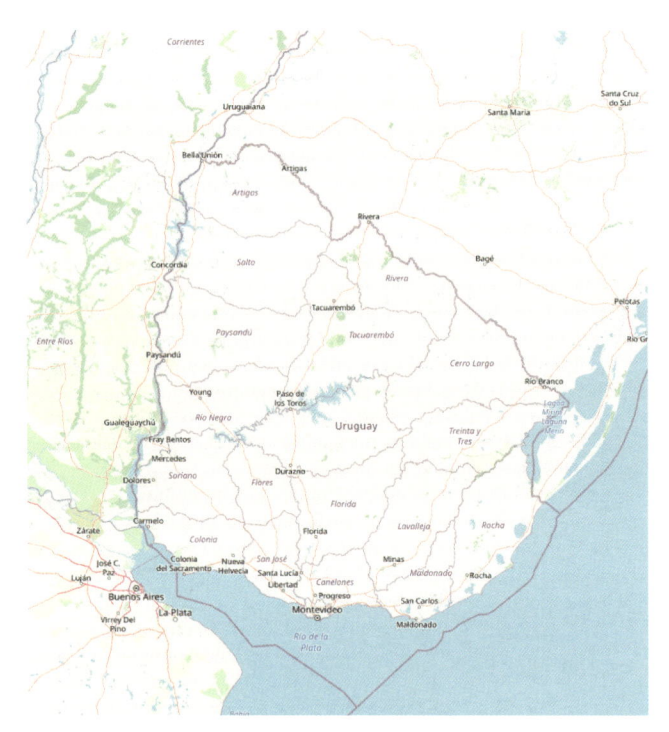

Navigational Map

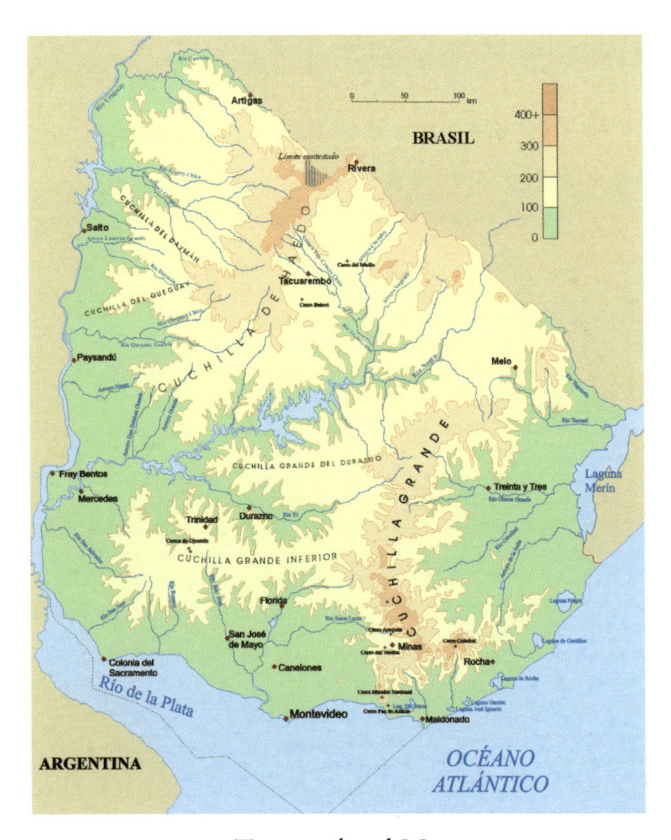

Topographical Map

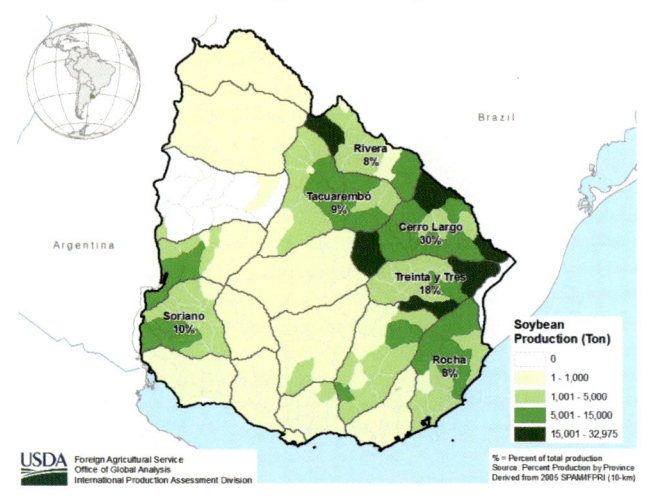

Thematic Map

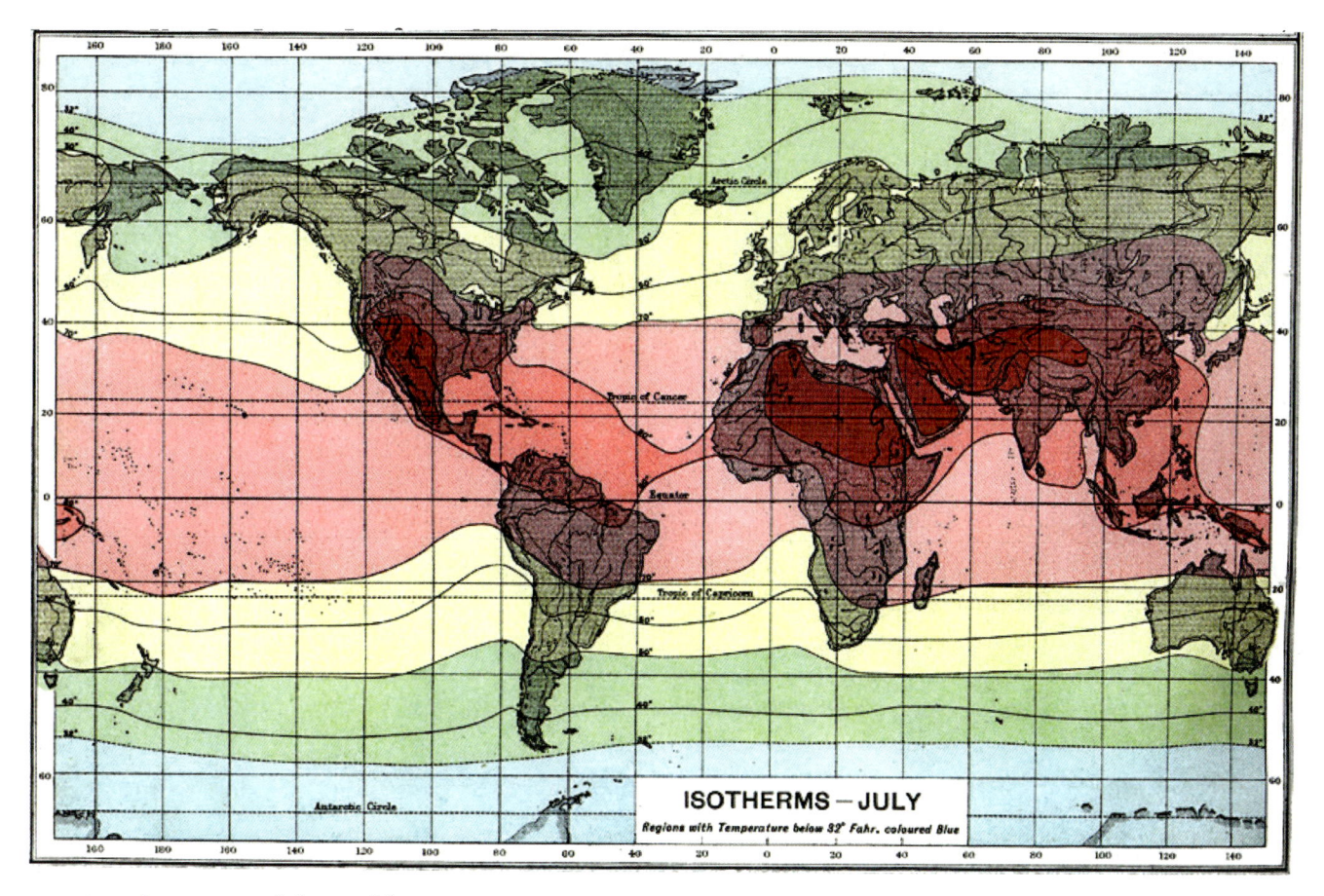

1915 isotherm map of the world

A physical or topographic map shows mountain ranges, rivers, and other elements of an area's terrain. A local map can be the map of a single state in the U.S., a county map, a city street map, or a neighborhood map.

Navigational maps help people get from one place to another. These include road maps, maps of bodies of water, and maps of a city's streets or transit system routes. Our digital and informational age has given us electronic navigational maps in our cars through GPS devices.

Thematic maps show information such as population distribution, where farmers grow certain agricultural crops, where significant oil reserves are located, and forested areas.

Weather maps include those that show current conditions, often generated with the help of weather radar, and those that show historical trends of temperatures and precipitation. Isotherms are lines that connect points that have the same temperature at a given time or the same average temperature over a period of time. Isobars are lines that connect points that have the same atmospheric pressure at a given time or the same average over a period of time.

Each of the maps we have described provides a great deal of information. Outline maps, by contrast, give a minimum of information, such as just the shapes of continents or just the political borders of states and countries. Students often complete assignments that involve adding more information to outline maps.

An atlas is a collection of maps. A U.S. road atlas might have a collection of state maps showing main highways and city maps showing major city streets. A world atlas will have maps of all the countries of the world. The same atlas might have more than one kind of map, such as topographic and thematic maps. Atlases are named for the mythical Greek god Atlas. According to Greek mythology, Atlas rebelled against Zeus. Zeus condemned Atlas to carry the sphere of the heavens (including the earth) on his shoulders.

Representation Versus Reality

Maps are invaluable tools for people in many walks of life. They are essential for industries that want to locate natural resources, for guiding military operations, for business and recreational travel, and for many other purposes. Maps represent reality for us, and we can hardly imagine getting along without them. However, maps have limitations. Just as we should understand that maps are essential, we should also recognize their limitations. Maps represent reality, but they are not reality.

The earth is a three-dimensional spheroid. The most accurate representation of the earth's surface is a globe, but even a globe has limitations. First, the earth is not a perfect sphere, so a globe will have some inaccuracies. Second, a globe is usually not large enough to show much detail. Even an accurate globe will not help someone drive across Germany from Berlin to Bonn. Third, a globe is not convenient to carry with you.

16th-century Engraving of Mercator by Frans Hogenberg

So people usually represent our spherical, three-dimensional earth with flat, two-dimensional maps. Projection is the process of transferring locations on the earth to the surface of a map. Because of this transfer, a map will always have distortions in portraying the way the earth really is. Various people have devised different methods of projection. Below are five of these methods.

Gerardus Mercator was a geographer who lived in what is now Belgium in the 1500s. In 1569 he published a world map that used his new method of projection. The Mercator Projection has probably been the most common method for drawing maps.

Mercator Projection. This projection is convenient in that it can portray the world on a rectangular sheet. Navigators at sea like this projection because a route that follows a constant direction on the compass appears as a straight line.

However, although the Mercator Projection renders distances at the equator accurately, it distorts distances as one nears the North and South Poles because the Mercator Projection stretches the surface areas near the poles to make the map rectangular. As a result, for instance, Greenland and Antarctica appear larger than they actually are. Africa appears to be smaller than it really is. A Mercator map shows Greenland and Africa as about the same size, when in reality Africa is about fourteen times larger than Greenland.

Eckert Projection. German geographer Max Eckert first proposed this projection in 1906. It renders the equator and the north-south meridian of longitude that is at the center of the map as straight lines, but it bows the other parallels of longitude outward from the center. Like the Mercator Projection, the Eckert Projection also distorts distances closer to the polar regions.

Robinson Projection. This method is similar in some ways to the Eckert Projection. It is visually appealing but it has distortions the farther one moves away from the center of the map. This projection

Lesson 7 - You Can Learn a Lot from a Map

Mercator Projection

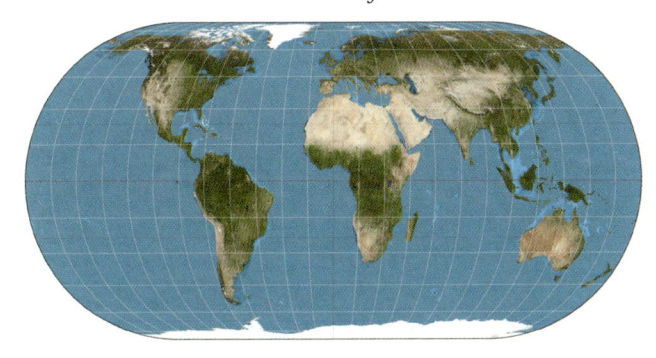

Eckert Projection

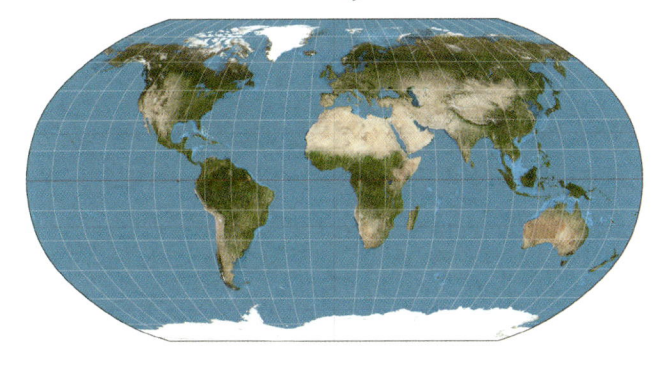

Robinson Projection

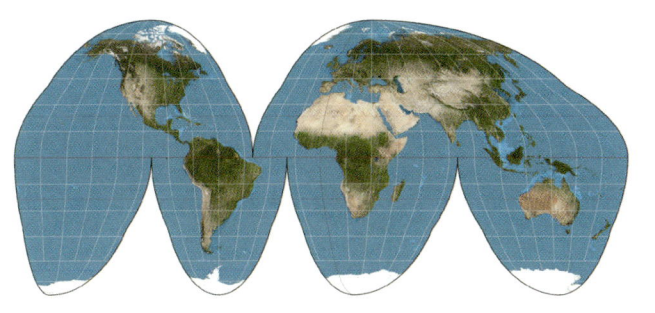

Goode Homolosine Projection

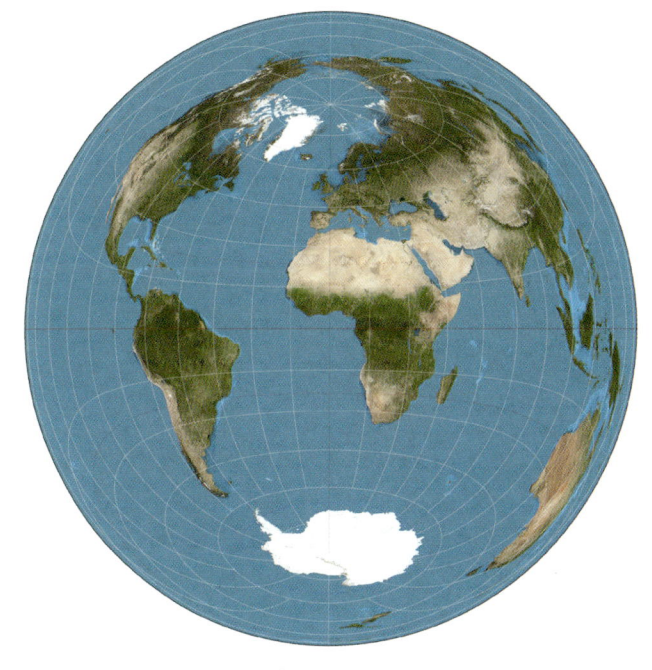

Lambert Projection

renders the North and South Poles as lines instead of points. Arthur Robinson was a twentieth-century American geographer who published this projection in 1963.

Goode Homolosine Projection. John Paul Goode, an American geographer, created this projection. Published in 1923, it shows the continents accurately but distorts the sizes and shapes of the oceans.

Lambert Projections. Swiss scientist Johann Lambert proposed several methods of map projection in 1772. One is the Lambert Azimuthal Equal-Area. It is useful for rendering the regions of the North and South Poles as circles. It is accurate at its center but distorts the land masses that are farther away from the pole.

The *Gazetteer* that is part of this curriculum uses the Universal Transverse Mercator Projection. This is a common modern projection. This method puts a cylinder around the earth's globe and slices it sixty times. The cartographer flattens each slice, which is 6 degrees wide, to produce a two dimensional map. The result is a map with less distortion than the original Mercator Projection.

Elements of a Map

Besides the method of projection used, a map has other features that the user should note.

Title. The title tells what the map represents, such as "The World" or "The U.S. Interstate System." It is important to note the title or subject of a map because you cannot get information from a map that it does not convey. You cannot use a geopolitical map of Africa, for instance, to locate fjords in Scandinavia.

Orientation. Orientation indicates the direction of north (and thus the other compass directions) on the map. North is usually at the top of a map, but not always. For instance, for a map of a state or a country to fit well on a page, north might have to be pointed toward the upper right or upper left corner. The direction of north is shown by the compass rose, which indicates the four cardinal directions (north, south, east, and west) and often some intermediate directions (northwest, northeast, and so forth).

Scale. The scale tells the user what length on the map equals a certain distance in the real world. For instance, one inch on the map might equal one hundred miles on the earth. Often a map will have a short line with numerical markings that accompanies the statement, "One inch equals approximately x miles."

Legend or Key. The legend (sometimes called the key) provides the meaning of the symbols used on the map. The key will explain, for instance, that an airplane stands for an airport, a tree indicates the location of a state park, a star shows the city that is a state or national capital, and so forth. On a thematic map, the legend will tell which colors indicate levels of population density or where farmers grow particular crops.

Grid. A grid is a system of equally-spaced, intersecting horizontal and vertical lines that create squares which are used to locate places on a map. The squares might be labeled "A, B, C," etc. across the top and "1, 2, 3," etc. down one side. Hence each square has a name, such as A1 or E5. If you are looking for a specific place, you can probably find it in a listing of places in a corner of the map or in the index at the back of the atlas. The index will say something like, "Mumbai...C3." You can find that square in the grid, and you will find Mumbai inside it. This makes finding a particular location on a map easier than if you just start scanning the map and hope that you run across what you are looking for.

Date and Author. The date of a map is important information because our world changes (see Lesson 4), and older maps might not accurately reflect the current situation. For example, maps published before 1991 will not show the countries that came into existence after the breakup of the Soviet Union. An older highway map will not show new roads that have been built since its date of publication. A city map from several years ago will not show new streets, housing developments, and industrial zones.

Lesson 7 - You Can Learn a Lot from a Map

The authorship of a map is important to know because the creator of a map can have an agenda that leads to distorted information. For instance, maps in textbooks that some Palestinian schools in the Middle East have used have not clearly identified the state of Israel.

In the next lesson we will discuss more elements of a map. We will also note what a map is not.

The psalmist praised the wonders of God's creation, which honor their Creator and which maps attempt to illustrate.

For the Lord is a great God
And a great King above all gods,
In whose hand are the depths of the earth,
The peaks of the mountains are His also.
The sea is His, for it was He who made it,
And His hands formed the dry land.
Psalm 95:3-5

Assignments for Lesson 7

Worldview Copy this question in your notebook and write your answer: How did the world come into existence?

Project Continue working on your project.

Literature Continue reading *Know Why You Believe*.

Student Review Answer the questions for Lesson 7.

Standing on the Prime Meridian

8 What a Map Is and What It Is Not

Question: *How could five hundred people lose their birthdays?*
Answer: *By government decree.*

An international conference in 1884 established the meridian of longitude that runs through Greenwich, England, as the Prime Meridian, designated as zero degrees longitude. That was an important decision, but one small matter remained unaddressed:

Where on the earth does a day start?

By custom and general agreement (but not by any official action), the meridian at 180 degrees longitude, in the Pacific Ocean on the opposite side of the world from England, became the International Date Line (IDL). A day starts there. When the time is 1:00 a.m. Friday in the time zone just to the east of the IDL, it is midnight on Saturday at the IDL.

Countries can declare their time to be whatever they want it to be, but countries generally follow the world time zones. In 1892 the island nation of Samoa moved the portion of the IDL nearest to it to the west to align its days with the United States. In more recent years, however, Samoa has developed closer ties with Australia and New Zealand, which lie to the west of the line. So in 2011 the Samoan government decided to move the IDL to its east so it could be on the same day as its largest trading partners. This meant that Samoa lost a day on the calendar. When December 29, 2011, ended in Samoa, the date became December 31, 2011. Samoans who had a December 30 birthday had to figure out another way to celebrate it that year.

Countries have nudged the IDL east or west at various times so that they would not have two days happening within their borders at the same time. As a result, the IDL does not coincide with the 180th degree of longitude from the North Pole to the South Pole. See the illustration on the next page.

All this leads us into a discussion of latitude and longitude and other matters regarding maps.

Latitude and Longitude

Geographers have developed a system of imaginary lines on the surface of the earth to determine the exact location of any point on the earth. The system of latitude and longitude is based on the fact that the distance around a complete circle can be measured as 360 degrees. Each degree of latitude or longitude is divided into sixty minutes, and each minute is divided into sixty seconds. As you read the following discussion of latitude and longitude, consult the globe map on page 50.

Lesson 8 - What a Map Is and What It Is Not

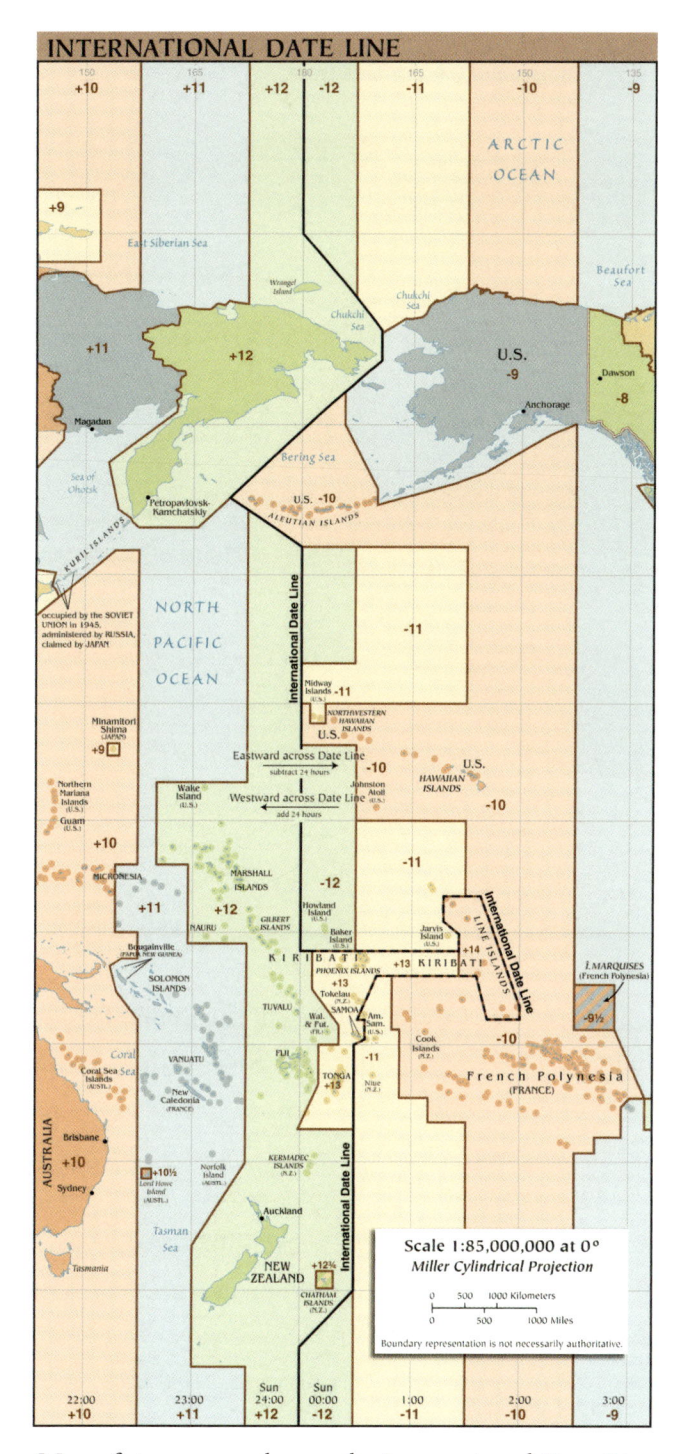

Map of time zones close to the International Date Line

points are parallels of latitude, lines parallel to the equator, which are given increasingly larger numbers as one moves north or south from the equator.

Circles of Latitude. In addition to the equator, geographers have defined four other major circles of latitude. The equator always exists at 0° latitude. Because of variations in the tilt of the earth's axis, the latitude of the other four lines changes slightly from year to year.

The Tropic of Cancer is the parallel of latitude about 23 degrees and 26 minutes north of the equator. As the earth orbits the sun, the sun shines directly down on this line on the first day of summer in the northern hemisphere (around the 21st of June). The Tropic of Capricorn is about 23 degrees and 26 minutes south of the equator. The sun shines directly down on this line on the first day of summer in the southern hemisphere (around December 21). The region between the Tropic of Cancer and the Tropic of Capricorn is the tropical zone on the earth.

The Arctic Circle lies about 66 degrees, 33 minutes latitude north of the equator. From there north, the sun does not set on the longest day of the year, around June 21, and does not rise at all there on December 21. The area between the Arctic Circle and the Tropic of Cancer is called the northern temperate zone. The area north of the Arctic Circle is considered the Arctic.

The Antarctic Circle lies about 66 degrees, 33 minutes latitude south of the equator. The sun never sets from there south on December 21 and never rises on June 21. The area between the Tropic of Capricorn and the Antarctic Circle is the southern temperate zone. South of the Antarctic Circle is the Antarctic. Almost all of Antarctica lies within the Antarctic.

Longitude. Meridians of longitude run north and south around the globe. They converge at the poles. The numerical designation of these meridians has to start somewhere. As we mentioned above, the 1884 international conference decided that the meridian of longitude that passed through the Royal Observatory at Greenwich, England, a suburb

Latitude. The equator is a line that runs east to west around the earth that is equidistant from the North and South Poles. It has the designation of zero degrees latitude. The North Pole is one-fourth of the way around the earth from the equator, so it is designated as 90 degrees north latitude. The South Pole is at 90 degrees south latitude. Between these

North Pole N 90°

- Arctic Circle
- Prime Meridian
- Tropic of Cancer
- Equator
- Tropic of Capricorn
- Antarctic Circle

Northern Hemisphere
Equator
Southern Hemisphere

Western Hemisphere Eastern Hemisphere

Example: Buenos Aires
34° 36′ 13″ S and 58° 22′ 54″ W

**-90°
South Pole S**

Latitude and Longitude

---

of London, would be the Prime Meridian, with a designation of 0 degrees. East longitude increases in number as one proceeds east of Greenwich, while west longitude increases in number as one proceeds west of Greenwich. The numbers of both east and west longitude increase until they reach the 180th meridian, which is the other side of the world from Greenwich and approximately designates the IDL.

Any spot on earth can be designated by latitude and longitude coordinates. Many geographers today use decimal degrees (DD) instead of degrees, minutes, and seconds (DMS). Latitude is always given first. For instance, the coordinates of a point in Buenos Aires are:

34 degrees, 36 minutes, 13 seconds south latitude DMS (or -34.6037 DD)

58 degrees, 22 minutes, 54 seconds west longitude DMS (or -58.3816 DD)

As a trick to help you remember which lines are latitude and which are longitude, think of east-to-west LAT-titude as the LATERAL (side to side) rungs on a LADDER. A lateral pass in football is one that goes to the side. Think of north-to-south LONGitude as the LONG legs of a ladder that go up and down.

Hemispheres. The equator and the Prime Meridian-IDL divide the earth into hemispheres (halves of a sphere) in two different ways. North of the equator is the Northern Hemisphere, while south of it is the Southern Hemisphere. The area from the Prime Meridian east to the IDL is the Eastern Hemisphere; the area from the Prime Meridian west to the IDL is the Western Hemisphere.

Of course, lines of latitude and longitude are not actually on the surface of the earth. They are used on maps and in navigation to determine location. If a ship radioed that it was marooned about 750 miles west of San Francisco, that information would not be precise enough for rescuers to know where they should go. Latitude and longitude coordinates identify a place more precisely.

Lesson 8 - What a Map Is and What It Is Not

What a Map Does Not Show

A map shows us many things and makes many things about the area on the map clearer to us. However, there is much vital information that a map does not and, in some cases, cannot show.

It cannot tell you why some neighboring countries are friends and some are enemies, nor what connections countries distant from each other have.

Unless the map is topographic it won't show you the height of Andes Mountains along the western coast of South America or that the Dead Sea is below sea level.

It will not show you how the maps of countries used to be drawn fifty or a hundred years ago.

Unless it is a population map, it will not show you that Russia is huge but that its population is not evenly distributed across its land area.

It will not show you how much of the world's ocean shipping passes through the Strait of Malacca in Malaysia.

Besides the distortions that occur as a result of projection, flat maps have other limitations that we need to keep in mind. Maps can show rivers, but they do not show the direction in which a river flows. A map might use the same size dot to indicate cities that have very different sizes and shapes.

Because humans make maps, some can have inaccurate information. An incorrectly-drawn map intended for military purposes might fail to show an important bridge or wrongly show where two roads intersect. Such errors on a typical road map would be frustrating; on a military map, it could be disastrous. It is also good to remember that the larger the area that a map shows, the less detail the map will be able to include. However, even with these limitations maps are an important tool in geography and an important part of everyday life.

You cannot understand the physical and political world without a map—actually several maps. And you cannot appreciate what the map reveals without an understanding of the human dynamics that are taking place on the map.

Understanding our world has a geographic component and a human component. That is why this curriculum is about human geography.

The island of Taveuni in Fiji lies on the 180th meridian of longitude. Though the International Date Line officially passes to the east, tourists can imagine themselves standing with one foot in today and one foot in yesterday.

Science and Art

Mapmaking is a science. It involves the collection and application of a huge amount of carefully gathered scientific information. Mapmaking is also an art. How clearly and accurately a map conveys information affects how well it helps, guides, and teaches those who use it.

Maps are the result of the work of creative people made in the image of God. Maps are remarkable inventions that illustrate the amazing world that God created and sustains.

It is good for us to use maps to know the geography of the earth and the nations that live on it. It is even more important that the nations know the God who created them.

God be gracious to us and bless us,
And cause His face to shine upon us—
That Your way may be known on the earth,
Your salvation among all nations.
Psalm 67:1-2

Assignments for Lesson 8

Worldview — Copy this question in your notebook and write your answer: How does the world operate?

Project — Continue working on your project.

Literature — Continue reading *Know Why You Believe*.

Student Review — Answer the questions for Lesson 8.

Port of Cotonou, Benin

9 The Business of Geography

Look again at the map of the world at the front of the *Gazetteer*. Two of the greatest physical safeguards for the United States are the Atlantic and Pacific Oceans. France and Germany can't choose whether they are going to be involved in European affairs because they are part of Europe. But the United States has been able to decide to take part or not take part in matters elsewhere on the globe because we are separated from Europe and Asia by those oceans.

Look at Russia, the largest country in the world. For all its size, Russia does not have a year-round port on either the Atlantic or Pacific Ocean. This explains its obsession with having access to warm-water ports, such as on the Black Sea. Three-fourths of the Russian population lives in the European part of the country, while three-fourths of its land area is in the Asian part of the country. The challenges of governing a country with this geographic makeup are tremendous.

Look at Africa. The continent has few natural harbors. The rivers that flow to the sea descend from a central highland and have falls that are impassable for trading ships. Southern Africa is a long way from the traditional trade routes between North America and Asia and North America and Europe. The geography of Africa made trade and interchange with other cultures difficult for centuries.

A map shows what locations are close to each other and which ones are far apart. Physical maps show mountain ranges that have been barriers to travel, trade, and cultural interchange. Maps show where major cities such as New York and Shanghai lie.

Maps tell us an abundance of information about the geographic, political, and demographic features of the earth. And that's only the beginning.

The Story of a Map Company

It started with a print shop. The business grew by being in the right place at the right time.

William Rand opened a print shop in the Loop area of downtown Chicago in 1856. Rand offered to do any and all kinds of printing for the public. Two years later, Rand hired an immigrant from Ireland, Andrew McNally, at the high rate of $9 per week.

In 1868 the two men reorganized their business as the partnership Rand McNally and Company. That same year, they took over the print shop work of the *Chicago Tribune* newspaper. But the big step for Rand and McNally that year was their receiving the contract to print the tickets and timetables for the rapidly expanding railroad industry in Chicago. Chicago was the key railroad hub for the Midwest. The Transcontinental Railroad would be completed the next year, and the demand for printing by railroad companies became tremendous. Rand McNally published the first *Western Railway Guide* in 1869. The next year, Rand McNally expanded its publishing with business guides, an illustrated newspaper, and additional railroad guides.

Then in 1871, the company was in the wrong place at the wrong time when the Great Chicago Fire destroyed the city. The two business owners did what they could to save the company. They rescued two ticket printing machines and buried them in the sand by Lake Michigan. Three days later, the company was back in business in a rented location.

Rand McNally achieved a milestone in 1872. The company published its first map in the December edition of the *Railway Guide*. A new printing method greatly reduced the cost of printing maps, and the company's future course was set. The two partners incorporated their business in 1873. William Rand was president, and Andrew McNally was vice president.

Public schools were becoming an increasingly important part of American life, and again Rand McNally was in the right place at the right time. In 1880 they began publishing geography textbooks, maps, and globes for schools.

The company experienced another major transition in 1899 when William Rand left to pursue other ventures. Andrew McNally became president, and the McNally family ran the company for the next century.

Then Came the Automobile

The growth industry during the first quarter of the 1900s was automobiles, and Rand McNally was there. They published a road map of New York City and vicinity in 1904. Three years later, Rand McNally took over publication of the Photo-Auto Guides from another company. These guides

Rand McNally published this guide to help visitors who came to Chicago for the 1893 World's Columbian Exposition from around the world.

combined maps with photographs of intersections overlaid with arrows to show correct turns to help drivers be familiar with those intersections and know what to do when they got there. Andrew McNally's grandson, Andrew McNally II, took the pictures for the Chicago-to-Milwaukee edition while on his honeymoon. Talk about loyalty to the business! (We don't know whether the new Mrs. McNally approved of this activity.)

As automobiling became a national pastime, drivers wanted maps to guide their travels. Most roads were known by names, such as Baker Road or Stubblefield Road. Some roads had long names, such as Lexington-to-Louisville Highway, that were

Lesson 9 - The Business of Geography

challenging to print—and read—on maps. Rand McNally began using geometric symbols for roads. On a 1917 map of Peoria, Illinois, the company began using numbers to designate main highways. This was a major step toward the numbering systems that state, federal, and Interstate highways use today. Again, right place, right time. (By the way, another American company, Thomas Brothers Maps in Oakland, California, developed the grid system for its maps that is still used today to make locating specific places on maps easier). The year 1924 saw the publication of the *Rand McNally Auto Chum*, which was the first edition of what would become the *Rand McNally Road Atlas*.

Rand McNally published the first edition of *Goode's World Atlas* in 1923. The editor was Dr. J. Paul Goode, a geography professor at the University of Chicago. It became the standard atlas for American high school and college students. Rand McNally has now published the 23rd edition of *Goode's World Atlas*.

The company's maps were in demand in some unexpected ways. In 1927 aviator Charles Lindbergh used Rand McNally railroad maps to help him navigate over land during his historic Transatlantic flight. In 1939 stores across the country sold out of Rand McNally's map of Europe within 24 hours of Germany's invasion of Poland.

Norwegian scientist and explorer Thor Heyerdahl developed a theory that South American natives traveled west across the Pacific Ocean and settled the Polynesian Islands. In 1947 Heyerdahl navigated a raft for three months from Peru to Polynesia. Rand McNally took a chance on publishing his account of his journey, *Kon-Tiki*, in 1948. It became an international best seller and is still in print today.

The company has pioneered several advances in printing, maps, and electronic navigation. Rand McNally developed the first pressure-sensitive tickets for the railroad and airline industries in 1958, eliminating the need for carbon copies. The first full-color Rand McNally Road Atlas appeared in 1960. The 1993 edition was the first Rand McNally product printed with all-digital technology.

During the 1980s, the company began developing electronic routing and mileage systems for truckers. Today Rand McNally is a leading producer of GPS tablet systems for the routing and log-keeping needs of the trucking industry. In the 1990s, the company developed trip planning and street navigation software for home computers. In the early 2000s, Rand McNally created GPS devices especially for RVs.

Rand McNally products for sale in 2019

In 1997 the McNally family ended its ownership of the company when they sold the business to a private investment firm. However, the company continues to develop and provide numerous map and travel-related products. These include framed and decorative wall maps, large maps that cover an entire wall, state atlases and gazetteers, fabric maps that don't require careful folding and that can be used to wipe up spills and clean your eyeglasses, street maps for selected American and foreign cities, pull-down maps for classrooms, globes (including illuminated models), several products for children including trip activity books and a kids' road atlas, and even a globe Christmas ornament. The company publishes a new edition of the *Rand McNally Road Atlas* (in several sizes and versions) every year. Randmcnally.com offers a blog with travel information.

And to think it all started with a print shop in downtown Chicago just before the Civil War.

In life, as in geography, we need someone to show us the way. Jesus does that for us.

Thomas said to Him, "Lord, we do not know where You are going, how do we know the way?" Jesus said to him, "I am the way, and the truth, and the life; no one comes to the Father but through Me."
John 14:5-6

Assignments for Lesson 9

Worldview — Complete this statement in your notebook: I am a Christian; therefore . . .

Project — Continue working on your project.

Literature — Continue reading *Know Why You Believe*.

Student Review — Answer the questions for Lesson 9.

Margaret Thatcher and Gerald Ford (1975)

10 — Recipe for a Worldview

Conservative British prime minister Margaret Thatcher once said, "The facts of life are conservative." She believed that the nature of the world and of human relations were the basis for her political stance. In her view, the world didn't operate by principles that changed with the whims of public opinion and human "progress." To Thatcher, certain truths were rock solid and unchangeable, recognized as true from ancient times. Such truths might include principles such as these:

The world doesn't owe you a living.

Prudence is a better policy than recklessness.

You reap what you sow.

Elements of a Worldview

Your worldview involves what you believe about the big issues of life and how you apply those beliefs to specific questions. The concept of worldview might seem huge and somewhat ill-defined to you, so it helps to break down your worldview into specific elements. Your worldview involves how you answer these questions:

1. What is the most basic, central, and important reality in the world?

Most people you know probably know that this basic reality is God. The majority of people through history have recognized a divine power. A related aspect of your answer to this question is determining what you believe about the attributes or characteristics of our heavenly Father whom you believe in.

2. How did the world come into existence?

The two most common answers to this question are that (1) God brought the world into existence or that (2) the world came about through purely materialistic forces.

People have also held other ideas about the creation of the world. Some people have believed that the world came into being as the result of a cosmic clash between gods or between the forces of good and evil.

3. How does the world operate?

Does the world operate on the basis of God's will? Natural law? Men's decisions? Happenstance? Luck?

What is the nature of the world in which we live? Is it a proving ground for heaven? Is it meaningless? Is it stacked against you? Is it a common playing field in which everyone lives, with each person

57

Khōn is a form of drama in Thailand that combines music, singing, dance, and ritual. It builds upon a literary tradition that reflects religious beliefs and cultural values.

---

having some set of advantages and disadvantages? Is it a world in which our personal decisions and actions are key to the kind of life we live?

Is ours a fallen world? People talk about "the fall of Adam" and about living in a "fallen world," but the Bible does not use either of those phrases. Certainly the sin of Adam and Eve had consequences for them and for all of mankind, as Genesis 3 and Romans 5 teach. In Romans 8, Paul talks about the world groaning and one day being set free from its slavery to corruption.

Even in a world where people sin, unredeemed parents love their children. Non-Christians pay their bills on time and obey traffic laws. Good happens even in a sinful world. What you think about our world naturally influences your worldview. What does "fallen world" mean? What is your understanding of the nature of our world? What are the consequences of that view? Why do you believe as you do?

4. **Is Satan real, and if so how does he operate?**

In other words, what are we up against? The Bible describes God as the ultimate force for good in the world and that Satan is the ultimate force for evil. However, the Bible does not describe God and Satan as having equal power. The Bible teaches that, in the current world, Satan is real and that his work affects people in real ways. The Bible also says God will defeat Satan in the end.

Many people in the world, even if they believe in God, believe that they also live subject to the actions of evil spirits. These people often believe that evil spirits are agents of Satan and that the actions of these spirits explain certain events.

5. **What is the meaning and purpose of life?**

Why are we here? Why are you here? Is the whole universe a mere accident? Have you been given life to live for yourself, or have you been given life to serve others and to contribute something good to the world around you? Are you here to honor God?

Paul said that the meaning and purpose of life centers in Christ:

Lesson 10 - Recipe for a Worldview

> For by Him [Christ] all things were created, both in the heavens and on earth, visible and invisible, whether thrones or dominions or rulers or authorities—all things have been created through Him and for Him (Colossians 1:16).

In Philippians 1:21, Paul wrote, "For to me, to live is Christ and to die is gain." Paul found his life, his purpose, and his goal in Christ.

It is important that you understand the meaning and purpose of life in general and of your life in particular. If you don't, you will have a hard time knowing whether what you are doing is what you should be doing. When you get to the end of your life, how will you know that you have followed the right path and have arrived at the right place?

Your worldview involves other questions that we will discuss in later lessons; for instance:

6. What is truth, and how do you know it?

7. What is faith?

8. How do you see yourself and others?

9. What is a Christian perspective on the environment?

10. What is the impact of geography on a person's worldview?

11. Where are we headed in our individual lives and as humanity in general?

So . . .

What is your worldview? What do you believe is true? What is the motivation that drives you? Since the vast majority of students who use this curriculum come from a Christian point of view, the question before you is, "How am I as a Christian supposed to think and live?" In other words, how do you complete this sentence: "I am a Christian; therefore . . ."? The way you answer that question will give you a window onto your worldview.

All these worldview questions fit together and have an impact on your decisions and actions. For instance, what you believe about where we came from influences the value you place on other people. How you believe the world operates influences the effort you will make to accomplish things in your life. Your belief about truth will have an impact on whether you act on consistent principles or whether you adapt your actions to given situations. What you believe about where we are headed will impact whether you believe you will be accountable for your actions.

The main purpose of our survey of worldviews in this curriculum is not to fill your head with a few ideas

This photo of Cairo shows pyramids in the background and a mosque in the foreground.

about several different religions and philosophies. The main purpose is to help you identify your own worldview and to help you see your place in the big picture of our world. We believe that the world has a purpose: to glorify God.

What is your place within that grand purpose? The psalmist David pondered man's place in the universe and expressed the amazement and joy that he felt on the basis of his worldview. It is a worldview that I share. David said in Psalm 8:

O Lord, our Lord, how majestic is Your name in all the earth,
Who have displayed Your splendor above the heavens!
From the mouth of infants and nursing babes You have established strength
Because of Your adversaries, to make the enemy and the revengeful cease.
When I consider Your heavens, the work of Your fingers,
The moon and the stars, which You have ordained;
What is man that You take thought of him,
And the son of man that You care for him?
Yet You have made him a little lower than God,
And You crown him with glory and majesty!
You make him to rule over the works of Your hands;
You have put all things under his feet,
All sheep and oxen, and also the beasts of the field,
The birds of the heavens and the fish of the sea,
Whatever passes through the paths of the seas.
O Lord, our Lord, how majestic is Your name in all the earth!
Psalm 8

Assignments for Lesson 10

Worldview   Write or recite the memory verse for this unit from memory.

Project   Finish your project for this unit.

Literature   Finish reading *Know Why You Believe*. Read the literary analysis and answer the questions in the *Student Review Book*.

Student Review   Answer the questions for Lesson 10.
Take the geography quiz for Unit 2 in the *Quiz and Exam Book*.

Old City, Jerusalem

3 The Middle East Part 1

The Middle East is a geographic area of the world that has drawn significant attention for decades. The process which created several countries in the Middle East involved attitudes that reflected Western colonialism. The Israeli-Palestinian conflict may be the most difficult geopolitical issue in the world today. The Kurds have a strong ethnic identity but do not have a home country. The worldview lesson is the first of several lessons that survey various faith systems; this lesson is about Judaism.

Lesson 11 - The Physical Geography of the Middle East
Lesson 12 - Drawing Lines: The Making of the Modern Middle East
Lesson 13 - The Toughest Geographic Issue
Lesson 14 - "We Have No Jam": The Saga of the Kurds
Lesson 15 - Faith System: Judaism

Memory Verse Memorize Genesis 21:12-13 by the end of the unit.

Books Used The Bible
Exploring World Geography Gazetteer
Blood Brothers

Project In this unit, we are only giving one option for the project. We believe that each student should complete this assignment.

Write a 250-300 word essay setting forth your worldview. Answer the questions in Lesson 10 the way they apply to you. Include anything else you believe is important for others to know about your worldview.

Literature *Blood Brothers* tells the story of the creation of the state of Israel and the displacement of the region's Palestinians through the life story of author Elias Chacour. Chacour's eyewitness perspective places the reader in the midst of history as it unfolds. More than an autobiography, *Blood Brothers* helps readers gain an accurate and well-rounded understanding of these controversial events and the continuing conflict in the region. The book is not angry or vindictive, but upholds the Christian message of forgiveness and respect for all people. Chacour's personal experience and sacrifice for what he believes present a perspective worth hearing for the thoughtful reader seeking to better understand Israelis, Palestinians, and all people.

Elias Chacour was born in 1939 into a Palestinian Christian family that had lived for generations in Biram, a village of Galilee. His family experienced violent eviction from their home when the state of Israel was created in 1948. Chacour became a priest in the Melkite Catholic Church and has spent his life serving in his home region. His message and life work are characterized by service, reconciliation, and cooperation. He founded Mar Elias University and schools, providing education for thousands of students of all faiths in an effort to help them live and work together in peace.

Plan to finish *Blood Brothers* by the end of Unit 4.

Farmland in Syria

11 The Physical Geography of the Middle East

No other region of the world brings together the factors of history, religion, culture, geography, politics, and natural resources in such a dramatic way as does the Middle East. The Middle East is where Asia, Africa, and Europe meet. In ancient times it was the region where large and influential civilizations and kingdoms arose. Today it is a place of much controversy and conflict.

What Is the "Middle East"?

The term Middle East refers to the region that extends roughly from the Black and Caspian Seas in the north to the southern tip of the Arabian peninsula.

In their definition of the Middle East, some geographers and observers include Iran and the countries along the Mediterranean coast of Africa since they are predominantly Muslim. The entire region is sometimes abbreviated MENA, for "the Middle East and Northern Africa."

Egypt and Iran have often been involved in Middle Eastern affairs. However, in this curriculum, we will study Egypt in our unit on North Africa and we will study Iran in our unit on Southern Asia.

Just as the Middle East is filled with controversy, the very term "Middle East" is a controversial one. In the nineteenth century, people in Europe and North America commonly divided "the East" or the Orient (the lands east of Europe) into three smaller regions: the Near East, the Middle East, and the Far East. The Near East referred to the former Ottoman Empire and the Balkan Peninsula. The Middle East referred to the area from the Persian Gulf to Southeast Asia. The Far East described the countries in Asia that bordered the Pacific Ocean.

A once-common term for the region that people rarely use today is the Levant. This term is from the French word for "rising." It refers to the lands in the direction of the rising of the sun from the perspective of Europe.

In the period before World War II, the British military began referring to the Near East and Middle East together as the Middle East. Today we see the Middle East as mostly what was once called the Near East, except for the Balkans. India and the countries around it are called Central Asia or Southern Asia. The Far East is now simply called Asia.

However, the term does betray an assumption of looking at the region from the West. The Middle East is east of what? East of where "we" are, of course. A more neutral term for the region is Southwest Asia.

The Fertile Crescent

The Fertile Crescent describes an arc of land from Iraq or Mesopotamia on the east across Syria (once Assyria) along the eastern coast of the Mediterranean Sea, and usually including Egypt. The Fertile Crescent lies south of the Armenian highlands and north of the Arabian Peninsula. It follows a major east-west travel route. In ancient times, people traveling between Canaan/Palestine and Mesopotamia and Persia followed this crescent-shaped route rather than going through the arid Arabian desert.

American scholar James Henry Breasted coined the term in 1916 to describe the area where land is comparatively more fertile than the desert areas nearby. It also refers to the cultural fertility of the region in bringing forth civilizations early in history that had (and still have) a major impact on the world. These civilizations include the Sumerians, Babylonians, Egyptians, Assyrians, Israelites, Phoenicians, and the empire of Alexander the Great and his successors.

Agricultural and cultural fertility influenced each other. The fertile land encouraged the growth of city-states, while the growth of technology—such as the Sumerian yoke, wheeled cart, irrigation, and mathematics—enabled better agriculture. The eastern end of the Fertile Crescent has also been called the Cradle of Civilization because from here arose some of the earliest great civilizations of the world, such as the Chaldeans and later the Babylonians.

The Middle East Today

The countries of the Middle East cover about 2.8 million square miles and contain about 425 million people. Three-fourths of the people are Arabs, speak Arabic, and practice Arabic customs. The term Arab refers to people whose ancestors were from the Arabian Peninsula. Other languages widely spoken in the region are Hebrew (especially in Israel), Kurdish, Persian, and Turkish.

Ninety percent of the people in the Middle East are Muslims. In Israel, 80 percent of the people are Jewish. Although Christianity began in the region, Christian groups are the decided minority and are mostly groups identified with particular nations, such as Coptic Christians in Egypt, the Maronite church in Lebanon, and Orthodox believers in Turkey and some other countries. There are other smaller groups, including Catholics, Protestants, and some that reject any denominational label.

The Middle East is a place of conflict. You might first think of the Arab-Israeli conflict, but that is only one of the continuing sources of tension. Political and social unrest have marked several Arab countries in recent years as many people have wanted to see a change in government and society. Terrorist attacks that can be traced to Middle Eastern countries have affected countries around the world. Iran and Iraq fought a long and costly war in the 1980s. The civil war in Syria has resulted in death, destruction,

This map of the Fertile Crescent highlights the locations of some ancient civilizations.

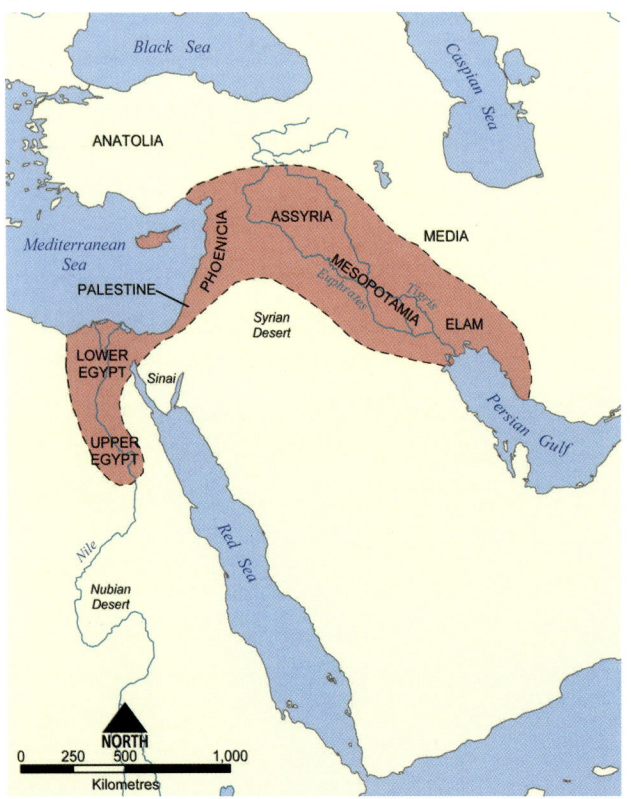

Syrian Refugees Seeking Asylum at the Turkish Border (2011)

instability, and tens of thousands of refugees who seek shelter in Europe and other countries, including the United States.

It is easy to think that the geographic characteristics of a particular region or country have always existed as they do today, but that is not the case. For instance, a key factor in the Middle East that impacts the entire world today is the presence of oil, but this resource played no role in the region in ancient times. Some of the political boundaries that exist between countries today are the result of modern diplomacy and do not reflect real divisions among ethnic groups. The cedar forests we associate with Lebanon have all but disappeared because of their being harvested by people over the centuries. Palestinians have dug tunnels for moving people and goods between the Sinai Peninsula and Gaza, with some extending into Israel. These have become a major security issue in the Palestinian-Israeli conflict.

The Geography of the Middle East

When you hear the term "the Middle East," you might think of one geographic term: desert. Much of the Middle East is desert, but the region is much more geographically complex than that. Through the region flow many rivers, including the Tigris, the Euphrates, and the Jordan. Seas that border or are within the region are the Mediterranean, the Red Sea, the Dead Sea, the Persian Gulf, the Caspian Sea, and the Black Sea. Mountain ranges include the Taurus, the Zagros, the Elburz, and the Caucasus. Many mountains rise 10,000 feet above sea level and more. Much of the Middle East is arid, but there are many areas that have productive agricultural activity. Among the crops produced are olives, spices, dates, figs, pomegranates, and coffee.

Most of the region receives little rainfall, and almost all of that usually comes in limited periods during the year. For instance, in Palestine the "early

The city of Nablus in Palestine is located between Mount Ebal and Mount Gerizim.

rains" usually begin in late October and help farmers plant their crops successfully. The "latter rains" that come in April mature the growing grain and help bring about a successful harvest (see Deuteronomy 11:14). Although rain is scarce, heavy dew forms along the seacoast and on the western slopes of mountains. The Bible often connects the dew and the rain (see, for instance, Deuteronomy 32:2, 2 Samuel 1:21, 1 Kings 17:1, and Job 38:28). Sea breezes during the day help farmers separate the chaff from the wheat.

Palestine/Canaan, Aram (Lebanon and Syria), and Jordan

The area that lies along the eastern coast of the Mediterranean Sea has four distinct geographic regions. Along the coast lie the Coastal Plains. In Palestine there were no good natural harbors in ancient times; Phoenicia to the north (modern Lebanon) had better harbors. However, Canaan had better farmland on its plains than did Phoenicia, where some mountain ranges extend to the coast. Because of this, the Phoenicians needed to trade with other countries to have the food they needed.

The next region to the east is made up of the Central Highlands. These extend from Syria to south of ancient Judea. Here we find Mt. Gerizim and Mt. Ebal, where the Israelites recited the Lord's blessings and curses when they entered the Promised Land (Deuteronomy 11:29). In this region lie the hills of Galilee, Samaria, and Judea. Here also is the city of Jerusalem. In Palestine the Jezreel Valley breaks up the highlands.

To the east of the Central Highlands is the Great Rift Valley, part of which is the Jordan River Valley. Geographers trace this valley as beginning in Turkey and continuing on to the continent of Africa all the way to Kenya. In this valley are the Sea of Galilee, about 700 feet below sea level, and the Dead Sea, the lowest point on the face of the earth at 1,242 feet below sea level.

The fourth region is a plateau east of the Jordan River, which includes the ancient Decapolis area as well as the modern country of Jordan

The following description demonstrates the variations among these four regions. From the Mediterranean coast to Jerusalem is only 30 miles, but the route ascends to 2,500 feet above sea level.

From Jerusalem east to Jericho is only 17 miles, but Jericho is 1,200 feet below sea level. Going east from Jericho to Amman, Jordan, 25 miles further, the traveler rises to 3,000 feet above sea level.

In the modern country of Lebanon, the Lebanon Mountains rise east of the coastal plains. A few miles to their east are the Anti-Lebanon mountains, which terminate in the south at Mt. Hermon. Between these ranges lies the Bekaa Valley, a fertile area where much vegetable farming takes place. The Anti-Lebanon range forms most of the border between Lebanon and Syria.

Arabian Peninsula

The Arabian Peninsula perhaps best fits the common perception of the Middle East as all desert. Ninety-five percent of the peninsula is desert. Geographers trace a large swath of desert starting at the Atlantic coast in North Africa, crossing Arabia, and continuing into the Gobi Desert in Mongolia.

The Al-Rab' al-Khali (Arabic for "empty quarter") is the largest contiguous sand desert in the world. It covers about 250,000 square miles, or about the size of Afghanistan, on the southern part of the Arabian peninsula. Few people live in the Empty Quarter. By comparison, Texas is about 268,000 square miles in size. It has large areas where no one lives, but it still contains over 27 million people.

The Empty Quarter covers about one-fourth of the land area of Saudi Arabia; but it also extends into Yemen, Oman, and the United Arab Emirates. The elevation in this desert varies from about 2,000 feet above sea level in the west, where the sand is fine and soft, to about 600 feet elevation in the east. There the desert has sand dunes, sand sheets (gently undulating areas of sand), and salt flats (large flat areas of salt).

The region has seen significant economic development because petroleum explorers have discovered major oil fields underneath it. Middle East desert and Middle East oil come together in the Empty Quarter.

The Rub' al-Khali ("Empty Quarter") Desert in Oman

The Gulf of Aqaba is to the east of the Sinai peninsula. It touches Egypt, Israel, Jordan, and Saudi Arabia.

Iraq

The term Mesopotamia is from the Greek and means "between the rivers." Most of the region of ancient Mesopotamia is now in Iraq, though some of it lies in modern Turkey and Syria.

The Tigris and Euphrates Rivers are the principal features of Iraq. The Euphrates is the longest river in Asia, extending 1,800 miles from its source in the Armenian highlands in Turkey. It cuts through the Taurus Mountains in Turkey on its way to the sea. The Tigris also begins in the Armenian highlands. It is 1,150 miles in length and carries more water than the Euphrates, although the Euphrates is more erratic in reaching flood stage. The two rivers join about sixty miles above the Persian Gulf to form the Shatt-el-Arab, which flows into the gulf.

Temperatures can exceed 120° F in the summer. Since the country receives only about eight inches of rain per year, and that is only during the rainy season, irrigation from the rivers is vital to agriculture. Iraq is a major producer of dates.

The Bible frequently uses geographic features, such as those in the Middle East, to convey its message.

[Brothers dwelling together in unity] is like the dew of Hermon
Coming down upon the mountains of Zion;
For there the Lord commanded the blessing—life forever.
Psalm 133:3

Assignments for Lesson 11

Gazetteer Study the entries for the Middle East and Israel (pages 1 and 8).
Read the photo essay on Cappadocia (pages 249-251).

Worldview Copy this question in your notebook and write your answer: What is your understanding of God's relationship with the nation of Israel, including the Jews of our day?

Project Start working on your worldview essay. Plan to finish it by the end of this unit.

Literature Begin reading *Blood Brothers*. Plan to finish by the end of Unit 4.

Student Review Answer the questions for Lesson 11.

Highway Sign in Jordan

12 Drawing Lines: The Making of the Modern Middle East

The Islamic State terrorist stared into the camera for the propaganda video and jubilantly exclaimed, "We are putting the last nail in Sykes-Picot."

Who—or what—is—or was—Sykes-Picot, and why does the Islamic State want it—or him—dead?

The answer lies in human geography, history, and Middle Eastern ethnic politics.

Look at the map of the Middle East on page 1 in the *Gazetteer*. Do you notice the many straight borders that separate countries? Did you ever wonder how borders could be so straight in a land of varied geographic features and diverse ethnic groups? You're about to find out.

The Background

In 1916 World War I was raging in Europe. Britain, France, and Russia were arrayed against Germany, Austria-Hungary, and the Ottoman Empire. The Muslim Ottoman Empire had ruled the Middle East since 1453. The Empire held Syria, Iraq, Lebanon, and Palestine. However, by 1916 its control over these areas was weak.

Great Britain, France, Germany, and other European countries had pursued the policy of establishing colonies in many parts of the world. European governments saw this practice as a way to increase their economic power and diplomatic prestige in the world. In other words, the more geography they controlled, the greater their power and prestige.

Great Britain and France had pursued their interests in the Middle East for many years. France had investments in Syria. Great Britain wanted to have secure, reliable access to India through the Suez Canal and the Persian Gulf. Now they saw practical reasons for exerting their influence. One reason was that war was taking place, and they wanted to defeat the Ottomans. Another reason was that, as the industrial and automotive age developed, oil from the Middle East was becoming increasingly valuable to the industrialized countries of Europe.

The Deal

In 1915 Great Britain made a promise to the Arab peoples: help us defeat the Ottomans, and we will support your desires for an independent country. The Arabs fought alongside the British with this understanding.

Sir Mark Sykes, 6th Baronet, was a British army veteran, writer, member of Parliament, and

Lesson 12 - Drawing Lines: The Making of the Modern Middle East

diplomat. Francois Marie Denis Georges-Picot was a French attorney and diplomat.

Sykes and Picot (with a representative from Russia present as well) began meeting secretly to decide how they might divide up a defeated Ottoman Empire in order to strengthen and secure their respective nations' influence in the Middle East and to maximize what they could get out of the region. The two men took a marker and a ruler and drew lines on a map to divide up the region between them. They signed their names on the map, which is pictured below.

France took the Mediterranean coastal region of Lebanon, the area that became modern Syria, and what became northern Iraq (colored in blue). Great Britain claimed a small section of the Mediterranean coast to have port access and the southeastern part of ancient Mesopotamia (colored in red). Arabs were to form a single state or confederation that encompassed lands under the "protection" of France (marked with an A) and Great Britain (marked with a B). The diplomats believed that Palestine (colored in yellow) could best be governed as an international area because of the conflicting interests of Jews, Christians, and Muslims in the land.

Arab Soldiers (1918)

Because of its alliance with France and Britain in the war, Russia received Istanbul and eastern Turkey and the area of Armenia that lay between the Black and Caspian Seas. France and Britain had agreed in May of 1915 to give Russia control of Istanbul (formerly known as Constantinople), something that Russia had long sought.

Those lines on a map, and all they represented, have influenced what has happened in the Middle East for the century since Sykes and Picot met.

The Issues

So what's wrong with this picture—or, more accurately, this map?

First, it was an agreement by European governments that decided the destiny of a large part of the world's geography and people without any opportunity for the people affected to express their desires.

Second, the diplomats conducted their negotiations in secret. This is how countries often carried on diplomacy in those days. Two or more countries would make an agreement in secret, then when they thought it would be advantageous, they would reveal the agreement, act on the basis of it, and expect other countries to respect it. This practice understandably complicated international relations. United States President Woodrow Wilson strongly opposed this practice, and he wanted the peace settlement that followed the war to insure "open covenants openly arrived at."

Third, the Sykes-Picot agreement appeared to violate Great Britain's commitment to the Arabs to support their independence and freedom from

Kurdish cavalry fought with the Ottomans against the Russians during World War I.

foreign rule. France and Great Britain saw the lands they divided as protectorates. This meant that they committed themselves to protecting the areas involved. But protectorates were not really about protecting independent countries from invasion. Instead, the protecting nations would call the shots within the areas they protected. Under the Sykes-Picot agreement, the Arab peoples would be dependent on Great Britain and France, whether they liked it or not. When the Arabs learned of the agreement, they felt betrayed.

Fourth, the lines that Sykes and Picot drew took little or no account of the geography of the region or the ethnic groups, tribes, and families who lived there. The Kurds were one notable victim of the agreement. It is a major reason why the Kurds are divided today among Turkey, Syria, Iraq, and Armenia instead of having their own unified country. As it is, they are minorities in those countries and have had little say in their own government.

And Then…

In 1917 the British foreign minister, Arthur Balfour, wrote a letter in which he stated British support for a homeland (not a state, specifically, but a homeland) for Jews in Palestine. This was a position seemingly at variance with Sykes-Picot and one that further infuriated Arabs, especially those who lived in Palestine.

In Russia in 1917, a Communist revolution overthrew the government of the tsar and created a new government. This new government surrendered to Germany to end its participation in the war. As the Communists gained control of the files of the former czarist government, they discovered and made public the Sykes-Picot agreement.

The Conference of San Remo, Italy, in 1920 established borders that divided up Arab lands into protectorates and led to the modern borders of Iraq, Israel, and the Palestinian territories.

```
                                    Foreign Office,
                                      November 2nd, 1917.

Dear Lord Rothschild,
          I have much pleasure in conveying to you, on
behalf of His Majesty's Government, the following
declaration of sympathy with Jewish Zionist aspirations
which has been submitted to, and approved by, the Cabinet

     "His Majesty's Government view with favour the
     establishment in Palestine of a national home for the
     Jewish people, and will use their best endeavours to
     facilitate the achievement of this object, it being
     clearly understood that nothing shall be done which
     may prejudice the civil and religious rights of
     existing non-Jewish communities in Palestine, or the
     rights and political status enjoyed by Jews in any
     other country"

     I should be grateful if you would bring this
declaration to the knowledge of the Zionist Federation.
```

This is the letter Arthur Balfour sent to Lord Rothschild, representative of the Zionist Federation of Great Britain and Ireland.

Sir Mark Sykes was thirty-six years old in May of 1916. He died in 1919, one month shy of his fortieth birthday. Francois Georges-Picot was forty-five in 1916; he died in 1951 at the age of eighty.

Today

The radical Islamic State (IS) has used Sykes-Picot as the target of its wrath. Citing the agreement's unjustified dividing lines, the IS used the agreement to blame the West for imperialism in what it claimed was its own land. As the IS sought to spread its control over Syria and Iraq and to create a unified Islamic state, it rejoiced in proclaiming that "Sykes-Picot is dead."

However, Sykes-Picot did not establish the borders of individual countries. And one historian of the Middle East has noted that the lines of division that Sykes-Picot established are close to what the IS recognized. Other, more complicated local conflicts among Muslims had much to do with the placing of national borders. The lines that Sykes-Picot established are still in place and people have accepted them. Redrawing the map of the Middle East without addressing current simmering issues would only change and not remove conflict in the Middle East. Still, in 2002 British foreign minister Jack Straw admitted the failure of Sykes-Picot in ending conflict in the region.

People choosing the areas they want in the Middle East is something that has taken place for a long time, as we see in Genesis when Lot separated from his uncle Abram:

So Lot chose for himself all the valley of the Jordan, and Lot journeyed eastward. Thus they separated from each other.
Genesis 13:11

Assignments for Lesson 12

Gazetteer Read the entries for Cyprus, Jordan, Lebanon, and Syria (pages 5, 9, 11, and 15).

Worldview Copy this question in your notebook and write your answer: Why do you think the Jews have been persecuted in so many times and places?

Project Continue working on your project.

Literature Continue reading *Blood Brothers*.

Student Review Answer the questions for Lesson 12.

Erez Border Crossing Between Gaza and Israel

13 The Toughest Geographic Issue

It's a small area, about the size of New Jersey. God described it as a land flowing with milk and honey. In ancient times it was on a major travel route, and as such it was a frequent target for foreign domination. Today it is the location of perhaps the most difficult and intractable geographic issue in the world.

The issue centers on the fact that two peoples, the Israelis and the Palestinians—who have a history of not liking each other—both claim the same land. Who has the right to own it? This is the crucial question regarding the area we today call Israel or Palestine.

The question has no easy answer. Many proposed solutions have merit but also have difficulties and inconsistencies. The rival claims and the difficulty of their resolution are why, over seventy years after the founding of the modern state of Israel, such a high level of tension exists in the region.

Opinions about the Israeli-Palestinian issue differ greatly, and people hold their opinions with deep emotions. Even finding a source of information about the issue that does not clearly favor one side or the other is difficult. Any perspective on the issue will likely meet with strenuous opposition from those who disagree with it and who assert that the perspective is wrong or prejudiced or doesn't do justice to the real situation.

A Brief History of the Region

When God called Abraham and promised to give him and his descendants the land called Canaan (Genesis 17:8), pagan Canaanite nations were already living there. Abraham and his descendants lived on the land in generally peaceful relations with the Canaanites, although there were occasional times of conflict.

Hundreds of years later, God led the Israelites out of slavery in Egypt by means of the Exodus and into the land of Canaan. The Israelites defeated enough of the Canaanite nations to take possession of the land, although some Canaanites continued to live there. Thus Canaan became Israel.

God's promise to the Israelites that they would dwell in that land was conditional on Israel remaining faithful to God (Deuteronomy 28:58-67). Since the Israelites were unfaithful to God, God sent invaders who carried them into captivity. Following the reign of Solomon, Israel divided into the Northern Kingdom (called Israel) and the Southern Kingdom (called Judah). Assyria conquered the Northern Kingdom, carried many

Caesarea on the coast of Israel was originally a Phoenician naval station. Herod the Great oversaw major construction during the Roman era. The ruins of his palace are shown in the foreground. In the background are the remains of fortifications built by the Crusaders.

of the Israelites into captivity elsewhere, and sent in other pagans loyal to Assyria to repopulate the land. The Samaritan people of New Testament times were descendants of the intermarriages of these transplanted people and the people of Israel who remained on the land.

Later, the Babylonians, who had conquered the Assyrians, carried into captivity thousands of Jews from the Southern Kingdom of Judah (the term Jews comes from the word Judah). The Persians, who later defeated the Babylonians and took over their empire, allowed the Jews to return to their homeland, although they remained under Persian rule. Authority over the land of Israel later passed to the Greeks, and eventually to the Romans. In the second century AD, after the Jews had twice attempted to rebel against Roman rule, the Roman emperor Hadrian changed the name of the province from Judea to Palestina, a name based on the term Philistines, in an attempt to eradicate any reference to the Jews. Thus Israel became Palestine.

Rule from Rome gave way to rule from Constantinople, the capital of the Eastern Roman Empire. Then Muslims captured Palestine in 634 and became the predominant people group there. Medieval Crusaders put it under Christian rule for a time a few centuries later, but then they lost control of it to the Muslims again. The Muslim Ottoman Empire, based in Constantinople (modern Istanbul), assumed authority in 1516. There the matter remained for centuries. A small number of Christians and Jews lived in Palestine along with the majority Muslims. Some Jews who lived elsewhere bought land in Palestine and moved there. Despite their religious differences, the Christians, Jews, and Muslims in Palestine generally lived peacefully with one another. The Ottomans allowed non-Muslims to live in Palestine unhindered as long as they paid a tax.

The Rise of Zionism

Jews were the target of discrimination in Europe for centuries. After a period of increasing hostility toward Jews in Europe in the late 1800s, many European Jews expressed a longing for a separate homeland where Jews could live in peace and security. Jews discussed several places in the world as possible locations, but the one that most Jews vastly preferred was Palestine, their historical

Lesson 13 - The Toughest Geographic Issue

homeland from centuries before. Mt. Zion is the historic location of the temple in Jerusalem. The Jewish desire for a homeland came to be called the Zionist movement, even though the temple had long since been destroyed and Muslim structures stood on the site. Many Jews bought land in Palestine in the late 1800s and early 1900s, moved there, and brought other Jews with them. This led to a rising level of uneasiness among the Muslim Arab residents of Palestine, who began to feel threatened.

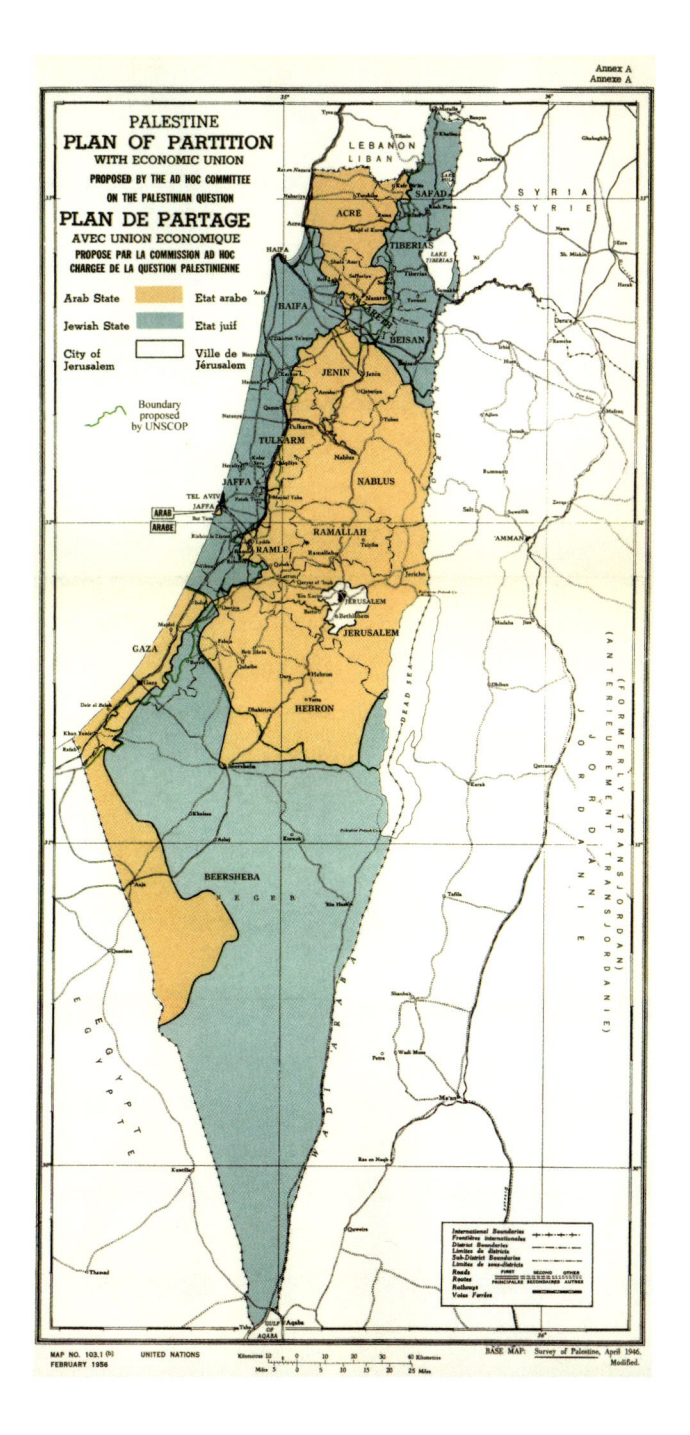

During World War I British troops liberated Palestine from Ottoman control. Later during the war, Great Britain stated its support for the creation of a Jewish homeland in Palestine. After the war, the League of Nations gave Great Britain the mandate to oversee Palestine.

The Nazi Holocaust in the 1930s and 1940s intensified the desire of many Jews for a secure homeland. Some Jews were frustrated by British oversight of Palestine and wanted to seize control of the land. Some Arabs resented the Jews' desires. Jewish groups used violence to discourage a continued British presence, and Arab groups used violence to try to stop the Jews.

Following World War II, Britain began withdrawing from its longtime role as world policeman. Because of the increasing level of violence against the British presence in Palestine, Great Britain announced that it could no longer fulfill its mandated oversight of Palestine. In 1947 the United Nations (UN) passed Resolution 181, which called for the partitioning of Palestine into a Jewish state and an Arab state with Jerusalem under international control.

Jews supported this resolution because they would get more than they previously had and because it fulfilled their long-held dream of a homeland. They began working toward this end. Hundreds of thousands of Jews moved to Palestine.

On the other hand, Arabs in Palestine and neighboring countries rejected the partition plan because they believed it would leave them with less than they had. The land in question was land on which the Palestinians and their ancestors had lived for centuries. They did not believe it was the United Nations' land to give. They believed it to be theirs, and now they feared that a major part of it was in danger of being taken away.

This map of the 1947 United Nations Plan of Partition outlines the Arab State in orange and the Jewish State in blue-green.

Statehood and Continued Conflict

Great Britain withdrew from its mandate role on May 14, 1948. That same day, with emotional support from the Zionist movement and political support from the United Nations, Israel declared itself to be a nation. The United States was the first country to extend diplomatic recognition to Israel. Hundreds of thousands of Palestinian Arab Muslims fled to nearby Arab countries as refugees, not wanting to live in a Jewish-controlled state. In pursuit of their goal of nationhood, the Jewish military forced some Palestinians from their homes. Caring for refugees created a hardship on those other countries. In some places the Palestinians were not wanted. Some Palestinian families have now lived in refugee camps for generations.

Several Arab countries launched a military attack on Israel, but Israeli defense forces defeated the attack. Other Arab-initiated wars took place in 1956, 1967, and 1973; but Israel defeated all of these assaults as well. The Arab states in the region announced their desire to drive Israel into the Mediterranean Sea (i.e., destroy it) and take back the land of Palestine.

Israel seized the Golan Heights, an area that many people understood to be Palestinian territory, in 1967 and annexed the area in 1981, both as a security measure and to provide housing for the

These Israelis are waiting in a bomb shelter during the 1967 Six-Day War.

These Egyptian vehicles were destroyed in the Sinai during the 1956 war.

growing Israeli population. Israel took control of Jerusalem and declared it to be their national capital (the capital had been Tel Aviv). Moreover, Israel exerted control over Palestinian territories in Gaza and the West Bank of the Jordan. Israeli government policy required security clearance for Palestinians to leave and enter what they believed to be their own land. Thus Israel did not remain within the boundaries that the original United Nations resolution proposed. Israel also did not accept international control of Jerusalem.

Palestinians said that Israel's actions robbed them of their freedom, but at the same time Palestinian leaders maintained their position of wanting to annihilate Israel. Palestinians formed such groups as the Palestinian Liberation Organization (PLO) to exert Palestinian autonomy and to fight against Israel. Palestinians began to launch terrorist attacks on Israel, and other Arab countries supported this policy.

Beyond attacking Israel itself, the policy of terrorism took aim at Israel's allies such as the United States and Great Britain. As a result, terrorist attacks have spread around the world, not just in Israel against Jews. The goal of this terror campaign is to discourage support of Israel in order to enable a Palestinian takeover of the land of Israel. However, Muslims who oppose Israel do not speak with a unified voice. They have divided into several groups with competing leaders and different specific agendas.

Lesson 13 - The Toughest Geographic Issue

U.S. President Jimmy Carter facilitated a meeting between Egyptian President Anwar Sadat (left) and Israeli Prime Minister Menachem Begin (right) at Camp David in 1978.

The Middle East region has seen a few steps toward resolution of the conflict there. Egypt and Israel signed a treaty of peace in 1979, and Israel and Jordan signed a peace treaty in 1994. In 2020 the United Arab Emirates, Bahrain, and Sudan agreed to establish diplomatic relations with Israel and work together in various ways.

However, conflict and complicated situations still abound. For instance, unrest in Lebanon has involved militant Islamic groups there and has led to attacks on Israel, Israeli attacks on certain groups in Lebanon, and even an armed Israeli presence on Lebanese soil for a time in an attempt to stop the attacks. In addition, Palestinians have dug numerous tunnels into southern Israel trying to smuggle weapons and attackers into Israel.

After years of sporadic attacks, an agreement between Israel and the PLO in 1994 called for Israel to return some lands in the West Bank to Palestinians and for the PLO to agree no longer to say that its goal was the destruction of Israel, in other words admitting that Israel had a right to exist.

The Palestinian Authority was created to administer Gaza and sections of the West Bank as Palestinian territory. However, not all Arabs were pleased with these developments; some militant Muslims denounced the agreement.

Although war cost Israel and the Palestinians and other Arabs dearly, the pursuit of peace has had a price also. In 1981 Egyptians who opposed the establishment of peace with Israel assassinated Egyptian President Anwar Sadat, who had concluded the peace agreement with Israel. In 1995 an Israeli assassinated Israeli prime minister Yitzhak Rabin because he thought Rabin was making too many concessions to the Palestinians.

Modern Israel

Israel is economically advanced, well educated, and highly urbanized (92% of the population lives in urban areas). It has a stable government and a strong military.

Israel has few natural resources. Even though it has many modern industries, it imports much more than it exports. About 20% of the land is suitable for farming, and about 40% of that is irrigated. Much of the farming takes place on kibbutzes or cooperative farms.

Israel has a policy of accepting Jews from anywhere in the world, so its citizens have a diversity of cultural backgrounds. Israel has a policy of freedom of religion. The country maintains religious courts that address issues such as marriage conflicts, divorce, and inheritance. Some religious sects have their own courts.

About 80% of Israel's 8.3 million population is Jewish. Of the remaining 20%, 15% are Arab Muslims and 2% are Arab Christians. Twenty-eight percent of Israelis are under 15 years of age. Israel has two official languages, Hebrew and Arabic. English is a common second language. The country operates two school systems, one in Hebrew and one in Arabic.

The government is a parliamentary democracy. The legislature is the one-house Knesset. Israel has several political parties that represent a wide spectrum of views. Citizens vote in parliamentary elections by choosing a party, not by voting for individual candidates. The leader of the majority party in the Knesset becomes the prime minister. However, with so many parties, a single party often does not have a majority by itself. In such cases, a coalition of parties form a government. The Labor Party is a moderate party that favors more negotiations with the Palestinians and Arabs and a peaceful settlement to the conflict with the Palestinians. The Likud bloc is a coalition of parties that takes a harder stance against the Palestinians.

The United States has long been Israel's strongest ally nation. About 30 countries in the world do not extend diplomatic recognition to Israel. Most of these are Arab countries in the Middle East.

We must not think that the modern political state of Israel is the embodiment of the nation of Israel that we read about in the Bible. They have the

Meeting of the Knesset (2019)

same name and exist on much of the same land, but they are not the same.

Modern Israel does not follow the Law of Moses, as Biblical Israel was supposed to do, nor does it seek to do so. The Knesset passes laws for the modern state of Israel. Modern Israel does not worship at the Temple in Jerusalem, it does not recognize the Levitical priesthood, nor does it practice animal sacrifices as Israel in the Bible did. Modern Israel does not have the tribal identity that ancient Israel did with its twelve tribes. The land that modern Israel occupies is not exactly the same as the land to which God led the Israelites after the Exodus. Many people in modern Israel understand it to be a secular state and do not want religious ideas and leaders to have a role in public policy. Modern Israel makes no claim to be ancient Israel. And yet, for some Jews in modern Israel, the country does embody their hopes and dreams based on God's promises to Biblical Israel.

The Palestinians and the State of Palestine

Ishmael, the son of Abraham by Hagar, is generally thought to be one ancestor of the Arab people. A large majority of Palestinians are Muslims, speak Arabic, and take part in Arab culture. However, their ethnic background is not simply Arab, although that is part of their story. Some believe they are descendants of ancient Philistines and Canaanites.

Lesson 13 - The Toughest Geographic Issue

The world population of Palestinians may be as high as 12 million, although half of their number live outside of Palestine.

The West Bank is the region that lies to the west of the Jordan River. In the original UN proposal, the area was to belong to the Palestinians. The country of Jordan (once called Transjordan because it lies east of the Jordan River) administered it. Israel took control of the area in the 1967 war. Many in Israel think of the West Bank as "Judea and Samaria," the names of parts of the West Bank in ancient times. Once Israel established control over the West Bank, the Israeli government began allowing Jewish settlers to build homes in the West Bank, a move which the Palestinians opposed and the majority of other nations believed to be illegal. Israel now provides defense forces for parts of the West Bank and allows Palestinian forces to operate in other parts. The West Bank population includes about 2.2 million Palestinians and about 400,000 Israelis. Israelis have also been moving into East Jerusalem, formerly an Arab area, to increase their numbers there.

In 1988 Palestinians declared the existence of the State of Palestine, which consists of Gaza (also called the Gaza Strip) and the West Bank. These two areas are not connected. The total size of the State of Palestine is about one-third that of Israel, or about half the size of Connecticut. About 137 of the almost 200 UN member nations have extended diplomatic recognition to Palestine. The United Nations has only granted it the status of a non-member observer state, but the term does imply recognition of it as a state.

Is There a Solution?

The Israeli-Palestinian conflict centers on sharply differing religious and cultural views about a section of geography. Here are three possible scenarios for resolving this conflict.

1) The Palestinian Arabs should own the land because they had come to possess it centuries ago. Their claim rests on historic ownership. Some Palestinians and their supporters stake their claim to ownership of the land on the assertion that the Palestinians are descendants of the Canaanites who lived on the land prior to the Israelites settling there after the Exodus. There is as yet no convincing proof to support this claim.

2) The Jews of Israel should own at least some of the land because of the 1947 United Nations resolution supporting the creation of the state of Israel. Some people believe that God giving the land to descendants of Abraham supports modern Jewish ownership of the land.

3) The Palestinian Arabs and the Jews should own different parts of the land because of their rival claims and because the 1947 UN resolution called for partition of the land into a Jewish state and a Palestinian state, and for the designation of Jerusalem as a neutral city under international oversight.

Jews have been victims of persecution for centuries. Should the world sit by and let a people be destroyed because others hate them?

And yet, Palestinian Arabs have a claim as well. Suppose the UN declared that some section of the United States should be carved out and given to a foreign group as their national homeland, or perhaps to native nations who had once lived there. Would we as Americans accept that decision and give up our claim to that land? What should be done about the claims of the Palestinians?

Ethiopia recognized the State of Palestine in 1989. This photo was taken in Addis Ababa, Ethiopia, in 2007.

How can our world, and especially the people who live in modern Palestine, move forward? Do the people there want to move forward? How far back in history must we go to seek settlement of claims and resolution of attacks? How can neighboring people groups with profound differences live in peace? Must people take the lives of others to express their claim of ownership of a piece of geography?

Many people have claimed and fought over the piece of geography in the Middle East that we have known as Canaan, Israel, and Palestine. When and how can the fighting cease?

Genesis describes how Israel and the Arab people came about and the difficult relationship between them. God promised to bless them both, although He gave His covenant promise to Isaac.

But God said to Abraham, "Do not be distressed because of the lad [Ishmael] and your maid; whatever Sarah tells you, listen to her, for through Isaac your descendants shall be named. And of the son of the maid I will make a nation also, because he is your descendant."
Genesis 21:12-13

Assignments for Lesson 13

Gazetteer Read the entries for Azerbaijan, Iraq, and Kuwait (pages 3, 7, and 10).

Worldview Copy this question in your notebook and write your answer: What are some ways in which you respect the Jewish people?

Project Continue working on your project.

Literature Continue reading *Blood Brothers*.

Student Review Answer the questions for Lesson 13.

Holding the Kurdish Flag in Northern Iraq (2019)

14 "We Have No Jam": The Saga of the Kurds

Adnan Hassan of Syria saw Islamic State (IS) fighters wipe out his hometown and kill ten members of his family. Then a coalition of forces pushed the IS out of the region, and Hassan and his people were able to breathe the air of freedom. They made plans to open a university and, for the first time in Syria, to teach their language. However, later fighting, as well as changes in the coalition, brought Hassan and his fellows back to a sense of uncertainty.

Raz Razool was a college student in northern Iraq. With protection from a favorable Iraqi government and Western allies, she and others began working to better the lives of her people. They had a long way to go. Four years earlier the regime of Saddam Hussein committed horrendous ethnic cleansing on her people. Two years after that, one and a half million of her people fled the country to try to survive during a harsh winter. Still, Razool had reason to hope for better days. Then IS fighters entered her land and uncertainty arose again.

Adnan Hassan is Syrian and Raz Razool is Iraqi, but they have one important trait in common: they are Kurds. As Kurds, they also share the uncertainty and difficulty that their people have faced for centuries. However, amid that uncertainty they have one geographic anchor to which they cling: the mountains. The mountains are their home. The mountains are their friends. The mountains are their refuge.

About 25 to 35 million Kurds live in the broad region that many people call Kurdistan. The region extends from the Taurus Mountains in eastern Turkey, across portions of Syria, Iraq, and Armenia, into the Zagros Mountains of western Iran. See the map below.

83

Despite the region's commonly accepted name, there is no internationally accepted political entity called Kurdistan. The Kurds very much want to have their own country, but the international community does not recognize a nation called Kurdistan. The Kurds are the fourth largest ethnic group in the region, after the Arabs, Turks, and Persians. They are the largest ethnic group in the world that does not have a recognized national homeland.

A Rich but Tragic Heritage

The Kurds claim a rich heritage. Abraham probably passed through their lands on his journey from Ur to Canaan. The ancient Assyrian capital of Nineveh occupied much of the same location as today's city of Mosul in northern Iraq. Experts generally believe—as do the Kurds themselves—that they are descendants of the ancient civilization known as the Medes. Among those from various lands who were present on the Day of Pentecost at the pouring out of the Holy Spirit and the founding of the church were Jewish Medes (Acts 2:9), probably descendants of some who did not return to Canaan after the Babylonian captivity. According to strong tradition, the apostles Thomas and Thaddeus brought the gospel to the Kurds in the early years of the church, and the fellowships they established are still functioning today.

The Kurds for the most part converted to Islam after that religion developed in the region. Today the Kurds are predominantly Sunni Muslims; but they are Persians, not Arabs. The main Kurdish language group derives from Persian. The powerful Muslim leader Saladin (1137-1193) was Kurdish. As Sultan of Egypt, he pushed the Crusaders out of Jerusalem in 1187.

Kurds are more accepting of beliefs other than traditional Islam than are many Arab Muslims. For instance, many Muslim women in Kurdish areas do not wear a head scarf, and women play unusually prominent roles in Kurdish government, education, and society. Even today Kurdish Jews and Kurdish Christians generally live peacefully in areas that Kurdish Muslims dominate.

When Jews in the Kurdish region faced increasing persecution by the Iraqi government after World War II, and especially when Jews established modern Israel in 1948, large numbers of Jews left the Kurdish villages where they lived. In response, for decades Kurdish Muslims maintained the synagogues in dozens of villages as a memorial to the friendship these Jews and Muslims shared.

The Kurds do not want to dominate other people or lands, but they will fiercely defend their own. Despite this generally peaceful spirit, the Kurds have often been victims of discrimination, persecution, and even ethnic cleansing at the hands of the governments under which they have lived.

One main reason is that the Kurds do not want to go along with others who are politically or militarily stronger than they are at a given moment. They do not want to be Arabs; their roots are Persian. They

The Tomb of Saladin in Damascus, Syria

This bazaar is in Sanandaj, capital of the Kurdistan Province of Iran.

do not want to be Turks or Syrians or Iraqis; they are Kurds. The majority of Muslims in the world are Shi'ites, but the Kurds are predominantly Sunni. (You'll learn more about the difference in Lesson 25.) In addition, other nations of the world have ignored or refused to help the Kurds in their desire for a national homeland. International agreements have divided them among several other countries, either out of prejudice or out of an attempt to limit their power.

This pattern has given rise to a cynical saying among the Kurds. When a Kurdish shopkeeper does not have jam or another article that a customer wants, the shopkeeper will sometimes say, "We have jam, but we have no jam." In other words, there is jam in the world, but this shop does not have any. It is an expression of Kurdish being and not being at the same time. Applying this thought to their situation, a Kurd might say: "We are a people, but the world does not recognize us as a people. We have history, heritage, culture, language, ethnic identity—everything that would make us a people—but we are not officially a people. We are humans, but others often deny our human rights. We are Kurds, but there is no Kurdistan." This has been the saga of the Kurds for centuries.

A Pattern of Persecution

As mentioned above, the ethnic group that became the Kurds were major players in the ancient world. They were present in the time of Daniel and Esther. Some speculate that the magi who came to visit the infant Jesus might have been Zoroastrians. If they were, they would have been part of the religion that most of the Kurds followed before the coming of Islam. Kurdish lands once played a key role on the overland trade routes between Europe and Asia. However, with the development of sea trade around Africa, the Kurdish lands became what one observer called a "mountainous irrelevancy."

The Ottoman Turks invaded Kurdish lands in the 1500s. After enduring this foreign rule for centuries, some Kurds revolted against the Ottoman Empire in the 1880s; but they were not able to establish independence. Following World War I and the defeat of the Central Powers and their Ottoman Empire ally, United States President Woodrow Wilson pressed for the independence and autonomy of ethnic groups around the world. One provision in the proposed Treaty of Sevres between the Allies and Ottoman Turkey in 1920 called for an autonomous Kurdistan. However, Turkey rejected the treaty.

The 1923 Treaty of Lausanne that the parties did adopt, which officially ended the Ottoman Empire, made no mention of the Kurds. The Western allies preferred a strong new Turkey, which did not want to grant autonomy to the Kurds. Thus the opportunity for Kurdish independence faded away. Turkey persecuted the Kurds and outlawed the use of the Kurdish language and other elements of Kurdish culture. Turkey even referred to the Kurds who lived it its eastern mountains as "mountain Turks," not Kurds.

Following World War II, the Soviet Union controlled northern Iran with approval of the other Allies. Iranian Kurds declared themselves to be the independent Mahabad Republic. It was the first declared Kurdish state in history. However, the Allies did not support it, since they wanted to encourage a unified Iran. The government of the Shah of Iran crushed the republic and persecuted the Kurdish people. Again, the decisions of other nations quashed Kurdish hopes for independence.

Later, in Iraq, the Baath Party's military dictatorship suppressed Kurdish rights in that country. After years of persecution, Saddam's hatred of the Kurds culminated in the Iraqi government's use of chemical weapons against the Kurdish city of Halabja in 1988. The attack killed thousands of people.

A Brighter Day

The region known as Kurdistan illustrates the importance of geography and the interaction of people and the land. This region has been associated

Newroz is a festival that coincides with the spring equinox. It has roots in ancient Iran and has become an important part of Kurdish culture. This 2010 photo shows Kurds celebrating in Turkey.

Lesson 14 - "We Have No Jam": The Saga of the Kurds

with the Kurds and their ancestors for millennia. The mountains and other features have defined their lives. In the Western efforts to create manageable nation states following the two world wars of the twentieth century, the Kurdish region was divided among the countries that came into existence. The governments of these nations have generally treated the Kurds harshly, attempting to make them conform to the larger whole instead of accepting them for who they are.

Despite the overall record of Kurdish persecution and disappointment during the past century, this lesson ends on a positive note. Ironically, it is in Iraq, where the Kurds have suffered the most, that the brightest prospects exist for them.

After the Persian Gulf War in 1991, the United States and its allies established a no-fly zone in northern Iraq. This meant that Iraqi planes could not fly over Kurdish areas, and it gave the Kurds a measure of peace and security. In 2003 an American-led coalition removed the government of Saddam Hussein in Iraq. After a transition period, Jalal Talabani, a Kurd, became president of Iraq. In 2014 Fuad Masum succeeded Talabani, thus becoming the second Kurdish president of Iraq. The current government of Iraq is more favorable to the United States than in Saddam Hussein's day.

The Kurdish region in Iraq has gained a degree of autonomy. It has a functioning government, it has welcomed foreign investment, and business and tourism are flourishing. The invasion of IS forces brought a degree of uncertainty, but the IS later suffered significant defeats and became much weaker. Life for Kurds in other countries is still somewhat uncertain, and the Kurdish community has suffered internal division among those who have had varying visions over what they think Kurdish life should be like; but the Kurds have made progress toward living with the freedom they have long desired.

The prophet Isaiah spoke about conflict between the Medes and the Babylonians in ancient Mesopotamia.

Behold, I am going to stir up the Medes against them,
Who will not value silver or take pleasure in gold.
Isaiah 13:17

Assignments for Lesson 14

Geography — Complete the map activity for Unit 3 in the *Student Review* book.

Worldview — Copy this question in your notebook and write your answer: In what way or ways do you think the religion of Israel and first century Judaism were the foundation for Christianity?

Project — Continue working on your project.

Literature — Continue reading *Blood Brothers*.

Student Review — Answer the questions for Lesson 14.

Musmeah Yeshua Synagogue in Yangoon, Myanmar

15 Faith System: Judaism

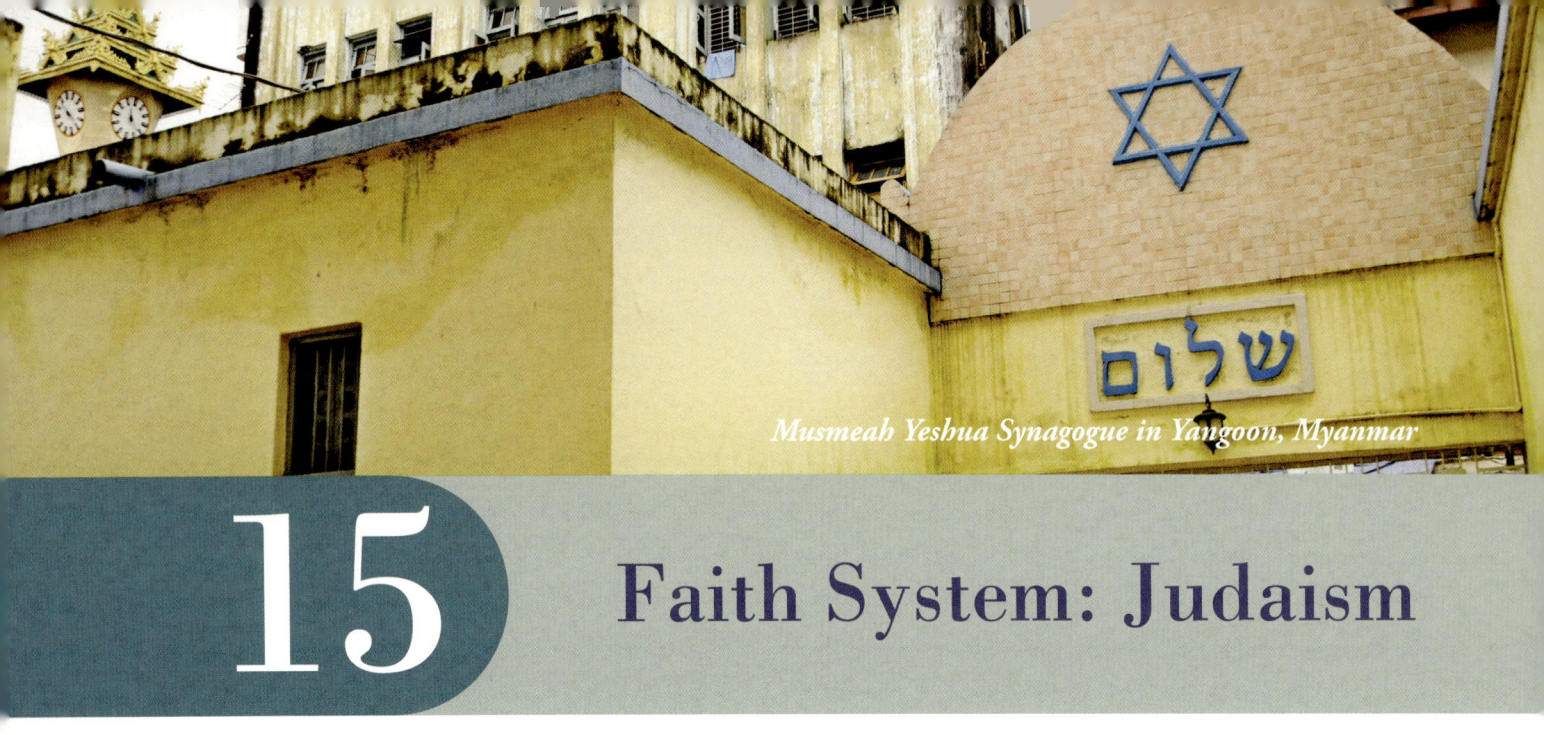

This is the first of the lessons that describe the most widely held worldviews today.

A chosen people.

A persecuted people.

A people with a rich heritage.

A people who were for a long time without a place.

The Jews.

The Jewish people have been on a long, complex journey that has brought them to who they are and where they are today. This journey has had a profound impact on their self identity and worldview. In this lesson we will give a quick survey of the history of the Jews, then discuss the main elements of how they see themselves and the world in which they live.

From Abraham to the Diaspora

God called Abraham to leave his father's country and family and go to a place He said He would give Abraham as his homeland. God promised to "make you a great nation" (Genesis 12:2), which He did through Isaac, Jacob, and their descendants.

Jacob's descendants became slaves in Egypt, but God brought them out from there through the Red Sea and led them to Himself. At Sinai God offered to make the people of Israel His unique, chosen nation, a relationship that He offered to seal by the covenant He proposed. The Israelites accepted this offer, and God gave them the Law by which they were to conduct themselves as His covenant people (Exodus 19:4-8).

Because of Israel's faithlessness, however, God made them wander in the desert for 40 years. Then the Lord led them across the Jordan River and into the land He had promised to them, the Promised Land of Canaan.

God richly blessed Israel in that land, but they became unfaithful to Him. As a result, God punished them. Assyria invaded about 722 BC and carried the ten northern tribes into captivity. We only have scattered bits of information about what happened to the people from this part of Israel (see, for instance, Luke 2:36). The tribe of Judah, along with the much smaller tribe of Benjamin, came to see themselves as the remnant of God's people.

But Judah and Benjamin were unfaithful also, and Babylonian invaders carried them into captivity around 606-586 BC. After 70 years in captivity, the new Persian rulers of Babylon allowed the people of Judah to return to their homeland. Some Jews chose to remain in Babylon, while others moved to Persia and a few took up residence elsewhere.

Lesson 15 - Faith System: Judaism

The Jews who returned to Canaan reestablished themselves as a nation, but they almost always lived under the yoke of a foreign power. Many Jews longed for the day when, as they believed, God would send His Anointed One (the Messiah) to gather His people together and make them free.

However, the Jews as a people rejected Jesus as their Messiah. Soon thereafter some Jews rebelled against their Roman overlords. Rome crushed their rebellion and scattered most Jews away from Canaan once again. Many Jews came to live in northern Africa and Europe. The Jews who lived away from Canaan were often called the Diaspora, which is Greek for dispersed ones.

European Judaism and Today

For centuries the Jews were associated with the land of Canaan, a location also called Israel or more recently Palestine. After the Roman dispersion, the majority of Jews did not live in Palestine. Eventually the majority lived in Europe.

The response by European ethnic and national groups to the Jews who lived among them varied considerably. In some places, such as Poland and Lithuania, people welcomed Jews. Residents in other places merely tolerated the Jews' presence and placed severe restrictions on them. For instance, in Venice Jews had to live in a designated area called a ghetto. In some cities officials locked down the ghetto at night out of fear that the Jews would commit crimes in the darkness.

Laws in many places forbade Jews from owning property. This is why Jews often became tailors, shopkeepers, and other artisans, as well as moneylenders, in order to support their families (at the time, the Catholic Church did not allow its members to loan money at interest, so many Christians borrowed money from Jews). At times Jews faced complete rejection. The rulers of some

Jews Praying in the Synagogue on Yom Kippur
by Maurycy Gottlieb (Polish, 1878)

countries, such as England in 1290 and Spain in 1492, ordered all Jews to leave their domains.

Mistrust, segregation, and discrimination gave way to outright persecution in some places. Many people hated Jews, whom they saw as "Christ-killers." These Gentiles, usually people professing to be Christians, resented the Jews' special dietary habits and other distinctive traditions. Jews became easy scapegoats when people wanted to find a reason for disasters and diseases such as the black plague. Physical attacks called pogroms, which included killing Jews and plundering their homes, took place from time to time in Eastern Europe and Western Russia. These pogroms took the lives of tens of thousands of Jews. The persecution of Jews reached its worst level in the twentieth century, when Nazi Germany took the lives of some six million Jews in the Holocaust.

For the most part, Jews simply wanted to live their lives peaceably and practice their religion as they saw fit. Jews who lived in Germany simply saw themselves as German Jews. Jews who lived in Poland saw themselves as Polish Jews. Jews who lived in Russia were Russian Jews. But because of the way people treated them, and because they knew their history, many Jews longed for a better and safer place to live. Perhaps, they hoped and prayed, the Messiah would come and lead them to such a place. It became a common practice to end the annual Passover feast with the phrase, "Next year in Jerusalem!"

Today the Jewish population of the world is estimated to be about 14-16 million, depending on whether the total includes people with one Jewish parent and people of Jewish ethnic descent who do not practice Judaism. Scholars who study population, called demographers, estimate that the Jewish world population is now about what it was before the Holocaust, which ended in 1945. Approximately 42% of the Jewish world population lives in modern Israel, and a slightly lower percentage lives in the United States. Millions of Jews emigrated from Europe to the United States in the late 1800s and early 1900s.

Since the 1800s, Jews have made significant accomplishments in many fields, including literature, entertainment, and business. A tiny minority of the world population, Jews have had a huge impact on world history and continue to have an outsized influence in the contemporary world.

Law and Tradition

The Law that God gave to Israel through Moses set forth requirements for what they were to sacrifice and when, regulations for the festivals the Israelites were to observe, rules to determine whether foods and other items were clean or unclean, laws regarding interpersonal relationships, and laws outlining personal responsibility.

A community known as the Cochin Jews lived on the southwestern coast of India for many centuries. Only a small number remain today.

However, the Law established more than a religion; for Israel it established a way of life. The people of Israel were not always faithful to God's commandments, but over the centuries and through many trials they repeatedly came back to the Law and saw it as their standard, the Word of God.

During the captivity in Babylon, the Jews were not able to conduct worship at the temple in Jerusalem as God commanded. As a replacement activity, the Jews began to gather in synagogues (the word is from the Greek for gathering together) for Sabbath-day worship, prayer, and study of the Scriptures. Many Jews devoted themselves to intense, on-going study of the Scriptures that God had given to them, the Scriptures Christians call the Old Testament. Synagogue worship and the study of the Scriptures continued when Jews were allowed to return to Israel and to rebuild the temple in Jerusalem. Then when the Romans defeated the Jews' revolt, the Romans destroyed the temple. Once again the scattered Jews could not make sacrifices at the place where God had chosen for His Name to dwell, so synagogue assemblies, the study of the Scriptures, and the observance of Jewish festivals and traditions became major parts of Jewish life in Europe.

Lesson 15 - Faith System: Judaism

A Divided People

As with just about every religion, the Jews have suffered divisions within their ranks. In the New Testament era, the Jews were divided among Sadducees, Pharisees, Essenes, and Zealots. Those groups faded away, but new divisions emerged after the New Testament period. Today Judaism is divided into three main groups.

Orthodox Jews hold most closely to the traditional interpretation of Scripture regarding dietary laws, Sabbath observance, and keeping Jewish festivals. The Orthodox are also dedicated to maintaining Jewish traditions. Hasidic Jews are the most conservative sect of Orthodox. The Hasidim insist on wearing certain clothes and hairstyles. They are deeply pious, often demonstrating this by an intense practice of prayer.

Conservative Jews are moderates. They keep many traditional elements and customs in Judaism but allow for some modern practices that they believe are more relevant expressions of abiding truth. Conservative Jews keep the Sabbath; but they allow some modifications of dietary laws, and they have accepted women rabbis.

Reform Jews are the progressives in Judaism. They have accepted many modern ideas and interpretations of Scripture and tradition. The most liberal among Reform Judaism are Reconstructionist Jews. This relatively small group arose in the United States in the twentieth century and sees Judaism as an evolving religious civilization. They believe Judaism is a creation of the Jewish people and not of God. In fact, they deny the existence of the supernatural and the idea that the Jews are the chosen people (since, in their view, there is no God to choose them).

Differences in practice arose among Jews in Europe depending on where they lived. Jews from the region of Germany were called the Ashkenazim (Ashkenaz is the Hebrew word for Germany). Jews who trace their lineage from the region of Spain are the Sephardim (Sepharad is the Hebrew word for Spain). Over the centuries different synagogue practices and differences in language arose in the two groups. About 80% of Jews today are Ashkenazim, but in Israel the numbers of Ashkenazi and Sephardic Jews are about equal.

Paramaribo, Suriname, has a Jewish synagogue next door to a Muslim mosque.

Orthodox Jewish Family in Jerusalem

The term Jew can be understood in two slightly different ways. It can refer to someone who is a member of the ethnic group, and it can refer to someone who practices the Jewish religion. Not all ethnic Jews practice the religion of Judaism, and not all practicing Jews are ethnically Jewish. However, Judaism is the religion of the great majority of the Jewish ethnic group. In addition, modern Judaism is not the same thing as the religion of Israel as set forth in the Law of Moses and the rest of the Old Testament and as practiced in the time of Jesus. Today's Judaism is a form of that religion. Most Jews today might see the two as the same; but they are not the same in every detail.

Jewish Festivals

Jews observe several special days and feasts during the year. Some of these are in the Law of Moses, while others developed later in Jewish history. Since the Jewish festivals are on a lunar calendar, the dates for these festivals vary somewhat from year to year. These annual festivals maintain and deepen the self-identity of the Jews.

Rosh Hashanah. The Jewish New Year (September or October of our calendar) celebrates God's creating the world.

Yom Kippur. The Day of Atonement, which comes ten days after Rosh Hashanah, is a time when the Jews fast and confess their sins from the previous year.

Passover and the Feast of Unleavened Bread. The Law called for Israel to make three pilgrimages every year to the place where God would cause His name to dwell. The pilgrimages to Jerusalem are not part of these observances today. The first was Pesach or Passover, in March or April, to remember the Lord's deliverance of the Israelites from Egypt and His passing over their houses when He struck down the firstborn in the Egyptians' houses. Jews have a special meal at Passover each year. Pesach also celebrates the barley harvest. The Feast of Unleavened Bread is a weeklong festival that immediately follows the one day of Passover.

Shavuot, Feasts of Weeks or First Fruits (the Greek term is Pentecost). The second feast of pilgrimage comes seven weeks after Passover. It commemorates the Lord giving the Law on Mount Sinai and also celebrates the first fruits of the land.

Lesson 15 - Faith System: Judaism

Sukkot, Feast of Ingathering or Harvest, or Tabernacles. The Feast of Tabernacles comes five days after Yom Kippur. It recalls the Israelites wandering in the wilderness. For this third pilgrimage feast, Jewish families build shelters (tents or tabernacles) in which they spend time during this festival to remember their ancestors' experience in the wilderness.

Jews also observe additional special days that the Law does not mention.

The book of Esther tells of the establishment of the feast of Purim in February or March to celebrate the deliverance of the Jews in Persia from the evil Haman.

The Feast of Hanukkah in December celebrates the cleansing and rededication of the temple in 165 BC by the Macabbeans after the pagan Hellenist rulers had defiled it.

Jews also observe the Ninth of Av in remembrance of the destruction of the temple by the Babylonians and later by the Romans.

During the Middle Ages Jews began practicing a special observance that recognizes a young man coming of age and responsibility. This is called a bar mitzvah (son of the Law). In 1922 a Jewish man gave such an observance for his daughter and began the practice of holding a bat mitzvah (daughter of the Law).

Essentials of the Jewish Worldview

Traditional Jewish beliefs include these elements:

- **The oneness of God, who revealed Himself as YHWH.** The name is traditionally pronounced "YAH-weh," although Jews usually do not pronounce it at all, for fear of using the Lord's name in vain. They substitute Adonai, the Lord, for the name.

- **The inspiration and authority of the Jewish Scriptures.**

- **The observance of ritual practices.** These include the circumcision of male babies on the eighth day after birth, observance of the Sabbath as holy, and keeping kosher (eating only clean foods prepared acceptably and not touching anything that would make one ceremonially unclean).

- **The identity of Israel as God's specially chosen, covenant people and of the Jews as the faithful remnant of that people.** The self-identity of Israel as a suffering, persecuted people is strong, and the evidence of history makes this viewpoint understandable. For instance, most Jews believe that the description of the Suffering Servant in Isaiah 53 applies to Israel as a whole, whereas Christian teaching interprets the passage as a prophecy of Jesus Christ.

This 19th-century manuscript is from the community of Kurdish Jews. It has poems and readings related to the celebration of Purim.

Jews who believe in God and the Jewish Scriptures believe that their people have a future hope with God in the coming of the Messiah. Even in the context of their being taken captive as punishment for their sins, the Lord told the Jews, "'For I know the plans that I have for you,' declares the Lord, 'plans for welfare and not for calamity to give you a future and a hope'" (Jeremiah 29:11). However, Jewish teaching does not include a strong, uniform belief regarding an afterlife. Opinions on this subject among Jewish scholars differ. Instead, Judaism emphasizes maintaining personal morality in this life in accordance with the Law and Jewish tradition.

The Old Testament and the religion of Judaism are the foundation for Christianity. Jews who become followers of Jesus as the Messiah call themselves completed Jews.

Then what advantage has the Jew? Or what is the benefit of circumcision? Great in every respect. First of all, that they were entrusted with the oracles of God. What then? If some did not believe, their unbelief will not nullify the faithfulness of God, will it? May it never be! Rather, let God be found true, though every man be found a liar, as it is written,
"That You may be justified in Your words,
And prevail when You are judged."
Romans 3:1-4

Assignments for Lesson 15

Worldview Write or recite the memory verse for this unit from memory.

Project Finish your project for this unit.

Literature Continue reading *Blood Brothers*.

Student Review Answer the questions for Lesson 15.
Take the quiz for Unit 3.

Al-Amin Mosque and Maronite Cathedral of St. George in Beirut, Lebanon

4 The Middle East Part 2

We begin by considering the human and geographic factors that define Armenia. We see one example of the significance geography has in warfare as we study the disastrous World War I campaign at Gallipoli that threatened the career of Winston Churchill. Saudi Arabia has long restricted the activities of women, but that is changing. Geography is about places, and we look at the importance of special places from the perspective of pilgrimages that people make. The worldview lesson summarizes Christianity and the Christian worldview.

Lesson 16 - Armenia and Its People
Lesson 17 - Churchill and Gallipoli
Lesson 18 - To Be a Woman in Saudi Arabia
Lesson 19 - Sacred Geography: The Meaning of Pilgrimages
Lesson 20 - Faith System: Christianity and the Christian Worldview

Memory Verse	Memorize John 8:31-32 by the end of the unit.
Books Used	The Bible
Exploring World Geography Gazetteer
Blood Brothers
Project (Choose One)	1) Write a 250-300 word essay on one of the following topics:
- What is a special place for you or your family? Why is it special? What does it mean to you to go there; or, if you haven't been, what would it mean to be able to go there? What if any is the spiritual significance of the place? (See Lesson 19.)
- Explain your understanding of what the New Testament says about what women can and cannot do, and why the New Testament says these things.

2) Draw a picture of a special place you like to visit.

3) Write a short story about someone making a pilgrimage. |

Rug Weaving in Armenia

16 Armenia and Its People

Twin sisters Sahkanush and Haykanush Stepayan learned the ancient Armenian art of rugmaking when they were teenagers. Now in their mid-twenties, they have devoted their lives to weaving woolen rugs by hand.

Armenian rugmakers sit at their looms for hours, weaving intricate patterns of colors in multiple symmetrical combinations interspersed with variations. They imagine designs in their minds and then bring those designs to reality with woolen threads, sometimes tying 144 knots per square inch. What art experts often call the Turkish knot today is really the Armenian knot.

The craft dates from antiquity. The ancient Greek historian Herodotus (400s BC) and the medieval Italian traveler Marco Polo (1254-1324), among many others, recorded their praise of Armenian rugs. Women have usually been the artisans. They have made rugs as gifts, as parts of their dowries when getting married, and as commemorations of special persons, events, or themes.

Interest in the art form declined for several years, but it is making a comeback as one generation teaches another. The handing down of skills and the shared understanding of the meaning of the designs enables Armenian communities to appreciate their past.

Rugmaking weaves through many of the factors that define Armenia.

Geography and People Define Armenia

Geography defines Armenia. It is a narrow, landlocked, north-to-south country on the wide isthmus between the Black Sea and the Caspian Sea. This region divides the Middle East from Europe. Armenia lies just south of the Caucasus Mountains. Many geographers place it in the Transcaucasia or Lesser Caucasus region. The countries of Azerbaijan, Iran, Turkey, and Georgia are Armenia's closest neighbors. Armenia is slightly smaller than Maryland.

As one source put it when describing Armenia, "There are no lowlands." Mountains, high valleys, and plateaus dominate the country. Average elevation is over 5,000 feet, and only ten percent of the land is below 3,300 feet. Mount Aragats at 13,418 feet is the highest point in the country. Another mountain, Mount Ararat (located just over the border in Turkey), is a cherished symbol of the Armenian people.

Extinct volcanoes are numerous. Many relatively short rivers line the country. Lake Sevan covers 525

Yerevan, Armenia, and Mount Ararat

square miles, or about five percent of the surface area of the country. Scientists have identified over 3,000 plant species in the country. Armenian rugmakers often weave representations of the mountains, flora, and fauna of their country into their creations.

People define Armenia. Of the three million people who live there (Maryland by comparison has about six million), some 98% are ethnic Armenians. About half the population lives in the part of Armenia that is below 3,300 feet of elevation, and about two-thirds of the people live in cities. The other third of the population work in agriculture in rural areas. The cities are generally prosperous and economically advanced.

Because of wars, displacement, and emigration, about as many ethnic Armenians live outside of the country as live in it. Some 1.5 million Armenians live in countries of the former Soviet Union, and another one million live in the United States. Glendale, California, is the home of the Armenian Rug Society.

The Nagorno-Karabakh region in the southern part of neighboring Azerbaijan is a scene of conflict because of the people who live there. Although a mountain range separates Nagorno from Armenia, a majority of its people are ethnically Armenian. Its residents have fought for independence from Azerbaijan, have declared their independence, and have even held their own elections; but no other country recognizes the region as independent. Most of the Armenians in Nagorno want to be either independent, an autonomous region, or part of Armenia. The governments of Armenia and Azerbaijan have been at a standoff for many years; but in 2008 the leaders of the two countries agreed to work for a solution to which all parties can agree. The fact that it hasn't happened at this time of this writing is an indication of the difficulty of the task.

Lesson 16 - Armenia and Its People

Faith and Language Define Armenia

Faith defines Armenia. The country was the first one in the world whose leader declared it to be a Christian nation. Tradition has it that the apostles Bartholomew and Thaddeus spread the gospel into the region in the first century. Church historians credit Saint Gregory the Illuminator with converting King Tiridates III to the Christian faith about 314 AD, whereupon the king declared Armenia to be a Christian nation. Today about 93% of Armenians are members of the Armenian Apostolic Church, which is a branch of the Orthodox Church. After Armenia became a Christian nation, rugweavers often included crosses and angels in their designs. A small percentage of Armenians are Catholics, and an even smaller number are Muslims.

Language defines Armenia. Armenian is part of the Indo-European language family, but it is distinct from other languages in significant ways. In the early 400s St. Mesrop Mashtots developed an alphabet of thirty-six letters (two more were added later) to translate the Hebrew Old Testament and the Greek New Testament into Armenian. Even with its distinctiveness, several variations and dialects of Armenian exist. Sometimes rugmakers have used Armenian to inscribe a saying, a prayer, or information about the creation of the rug into their works.

Originally built in the 4th century, the Cathedral of Etchmiadzin has long been a spiritual center for the Armenian people.

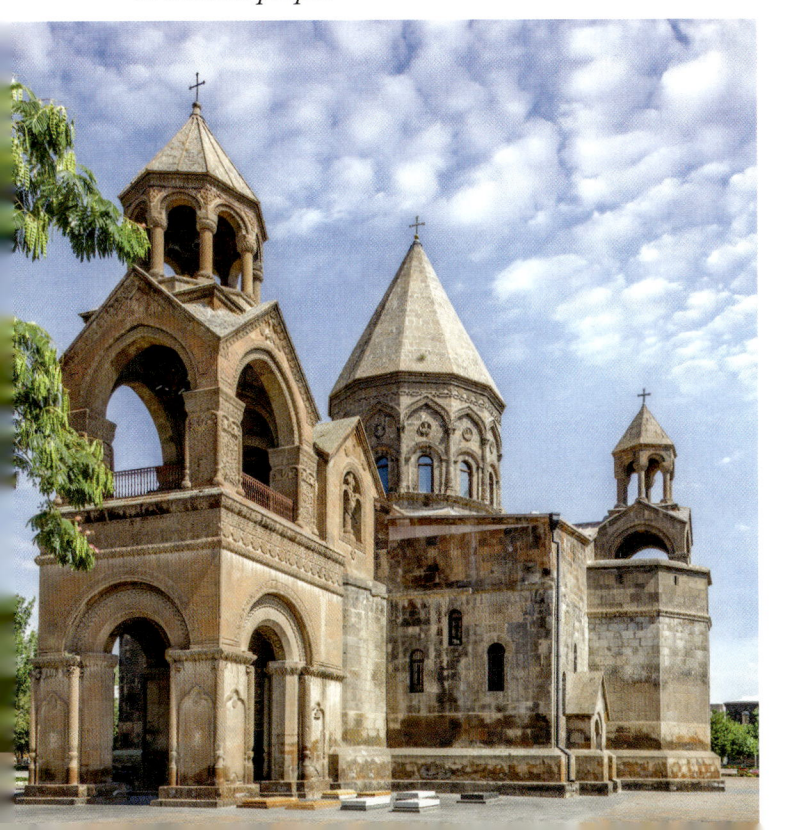

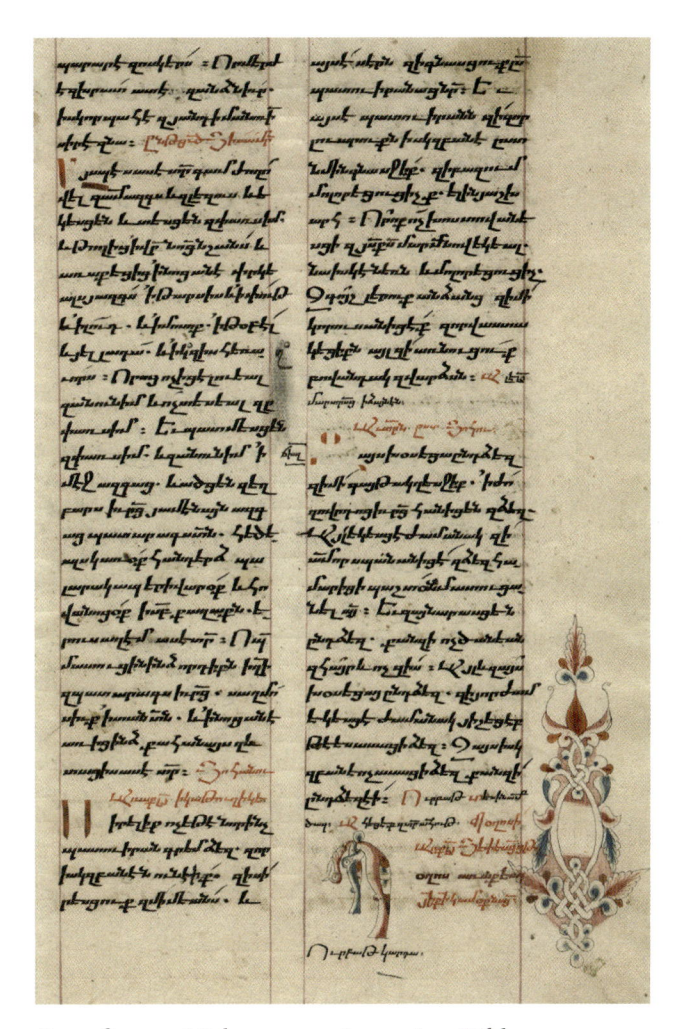

Page from a 12th-century Armenian Bible

The Arts Define Armenia

Obviously rugweaving is a major part of Armenian identity, but other artistic expressions flourish there also. Khachkars (which means "cross-stones") are elaborately-carved memorial stones that bear a cross along with other intricate designs. These

Khachkars at the Noraduz Cemetery

are usually burial makers but can serve as memorial stones for historic events as well. Khachkars are a distinctly Armenian art form. The most elaborate ones date from the 1200s.

Music defines Armenia. During the Middle Ages the Armenian Apostolic Church developed a distinctive style of music for its worship, and this unusual style is still used today. Traditional and popular Armenian music also have a style that is different from that of other countries.

History Defines Armenia

Many times in history, Armenians have been the victims of aggression and even genocide. Modern Armenia is just a small part of what was ancient Armenia. Much of its land was carved up and claimed by Turkey, Russia, and other countries. Many foreign powers have dominated the land at various times, including the Romans, Byzantium (Eastern Roman Empire), Arabs, Persians, and the Ottoman Empire. Shortly after Communists took over Russia, they took control of Armenia and made it one of the Socialist Republics in the U.S.S.R.

In the late 1800s the Ottoman Empire and the Kurds took the offensive to wipe out the Armenian people. The aggressors did not succeed, but they set the stage for one of the worst incidents of ethnic genocide in world history. During World War I, Ottoman soldiers forced large numbers of Armenians to leave their homes and move into the deserts of Syria. About one and a half million Armenians died from a lack of food and water or were killed by Ottoman, Kurdish, or Arab fighters. A memorial khachkar in Etchmiadzin, Armenia, pays tribute to the victims of this genocide.

During the Soviet era, the government forced rugmaking into factories and made it illegal to make rugs in homes, although some people continued to make them in their homes despite the law. As the Soviet Union was disintegrating, Armenia declared its independence in 1991. It is during this post-Soviet period that rugmaking has experienced a revival.

Armenia is a land of beauty and tragedy, a land of toil and great accomplishment. These themes of humanity weave through Armenian life and history as the threads of an Armenian rug weave their patterns. As Armenian rugweaving has endured, so the Armenian people have endured to the present day.

The Bible records the assassination of the Assyrian king Sennacherib. His killers escaped to "the land of Ararat" (in or near modern Armenia) after committing their foul deed.

Lesson 16 - Armenia and Its People

It came about as he was worshiping in the house of Nisroch his god, that Adrammelech and Sharezer killed him with the sword; and they escaped into the land of Ararat. And Esarhaddon his son became king in his place.
2 Kings 19:37

Assignments for Lesson 16

Gazetteer Read the entries for Armenia and Georgia (pages 2 and 6).

Worldview Read the poem "The Gods of the Copybook Headings" by Rudyard Kipling (pages 252-253).
Complete the assignment about the poem in the *Student Review Book* at Lesson 16.

Project Choose your project for this unit and start working on it. Plan to finish it by the end of this unit.

Literature Continue reading *Blood Brothers*. Plan to finish it by the end of this unit.

Student Review Answer the questions for Lesson 16.

Turkish Troops at Gallipoli (1915)

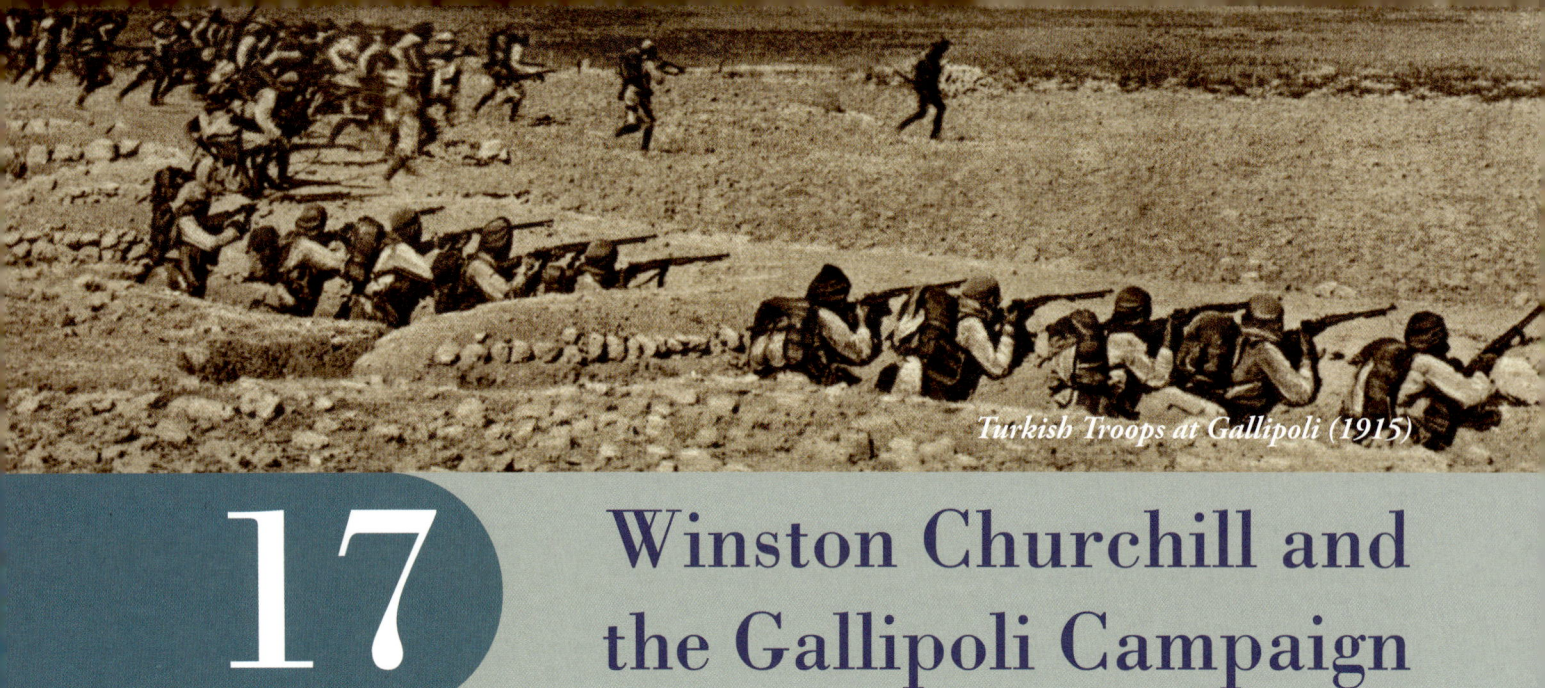

17 Winston Churchill and the Gallipoli Campaign

Winston Churchill needed a victory. As First Lord of the Admiralty of the United Kingdom (a position similar to the U.S. Secretary of the Navy), Churchill was involved in the military planning for the Great War (later known as World War I) that had engulfed Europe in 1914. By early 1915, he and the other British leaders had seen the war bog down to a trench war that for months had hardly moved along the entire length of the front in France. The Allies (primarily the British Empire, France, and Russia) felt the need for a decisive, dramatic blow against the Central Powers (primarily Germany and Austria-Hungary). The Ottoman Empire had been declining in strength for many years before the Great War began. Hoping to regain some of its power, the Ottoman Empire aligned with Germany.

As an ambitious politician, Churchill also needed a victory to advance his career. A demonstration of administrative skill in conducting the war would likely help his chances of becoming prime minister someday. To achieve this success, he needed to deal with the factor of geography, among other things, in a key area of the world.

Geography and War

Geography is a key factor in warfare. Military planners must consider the physical geography of the positions that each side holds, any geographic features that their forces would encounter in a movement, the ease or difficulty of maintaining supply lines, and possible escape routes for both armies. Success in battle requires different planning for the jungles of Vietnam than for the beaches of northern France.

Military leaders prefer to choose the landscape or seascape on which they fight, but they are not always able to do so. Armies prefer to hold higher ground because it is more defensible. The lay of the

Winston Churchill is depicted (standing at right) in this 1913 political cartoon related to conflict in the Balkans that preceded World War I.

102

Lesson 17 - Winston Churchill and the Gallipoli Campaign

land played a key role in the battle of Gettysburg during the Civil War. The battle of Chattanooga during the same war involved Confederate positions atop Lookout Mountain and Union troops scaling almost sheer cliffs to take the mountain. Invading armies have to consider what the terrain before them is like.

During World War II, fighting in northern Africa on sandy deserts required different equipment and strategy than did the Allied invasion of Italy, with its beaches and cliffs. The war in the Pacific involved naval battles and intense hand-to-hand combat as the Allies landed troops to take islands held by the Japanese. The Allies did not invade every island the Japanese held, but they "hopped" over some, leaving the enemy forces on those islands isolated and cut off from supplies.

The D-Day invasion in northern France involved intensive geographic planning. The Allies had to transport troops and equipment in Landing Ship Transports (LSTs) across the English Channel, facing enemy mines and gunfire as they did. In some places the Germans held elevated positions on cliffs along the beach, which Allied troops had to climb while Germans fired on them. Invasion planners had to wait for good weather to give them their best chance for success.

Weather is a significant factor in war. Moving men and material through mud is difficult. Until recent times, fighting generally ceased during winter because wet, snowy, and cold conditions made combat and movement difficult. Note the comment in 2 Samuel 11:1 that refers to "the spring, at the time when kings go out to battle." Napoleon's invasion of Russia in 1812 and Hitler's invasion of Russia in 1941 both failed in large part because of the harsh Russian winter and the overextension of supply lines.

Recent wars in Iraq and Afghanistan illustrate the significance of geography in warfare. Soldiers had to contend with vast sandy deserts in Iraq and mountains and caves in Afghanistan. The standard Army uniform changed from olive drab to desert camouflage to blend in with those settings.

A Country That Faces in Two Directions

We usually think of the Great War as occurring only in Europe, but the conflict was much wider in scope. In looking for a place where Allied troops could achieve a decisive victory, Winston Churchill looked to Turkey.

Turkey is where Asia and Europe meet. The Black Sea is to the north, the Mediterranean is to the south, Asia is to the east, and Europe and the Aegean Sea (an arm of the Mediterranean) are to the west. As a result, Turkey is economically and militarily significant. Turkey controls the water passage between the Black Sea and the Mediterranean Sea. This affects Russia, Eastern Europe, the Balkans, and the nations of the Middle East.

Bridge over the Bosphorus, Istanbul, Turkey

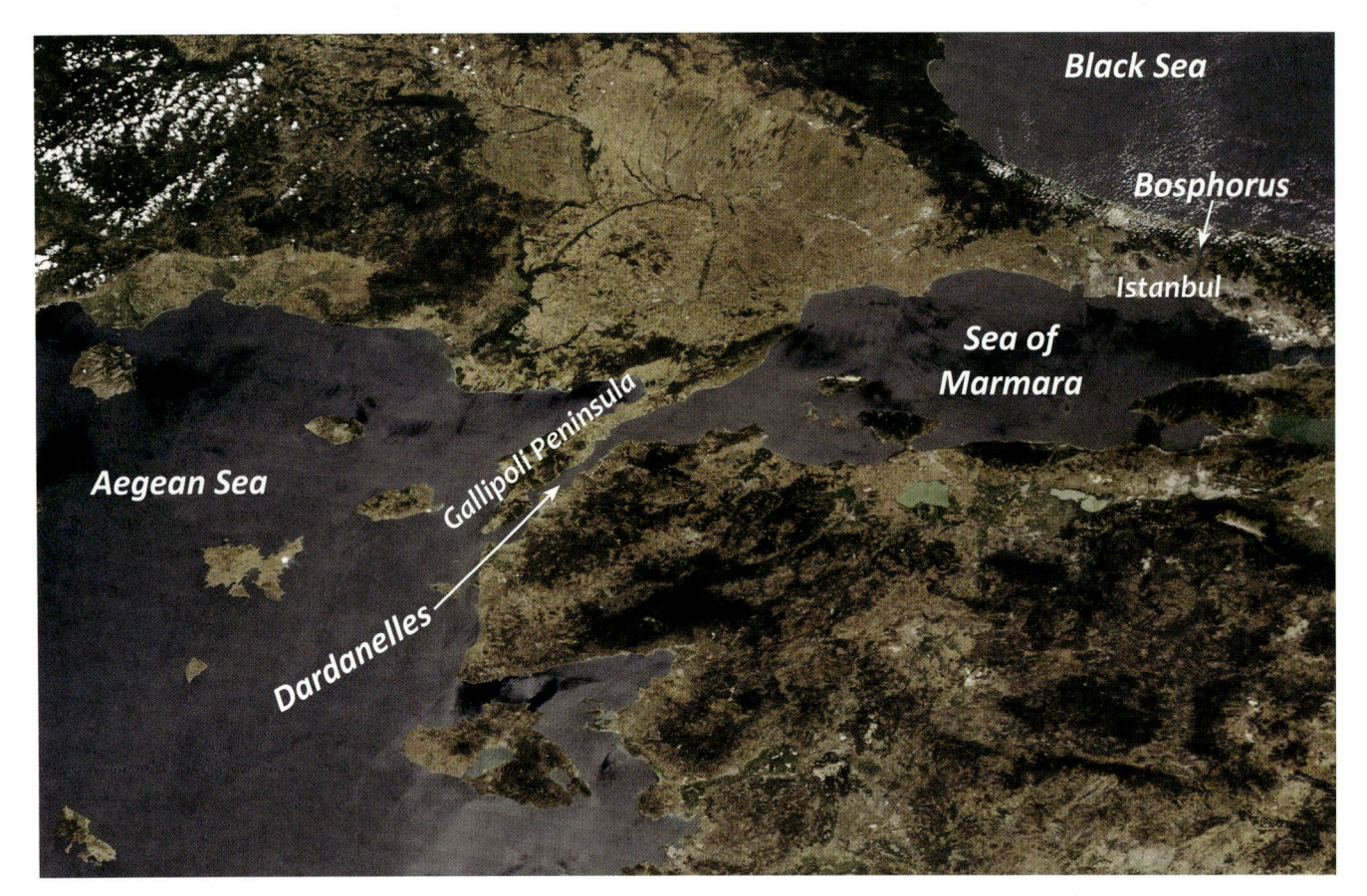

The Turkish Straits

Turkey is one of two countries located in both Asia and Europe. Russia is the other such country. The city that spans the two continents and that has been known successively as Byzantium, Constantinople, and (since 1923) Istanbul was the capital of the Byzantine and Ottoman Empires.

Ninety-seven percent of Turkey is in Asia. The other three percent, the region known as Thrace, is in Europe; but most of Istanbul is in Europe.

The Turkish Straits

The Dardanelles, the Sea of Marmara, and the Bosphorus, together known as the Turkish Straits, make up one of the busiest and most strategically located waterways in the world. The Straits separate Europe from Asia. They provide a water route between the Black Sea and the Mediterranean Sea.

The Bosphorus, to the northeast, is 19 miles long and varies between less than a half mile to 2.3 miles wide. The central part of Istanbul lies on either side.

The Sea of Marmara, which lies between the Bosphorus and the Dardanelles, covers 4,382 square miles, a little less than half the surface area of Lake Erie. The Dardanelles (once known as the Hellespont) to the southwest connects the Sea of Marmara with the Aegean Sea and then the Mediterranean. The Dardanelles is 38 miles long and varies from .75 mile to four miles wide.

The Turkish Straits hold strategic importance for Russia. Russia has no year-round, warm-water port that provides direct access to the world's seas for its naval and trading vessels. The country's best alternative is its Black Sea coast, which is why it has been interested in controlling the Crimean Peninsula for many years. But even with that port, Russia has to get its vessels through the Straits to get to the Mediterranean and then to the Atlantic. Russia is not able to do this if an enemy country controls the Straits. In 1915, at the beginning of World War I, the Ottoman Empire controlled the Turkish Straits and had the Russian navy bottled up on the Black Sea.

Lesson 17 - Winston Churchill and the Gallipoli Campaign

The Allied Strategy

In 1915 the British and Allied War planners, led by Winston Churchill, proposed a naval assault from the Mediterranean on the Gallipoli Peninsula on the northwest side of the Dardanelles. Their hope was to take the Dardanelles and then move on to capture Constantinople, drive the Ottomans out of the war, and allow the Russian navy to become more actively involved in the war. The Allies believed that taking Gallipoli and Constantinople would decisively weaken the Central Powers. They also assumed that their enemy Germany would have to send some of their troops fighting in Western Europe to help in the defense of Constantinople. Fewer German troops in Western Europe could turn the tide for the Allies in that region.

Gallipoli has long sandy beaches and high cliffs behind them. Ottoman forces held the area, but the Allies did not believe the defending forces would provide strong resistance to an attack. Churchill believed that taking Gallipoli would only require a few days.

An Allied fleet made up mostly of older vessels began an attack on Gallipoli in February of 1915, but the Ottoman forces held. Another naval assault in March again failed to dislodge the defenders. The Allies then decided to invade Gallipoli with ground troops and dispatched a 75,000-man force in April. Key units came from France and from ANZAC (Australia and New Zealand Army Corps). This was the largest military landing in history to that time. The Ottoman defenders numbered about 84,000 men. Some German troops took part in support of the Ottomans.

The Allies were able to secure small beachheads, but they underestimated the strength of Ottoman resistance and the difficulty of the terrain and could not move further inland. Just as was happening in Europe, the conflict bogged down, dragged on for months, and became a bloodbath. The troops had to contend with intense heat during the summer, flash floods, and freezing cold as winter came on.

As reinforcements arrived, a total of about 500,000 soldiers participated on each side. Estimates of casualties (killed, wounded, or stricken by illness) vary greatly, but some estimate the casualties to have been about 250,000 soldiers on each side. Perhaps 75,000 Ottoman soldiers and 45,000 Allied soldiers died. Many deaths were the result of disease because of the horrible health conditions that the soldiers had to endure. Allied leaders finally decided to withdraw, which they did in December of 1915 and January of 1916.

The Gallipoli campaign failed for several reasons. The Allies did not have sufficient knowledge of the terrain, and they underestimated Ottoman resistance. Planners did a poor job preparing for the attack. No element of surprise was possible with a fleet of slow-moving warships approaching from the Mediterranean. In addition, it is almost impossible to hold terrain without ground troops, whom the Allied planners did not bring into the campaign until too late. The troops that the Allies did put into action were inexperienced and had inadequate equipment and ammunition.

Australian Troops at Gallipoli

Consequences of Gallipoli

Winston Churchill resigned as First Lord of the Admiralty, although he became minister of munitions in July of 1917. Many people believed that the debacle would end his political career. He did recover enough credibility to serve as prime minister twice in future years, but criticism about his insistence on undertaking the Gallipoli campaign and his reluctance to end it haunted him for the rest of his life. The Allies were unable to strike a decisive blow in either Europe or the Turkish Straits, and the war continued with massive loss of life until November of 1918.

Though the Ottoman Empire won the battle of Gallipoli, it lost the war. The Allies defeated the Central Powers. The Ottoman Empire came to an end in 1923, but the country of Turkey arose from its ashes. The first president of Turkey was the Ottoman military commander at Gallipoli, Mustafa Kemal. He became known as Kemal Ataturk (which means Father of the Turks).

The Gallipoli campaign continues to have significance for participating countries. The Turks call it the Canakkale campaign, for a small town on the peninsula by that name. Turkey looks back on the battle as a source of national pride and identity. Turkey honors March 18, the date on which the Ottoman forces repulsed the last Allied naval assault, as Martyrs Day.

Although three bridges carry traffic across the Bosphorus, a bridge has never spanned the Dardanelles. A six-lane bridge, begun in 2017 and to be completed in 2023, will be the first. It will be the longest suspension bridge in the world at 6,637 feet. The Turks call it the Canakkale 1915 bridge. April 25, the date of the ground invasion, is ANZAC Day in Australia and New Zealand. About 50,000 ANZAC troops took part in the campaign, and 8,000 died.

British Troops from India at Gallipoli

One story from the Battle of Gallipoli tells of Turkish soldiers raising a white flag during a battle. A Turkish officer climbed out of his trench to pick up a wounded Australian officer. After carrying the man to the Australian trench, the Turkish officer returned to his trench and the battle resumed. This statue, designed by Turkish sculptor Tankut Öktem, was placed on the Gallipoli Peninsula in 1997.

Still Facing in Two Directions

Because of its geographic location, Turkey still feels pulled in two directions—toward Europe and the West on one hand and Russia on the other. When Ataturk took power, he moved the country strongly in a western direction and away from its Muslim past. Ataturk's government declared the country to be officially secular, although Islam was by far the most widely practiced religion.

Lesson 17 - Winston Churchill and the Gallipoli Campaign

The North Atlantic Treaty Organization (NATO) is a group of countries who agree to provide common defense of each other. Turkey became a member of NATO in 1952 and allowed the United States to base nuclear missiles there that were aimed at the Soviet Union. In the twenty-first century Turkey has taken steps to become a member of the European Union.

Although Turkey has sought closer ties with Europe and the United States, it has also made reconciling gestures toward Russia. While some Turkish people favor a secular government, other leaders have given renewed honor to some elements of Islamic culture and religion.

Istanbul is still an important city, but the city of Ankara, which is in the interior of the country, has been the capital of Turkey since 1923.

The Turkish Straits are one of the most important and potentially volatile transportation choke points in the world. The attention it receives demonstrates its significance to Turkey, Russia, and the rest of the world. Even over a century after it happened, the Gallipoli campaign stands as a stark reminder of how important this geographic location is.

How important are the Turkish Straits for the world? Napoleon is supposed to have said, "If the earth were a single state, Constantinople would be its capital."

Military planners need to make careful preparation before going into battle. Jesus mentioned this when talking about the even more important decision of following Him as a disciple.

Or what king, when he sets out to meet another king in battle, will not first sit down and consider whether he is strong enough with ten thousand men to encounter the one coming against him with twenty thousand? Or else, while the other is still far away, he sends a delegation and asks for terms of peace. So then, none of you can be My disciple who does not give up all his own possessions.
Luke 14:31-33

Assignments for Lesson 17

Gazetteer Read the entry for Turkey (page 16).

Worldview Copy this question in your notebook and write your answer: In practical terms, what is the significance of the fact that Jesus was fully God and fully man (John 1:1, 14)?

Project Continue working on your project.

Literature Continue reading *Blood Brothers*.

Student Review Answer the questions for Lesson 17.

18 To Be a Woman in Saudi Arabia

Road in Saudi Arabia

She drove a car.

That was revolutionary.

In 2011 Manal al-Sharif of Saudi Arabia posted a video on YouTube of herself driving a car. For that, she served time in prison.

Seven years later, Saudi Arabia changed its law to allow women to drive. To celebrate, Manal posted a picture of herself behind the wheel of a car.

Saudi Arabia was the last country in the world to allow women to drive. This change in policy reflects both the tight restrictions under which women live in that country and the gradual changes taking place in Saudi society.

Background

Historically the Arabian Peninsula consisted of many small Arab tribes each with its own leader or sheikh. In 1932, after many years of effort, the powerful sheikh Abdul al-Aziz al Saud gathered the other sheikhs in the central part of the peninsula under his leadership and founded Saudi Arabia. By law the king must always be one of his descendants.

For many years Saudi Arabia has been an ally of the United States, despite its opposition to Israel and despite its suspected support of Islamic terrorist

Manal al-Sharif received a bachelor degree in computer science from King Abdulaziz University in Jeddah, Saudi Arabia. She worked for the Saudi Arabian Oil Company for many years but lost her job because of her activism.

activities. The Saudis allowed the U.S. to station troops on its soil during the 1990-1991 Persian Gulf War that drove Iraqi forces out of Kuwait.

Saudi Arabia is the homeland of Muhammad. It is the place where he founded the Muslim religion, and it is the location of the two sites that Muslims consider most holy, the cities of Mecca and Medina. Today Saudi Arabia is officially a Muslim state. About 93% of the population are Muslims. A large majority are Sunni, while 10-15% are Shia. The royal family and the leading clerics are part of the Wahhabi sect, which holds to a strict literal reading of the Qur'an, the foundational book of Islam. The government follows sharia law, which is the application of Islamic religious principles to government. Evangelism and the public practice of any other religion are illegal. Non-Muslims may not be Saudi citizens.

The country is about one-fifth the size of the United States. Mostly desert, it is the largest country in the world that does not have a river. Saudi Arabia is believed to have about one-sixth of the world's known oil reserves. Oil has been the predominant part of its economy, but the country is seeking to diversify its economic activity.

Women's Rights—Or the Lack Thereof—in Saudi Arabia

As we consider the status of women in Saudi Arabia today, we should remember that it was only a century ago, in 1920, that all women in the United States received the right to vote with the ratification of the Nineteenth Amendment to the U.S. Constitution. The final ratification vote took place in the Tennessee state house of representatives, where the amendment passed by only one vote.

Historically, women throughout the world have not had the same rights as men. In Israel in Jesus' day, for instance, women were not able to own property and courts did not consider their testimony. The respect that Jesus accorded women was a radical change in that society. Even in the U.S., women could not serve on federal juries until 1957. Only in 1973 were women able to serve on juries in all fifty states.

Oil Facility at Shaybah, Saudi Arabia

As the rights of women have increased in many parts of the world, this trend has largely passed by Saudi Arabia because of its strict interpretation of sharia law. Every woman in Saudi Arabia must have a male guardian who must approve of every major decision in her life. This guardian is usually her father first and then her husband, but it can be a brother or even a son. Other Muslim countries practice the male guardian system to some degree.

Here are some examples of male guardianship as practiced in Saudi Arabia. A woman may not get married, obtain a passport, or travel out of the country without her guardian's approval. She must get his permission to work outside the home or open a bank account. Some variation in the level of restrictions does exist from place to place, but some Saudi hospitals require the consent of the guardian to admit a woman or to perform surgery on her. In some places a woman may not leave her home without the male guardian's permission.

Women in Saudi Arabia deal with a fairly high rate of domestic violence, but they cannot file for divorce. Until 2019 a man could divorce his wife without telling her. A 2019 law now requires the court to send her a text message telling her that her husband has divorced her and giving her the divorce certificate number and the name of the court. However, a man can divorce his wife orally, which results in no documentation she can keep. In keeping with sharia law, a Saudi man may have up to four wives provided that he treats them equally.

Saudi Arabia has granted limited political rights to women. They can vote and run for office in local elections. There are no national elections, but the king appoints a 150-member Consultative Council. In 2013 the monarch granted thirty seats on the Council to women. However, some women have been arrested for publicly advocating greater rights for women.

Market in Medina, Saudi Arabia

Is It a Big Deal That Saudi Women Can Drive?

The change seems small but not small. Saudi women have to take a driving course to obtain a driver's license, and they still have much larger and more fundamental restrictions that they have to deal with in their lives.

The new law is part of the trend of changes. More Saudi women are graduating from universities and entering professions than ever before. Ride-hailing services are recruiting women as drivers. One impetus for the new law is that the government wants to diversify the economy, so leaders expect that permitting women to drive will increase the number of women in the workforce. Another even greater change is that women do not need a guardian's approval to get a license or to drive alone anywhere in the country.

As with any group dealing with change, some Saudis—men and women—have conflicting ideas about it. Without a doubt, however, the changes are historic. As one female student in a driving school put it, "We're the ones that are going to tell our grandkids about it, that we lived through this."

The Arabian Peninsula has been a key region in the history of the world for thousands of years. The Bible tells us about the queen of Sheba, whom historians believe ruled in Arabia.

Now when the queen of Sheba heard about the fame of Solomon concerning the name of the Lord, she came to test him with difficult questions.
2 Kings 10:1

Assignments for Lesson 18

Gazetteer   Read the entries for Bahrain, Oman, Qatar, Saudi Arabia, United Arab Emirates, and Yemen (pages 4, 12, 13, 14, 17, and 18).

Worldview   Copy this question in your notebook and write your answer: Jesus demands that His followers die to themselves and give Him complete loyalty. In specific terms, how is this sometimes difficult for you?

Project   Continue working on your project.

Literature   Continue reading *Blood Brothers*.

Student Review   Answer the questions for Lesson 18.

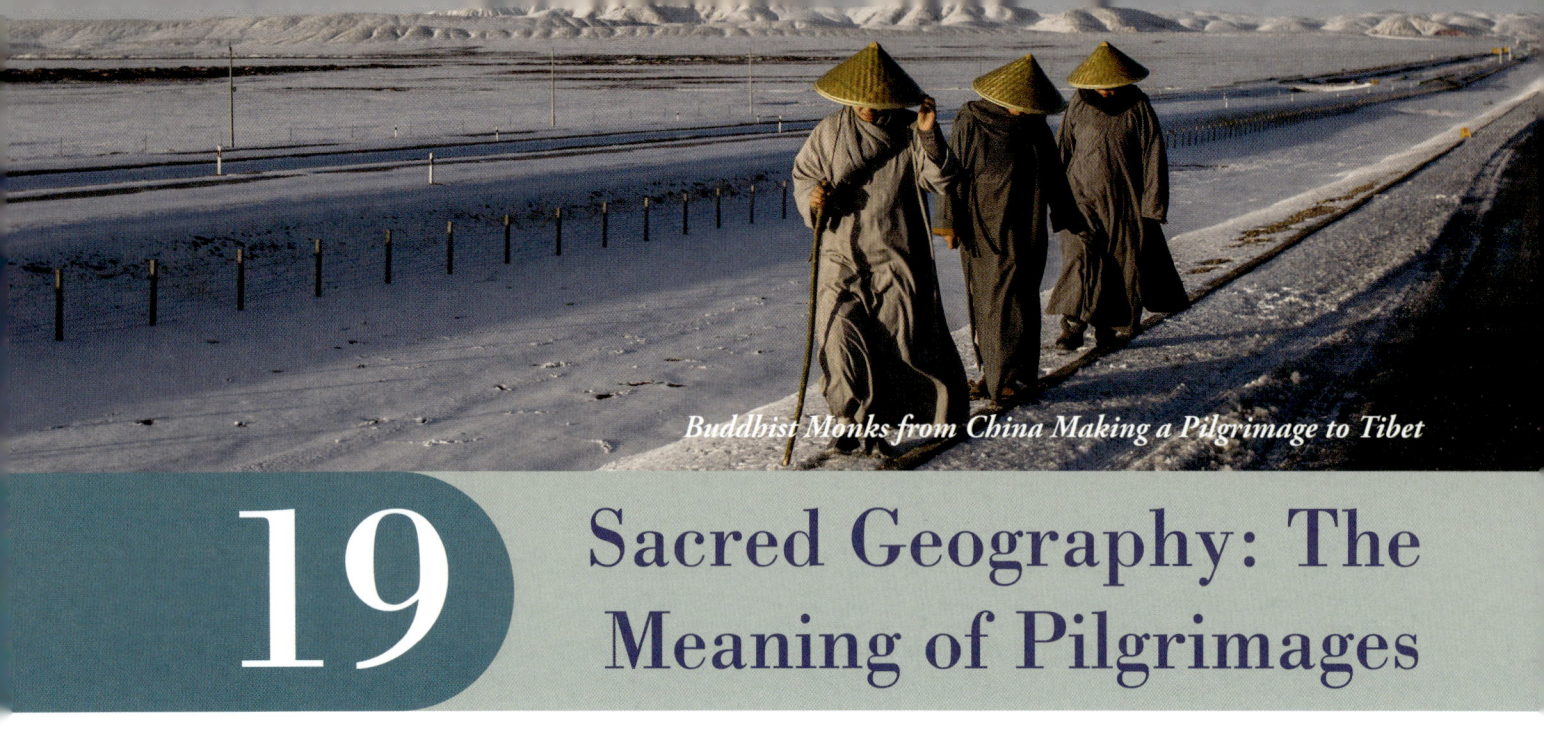

Buddhist Monks from China Making a Pilgrimage to Tibet

19 Sacred Geography: The Meaning of Pilgrimages

When I was growing up, every Mother's Day involved a ritual. Before church, we would go to my grandparents' house. In their backyard, which my grandfather had almost completely covered with flowers, my father would clip roses from a bush along the fence. We would wear those roses on the lapels of our suits in honor of our mothers when we went to church.

After church, we would go back to my grandparents' house and pick them up. My grandmother would laboriously climb down the steps. Ever since she had broken her hip, she could barely walk, even with a walker. We would drive out to the country to Godwin-Chappell Church. My grandfather's brother and his family would drive from Nashville, about an hour away, to meet us there.

The men would detach a couple of long shutters from a window and place them across chairs to make a table. The women would bring out all sorts of delicious food from the trunks of their cars. We set up folding chairs in circles. Someone would say an appropriate prayer, and then we enjoyed conversation and a delicious lunch. I especially remember the desserts.

After we had finished eating, the women would put away the serving dishes, the men would reattach the shutters, and we would dispose of the paper goods. While the adults talked, we children would go into the church building to see the beautiful dark-stained paneling, walk through the pews and around the pulpit area, leaf through the huge Bible that was open on the pulpit, and soak in the setting of this little country church where our ancestors had regularly gathered with their friends and relatives.

Then the adults would get from the cars the flower arrangements and trowels they had brought. We would go into the cemetery that lay up the hillside behind the church building to our ancestors' graves. The adults would place cut flowers in the dirt in front of the tombstones. Then we would drive a little further into the country to a small, overgrown family plot where more ancestors were buried and watch the adults place more flowers. The young people would wade in the cold water of Knob Creek, which we crossed to get to the more distant cemetery. Then we would say our goodbyes, exchange hugs and handshakes, and set off for our homes. The day's events were very different from how we usually spent our Sundays, but every year the day's events were almost exactly the same.

This annual pilgrimage did not directly have religious significance. Our faith tradition did not require it, it involved no solemn ceremony, nor did

112

Lesson 19 - Sacred Geography: The Meaning of Pilgrimages

we do it to obtain grace or a blessing. Yet in a sense it did have spiritual significance. It reconnected us with our family roots and with the walk of faith of those who had gone before. We traveled to a specific place to remember, and in a sense to pay homage to, people from earlier times who had an important role in our lives. We were blessed by doing so.

One more recollection conveys the personal and spiritual significance of these annual pilgrimages for me. On Friday, May 24, 1974, a young lady and I drove out to Godwin-Chappell. I got down on one knee on the front steps of that beautiful building in that beautiful and memory-rich setting and asked Charlene to marry me. I knew of no other place that had the same meaning for me in taking this step in my life.

This 18th-century painting from Peru depicts Roman Catholic pilgrims showing devotion to Our Lady of Cocharcas.

Pilgrims and Pilgrimages

A pilgrim is a traveler, someone who is on a journey. The King James Bible uses the word to describe the people of God as they live in this world (Hebrews 11:13, 1 Peter 2:11). A pilgrimage is a journey that a pilgrim takes, usually with a religious purpose. John Bunyan's *Pilgrim's Progress* is an allegory that describes one man's pilgrimage through life as he encounters people, places, and events that alternately strengthen his faith or challenge it.

The Jews traveled to Jerusalem three times per year to celebrate feasts as God commanded (Passover, Pentecost, and Booths). It was during one of these pilgrimages that Mary and Joseph became separated from Jesus (Luke 2:41-51). Bible scholars believe that pilgrims going to these feasts sang the Songs of Ascents (Psalms 120-134) as they traveled. Many modern Jews make a point of visiting the Wailing Wall in Jerusalem. This is the retaining wall that Herod the Great had built to support his remodeled temple. It is the only part of that construction project that still exists. Since Muslims possess the actual site of the Jewish temple, this is as close as Jews can get to the location of the temple that means so much to their history and faith.

In the early centuries of the church, many people wanted to visit the places associated with the ministry of Jesus (Bethlehem, Galilee, Golgotha, and so forth). The mother of Emperor Constantine traveled in Palestine and gave her official identification of the places that were significant in Bible history there. Centuries later, people also stared visiting the tombs of those the Roman Catholic Church designated as saints.

The Catholic Church has also identified certain places where purported miracles have taken place as appropriate sites for pilgrimages. A commonly reported occurrence has been a vision of Mary, the mother of Jesus. At some point Catholic teaching held that making such a pilgrimage could be a form of penance or a way to obtain forgiveness of sin. The New Testament nowhere enjoins Christians to make pilgrimages, but many Christians today fulfill a lifelong dream of visiting the Holy Land and report that the experience greatly enriches their faith. (And this is why Israel's Department of Tourism places advertisements in *Christianity Today*.)

Sometimes the goal of the journey is not the only part that a pilgrim considers sacred. A pilgrimage can be a journey that has several significant places along the route, such as a trip to visit locations that were important in the life of Patrick of Ireland.

Judaism and Christianity are not the only religions that have made provisions for pilgrimages. One of the five pillars of Islam is that a person make a pilgrimage to Mecca at least once in his life if he is able. Buddha's birthplace and the tree under which Buddhists believe that he gained enlightenment are common destinations for Buddhist pilgrims. Other religions have journeys and destinations that play important roles in the practice of those faiths.

Secular Pilgrimages

Pilgrimages express the importance of place. We carry our faith in our hearts everywhere we go every day, but we live out our faith in specific places: in the buildings and homes where we gather, and in places such as nursing homes and homeless shelters where we serve. Certain places have special meaning for us, such as where famous people lived or are buried and where special events have taken place. Martin Luther nailing his ninety-five theses for debate to a church door is not just a theological concept. It really happened at a specific time and place: the castle church in Wittenberg, Germany, on October 31, 1517. Travelers can go there and see the doorway (the actual door has been replaced). Nothing makes history come alive more than being at the actual places where historical events occurred.

Muslim Pilgrims in Mecca, Saudi Arabia

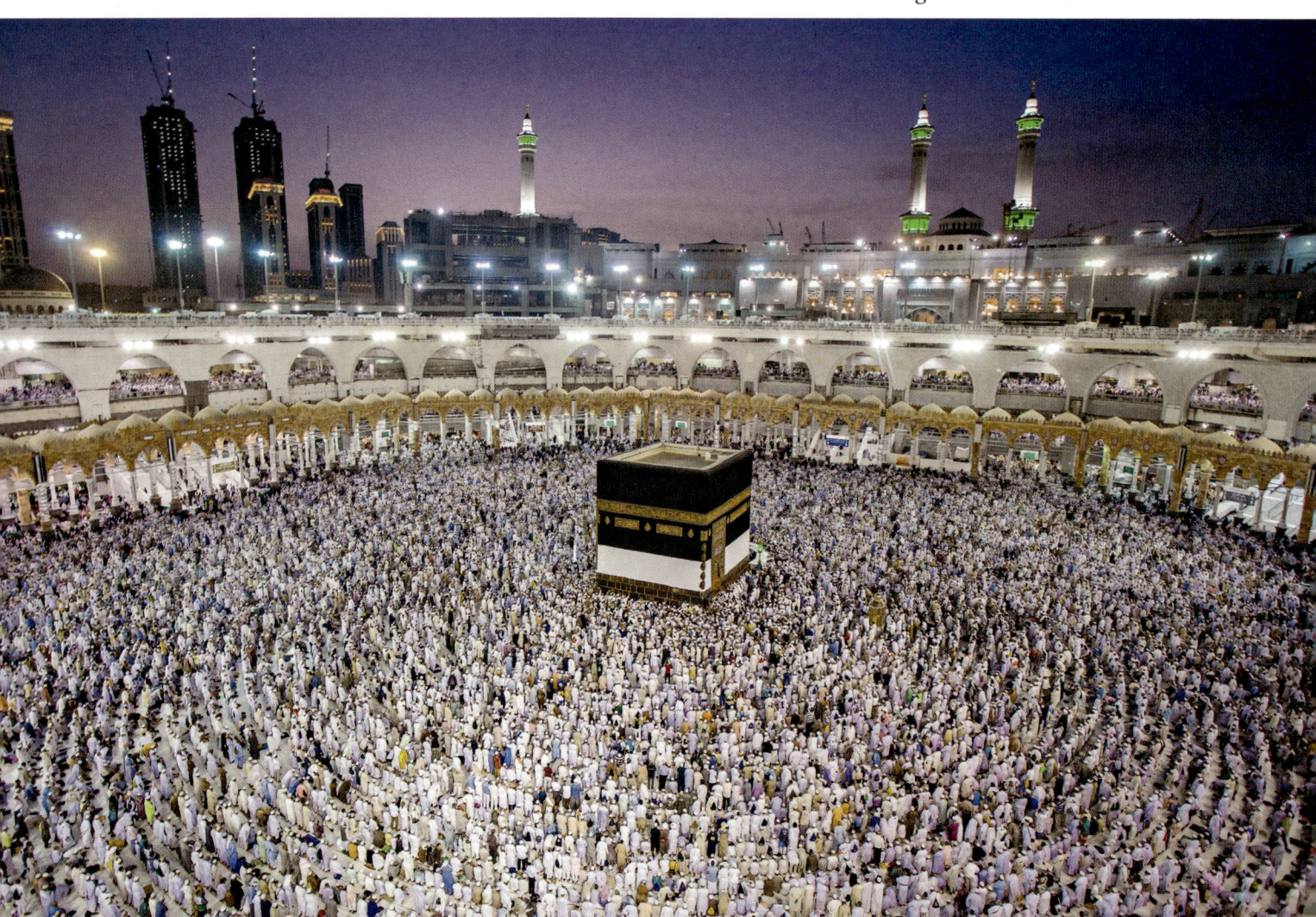

Certainly the religious meaning of pilgrim and pilgrimage is the most common usage of those terms, but they can have broader meaning. People sometimes use the word pilgrimage in a more everyday sense to describe a special trip, as in, "The businessman made his annual pilgrimage to the tax accountant to see how much he owed in taxes." Many Americans go on historic pilgrimages to see the places where events in American history occurred. We might consider trips to see historic sites in such places as Boston, Philadelphia, Gettysburg, Pennsylvania, and Washington, D.C., as historic pilgrimages. Many Turkish people go to the capital of Ankara to visit the tomb of their country's founder, Kemal Ataturk.

Another kind of pilgrimage could be going to natural features that many people consider to be quintessentially American, such as the Grand Canyon, Niagara Falls, and Yellowstone. Many American families sacrifice much and put out great effort to visit such places at least once during their children's years at home.

Pilgrimages can have personal meaning. For an adult, visiting the neighborhood where she grew up or campus of the college she attended can bring back many special memories. My father served in the U.S. Army in Europe during World War II. He crossed the English Channel the day after D-Day. About forty years after that event, he was able to visit the beaches of northern France in a tour group. When the French citizens who were there learned that he was a veteran of that dramatic landing, they gathered around him and sang to him. The event deeply affected him, and I almost come to tears myself writing about it now, over thirty years later.

A Personal Pilgrimage

My family and I do not give pilgrimages the religious significance that Catholics, Muslims, and followers of Eastern religions do; but I do believe I have a sense of what special—even sacred—places can mean.

My mother was born in Bristol, England. She met my dad at church when he was stationed in Bristol during preparations for the D-Day invasion. After spending the next several months in Europe following D-Day, he returned to England in April of 1945. He and Joan Clark were married on April 19.

Joan and Wesley Notgrass (1945)

When I was growing up, I heard and read a great deal about England. I came to appreciate such places as Stonehenge, Big Ben, the white cliffs of Dover, and other historically significant places there. My father taught me to love the novels of Charles Dickens, and I read several of them in my youth. In graduate school I came to appreciate the writings of C. S. Lewis. My mother died in 1975.

In 1997 my wife, our children, and I saved up, cashed in, and made the sacrifices necessary to travel to Great Britain. This was our first visit to the land of my mother's birth. We saw the famous and beautiful sights. We saw C. S. Lewis' house and the church he attended in Oxford. We toured Charles Dickens' home in London, and I bought a copy of David Copperfield at the gift shop there.

In addition, we met my uncle (Mother's brother) for the first time and I baptized him into Christ at his request. We saw Flax Bourton School that Mother attended (above left) and went to 36 Brighton Road where she was living when she met and married Dad (above right). I stood on the porch where Dad stood when he went to visit her. On this trip I renewed and deepened the connection with my mother that I had not had for over twenty years.

In terms of visiting places that are special—dare I say sacred?—to me, that trip was a pilgrimage.

Jesus, Mary, and Joseph made a pilgrimage to Jerusalem every year to celebrate the Passover.

Now His parents went to Jerusalem every year at the Feast of the Passover. And when He became twelve, they went up there according to the custom of the Feast. . . .
Luke 2:41-42

Assignments for Lesson 19

Geography   Complete the map activity for Unit 4 in the *Student Review* book.

Worldview   Copy this question in your notebook and write your answer: In practical terms, how does your relationship with God through Jesus Christ affect how you see yourself?

Project   Continue working on your project.

Literature   Continue reading *Blood Brothers*.

Student Review   Answer the questions for Lesson 19.

Borgund Stave Church, Norway

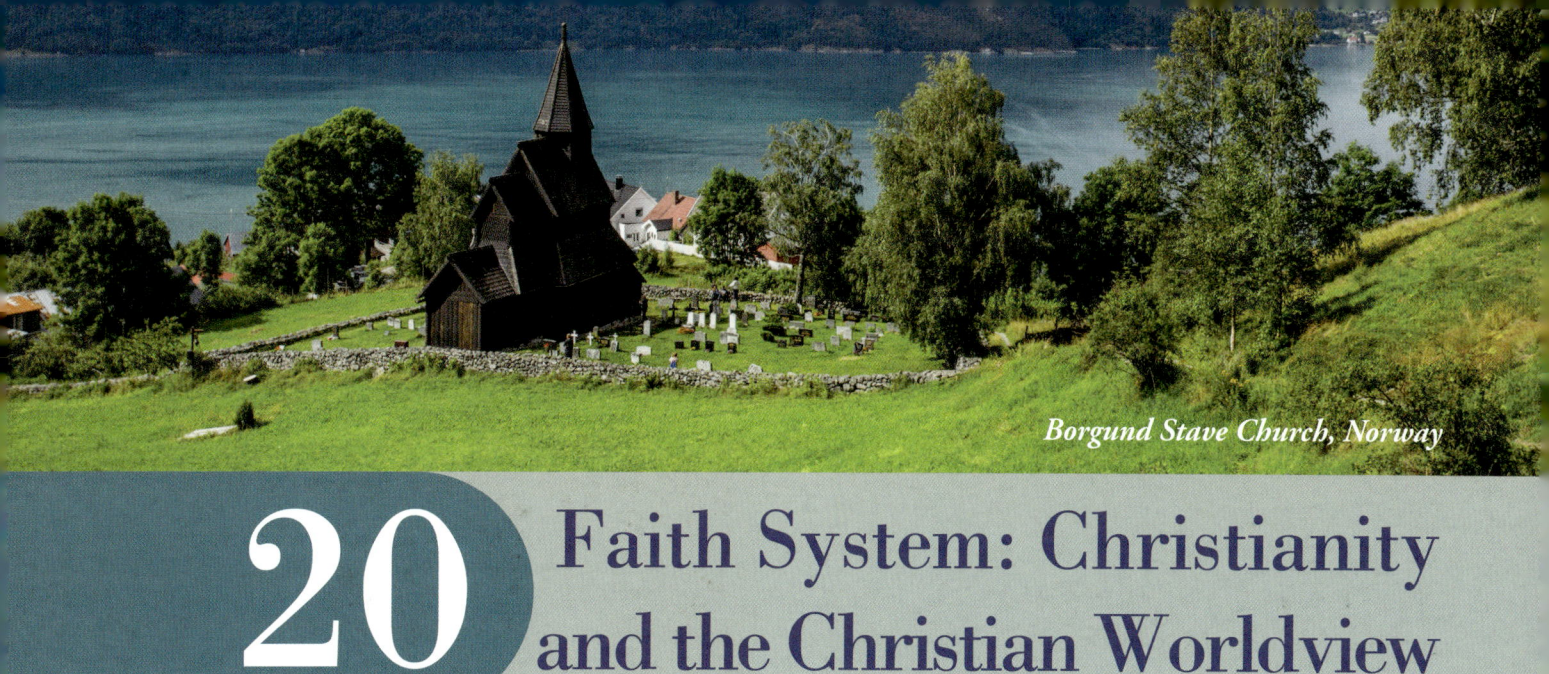

20 Faith System: Christianity and the Christian Worldview

The Christian faith stands or falls on Jesus Christ. The center of the faith is not the church, it is not the teachings of the church, nor is it the way you decide to define or describe Jesus. The foundation of Christianity is who Jesus is, as the New Testament presents Him. Anything else might have elements of Christianity, but it is not the basis of Biblical Christianity.

Who Jesus Is

The New Testament describes Jesus as the fulfillment of numerous prophecies of the Messiah in the Old Testament. He is the offspring of Eve who bruised (some translations say crushed) the head of Satan (Genesis 3:15). He is the One from Bethlehem who became ruler in Israel and was great "to the ends of the earth" (Micah 5:2-4). He was born the son of a virgin and was properly called Immanuel, which means God with us (Isaiah 7:14, Matthew 1:23). He is the suffering Servant of God who bore the sins of others (Isaiah 53). The accuracy of Jesus' fulfillment of these and all other prophecies of the Messiah goes beyond coincidence. His life proves that God brought about the coming of Christ and that Christ is who He said He is.

Jesus was fully God and fully man (John 1:1, 14). He lived without sin (Hebrews 4:15). He died on the cross to bear the sins of others (1 Peter 2:24). He arose from the dead and ascended into heaven (Matthew 28:6, Acts 1:9). Jesus will return one day to judge all men (Matthew 25:31-33).

Further Evidence for Christ

Another evidence regarding Jesus is the undeniable historical truth of his life. Without question Jesus lived, and a revolution in world history began with His life in Israel about 2,000 years ago. The world has known many influential leaders, but no other person has had the impact that Jesus has had.

The lives of the apostles and other early Christians provide further evidence regarding who Jesus is. At Passover one year, all of the disciples of Jesus fled and the Person they followed was crucified. Less than two months later, one of those who fled stood before his fellow Jews and proclaimed Jesus to be the long-awaited Messiah and Lord of the world. Before long, Saul of Tarsus, who had been a bitter enemy of the Way, began proclaiming Jesus as the Christ. Sabbath-observing Jews began meeting on the first day of the week to lift up Jesus among themselves.

Untold numbers of Christians laid down their lives rather than deny what they believed to be the truth of the ages: that Jesus was indeed Savior and Lord. These changed lives in the period immediately following Jesus' own life on earth are compelling evidence that Jesus was and is indeed the Christ.

Perhaps the strongest evidence for Christ is the empty tomb. The Jewish leaders knew exactly where Jesus was buried, but they never produced His body so as to crush the movement that claimed He was risen. The story of the resurrection is filled with surprising elements. For instance, women (whose testimony was not allowed in court) were the first witnesses of the empty tomb. Despite Jesus' predictions of His resurrection, apparently none of the apostles expected it. If the followers of Jesus had wanted to fabricate a story in an attempt to convince people, they could have come up with one that more people in their day would probably have believed. However, the story they told was true, and His followers had no other choice but to report it accurately.

Saint Sophia Cathedral, Harbin, China

Grytviken Church, South Georgia Island

The Call to Discipleship

Jesus did not call people to become adherents of a philosophy or a religion. He called people to die to themselves and to follow Him as His disciples (Matthew 11:28-30 and 16:24-27). Becoming a disciple means being born again (John 3:3-5). It means observing all that Jesus commanded (Matthew 28:19-20). It is a total trust in Jesus as Lord that commits all of a person's life to Him (Matthew 7:21, Acts 2:36-38).

All or Nothing

Jesus does not accept partial or divided loyalty. He is everything to you or He is nothing (Matthew 6:24, John 6:68-69). The claims that Jesus made about Himself and that the New Testament makes about Him are exclusive; they are all or nothing (Acts 4:12). He said that He and the Father (referring to God) are one (John 10:30). Hebrews 1:3 says that Jesus is the exact representation of God's nature. Jesus claimed to be able to forgive sins (Mark 2:5-10). He said that He was alive before Abraham was born (John 8:58).

Lesson 20 - Faith System: Christianity and the Christian Worldview

To say that Jesus was nothing more than a good moral teacher is to describe Him in a way that is inconsistent with what He said. He was not a good moral teacher if what He said about Himself and His relationship with God was mistaken, deceptive, or delusional. Because what He said is true, He is much more than a good moral teacher; He is Lord and God.

The Purpose of His Life

The central truths about Jesus that we mentioned earlier concern His birth, death, and resurrection. The gospel accounts in the New Testament have this emphasis. But the gospels also describe His life, His teachings, and His interactions with others. Why all this information about His life if all that matters are His birth, death, and resurrection?

The gospels tell about His life because through His life He taught His followers how to live. "The one who says he abides in Him ought himself to walk in the same manner as He walked" (1 John 2:6). "Christ also suffered for you, leaving you an example for you to follow in His steps" (1 Peter 2:21).

The Reason Christ is Central

The crucial question to answer is whether Christ is trustworthy. Because Christ is who He said He is, because His words are true, and because His life is consistent with what He said, then important consequences follow:

His teachings that reveal the nature of God let us know that we can believe God exists and that we can trust Him.

His dependence on Scripture as the inspired and authoritative Word of God teaches us the right attitude to have about Scripture.

His teachings about sin tell us that sin is real, that we can understand the true nature of sin, and that God has provided a way for us to deal with sin's presence and its penalty in our lives.

His instructions on how to live in this life will help us have a more successful life, as God defines success.

The Christian worldview includes the belief that God will triumph in the end, that Jesus will return and judge all and reign for eternity. His teachings about what happens after we die reveal to us the truth about the final things.

Church in Tabora, Tanzania

St. Andrew's Church, Chennai, India

The Church and Doctrine

After Jesus returned to heaven, God formed a fellowship of believers to communicate the truth of Christ to themselves and to others, to encourage each other in their walk, and to be the body of Christ. This fellowship is called the church.

The church is not the center of Christianity; the church is built on Christ, who is the center. Doctrine is not the center of Christianity; doctrine is based on Christ. Because Jesus is true, the church and doctrine flow from that truth. The center of Christianity is a Person and the relationship that individuals have with that Person who is Savior and Lord. Those people make up the church and believe the doctrines about Him that the New Testament teaches.

We need to understand and appreciate the significance of the fact that God's ultimate and most important word or message to mankind, His word on how to live in this life and how to live with Him forever, came not in the form of a scroll or a tablet but in the form of a person. God's truth comes to us in the form of a story. The central message of the universe, the foundation of hope that humans can have, is not primarily about power or dominion or intelligence or cleverness. God's central message includes all those things, but the heart of the message is suffering love. Christ showed this to us in clear and definite terms, and in this we can have a transformed life and the hope of redemption.

The Resulting Worldview

The reality of Christ as being who He says He is and the trustworthiness of His life and words lead to the Christian worldview. The Christian worldview grows out of the fact that God loves us.

"For God so loved the world, that He gave His only begotten Son, that whoever believes in Him should not perish, but have eternal life" (John 3:16). To be loved, to receive that love, and to understand the significance of that love is to be transformed. To be loved says that you matter and are worth the best and costliest that God can give, specifically the death of His Son. To view the world as one who is loved by the Maker and who is not insignificant or a piece of junk releases you from feeling a desperate need to make your mark or from wanting to trample on others out of a bitter heart. It frees you to love others as you are loved. "Beloved, if God so loved us, we also ought to love one another" (1 John 4:11).

The Christian worldview should affect how Christians view the world.

"By Him all things were created, both in the heavens and on earth, visible and invisible, whether thrones or dominions or rulers or authorities—all

Entoto Maryam Church, Addis Ababa, Ethiopia

Lesson 20 - Faith System: Christianity and the Christian Worldview

things have been created through Him and for Him. He is before all things, and in Him all things hold together" (Colossians 1:16-17). The world is from Him and for Him. This means that Christians should deal with the physical world—the environment, the things we value highly, and the everyday things we deal with—in a way that honors Him.

The Christian worldview informs us as to how we are to relate to other people.

"Therefore from now on we recognize no one according to the flesh" (2 Corinthians 5:16). This means that we see other people for who they are, as spiritual beings made in God's image whose lives matter to God and who will one day face their eternal destiny. We have the opportunity and responsibility to communicate to them their worth and their need to be reconciled to God.

A Minority Worldview

At some point in the history of the world, people began developing faith systems and worldviews that did not have God at the center. These faith systems grew, spread, and multiplied. At some later point, belief in the one true God became a minority view among the world's population. When God called Israel to Himself, most people in the world did not believe in Him. When Jesus came, most people in the world did not believe in God.

The second century Epistle to Diognetus is an explanation and defense of Christianity. His subject is not paganism, the dominant worldview of the day; nor is it Judaism, the exception to paganism that the Roman government permitted to exist without compromise with pagan beliefs. The writer calls Christians the "third race" or the "third way." It is not paganism or Judaism. The Christian worldview and lifestyle is distinct from those other two worldviews. The reason is not because Christians just wanted to be different. The reason Christianity is the third way is because of Jesus Christ. Those who were loyal to Him understood that they could not be loyal to any other way.

Our Lady of the Assumption Cathedral
Granada, Nicaragua

Today, even though Christianity has the largest number of adherents of any religion in the world, it is still a minority of the total world population. If you add up Muslims, Jews, Buddhists, Hindus, folk religions, and those who express no faith in God, the total is much greater than the number of people who say they are Christians. The general trend in our culture for many years has been a decline in the number of people who say they are Christians. Moreover, Christianity and confessing Christians have become a popular target for criticism and ridicule in the media.

All that is to say this: If you decide to follow Christ, you will likely get support from your family and immediate circle of acquaintances (although if

Singapore Life Church

those people are not Christians you might not get their support), but in the world in general you will likely be questioned, ridiculed, ignored, or worse. Choosing to follow Christ will not be popular with many people, either because they see it as an outdated and flaw-riddled belief system, or because your faith challenges them to think about their own lack of faith. A decision to adopt and follow the Christian way must come from a deep commitment to Christ as the Son of God, Savior of the world, and Lord, regardless of what others say or do.

[Jesus] said to them, "But who do you say that I am?"
(Matthew 16:15)

Assignments for Lesson 20

Worldview Write or recite the memory verse for this unit from memory.

Project Finish your project for this unit.

Literature Finish reading *Blood Brothers*. Read the literary analysis and answer the questions in the *Student Review*.

Student Review Answer the questions for Lesson 20.
Take the quiz for Unit 4.

Traditional Home in Algiers, Algeria

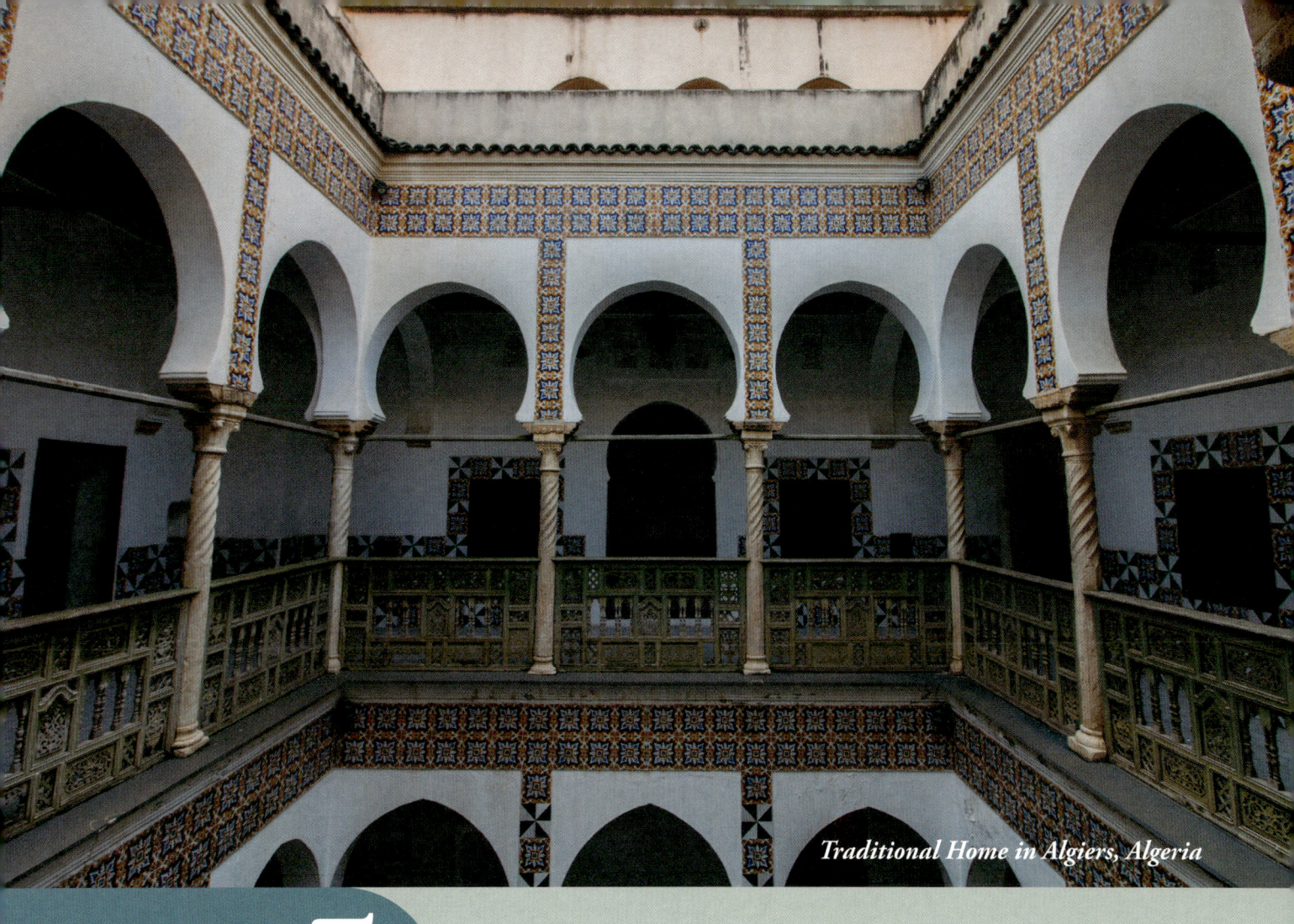

5 North Africa

The Arab Spring revolt that began in North Africa in 2010 spread to much of the Arab World and challenged the cultural and political status quo in many countries. The completion of the Suez Canal in 1869 was a geographic accomplishment that has had wide economic and political consequences even into contemporary times. The Berbers are a North African ethnic group who have had impact in the Mediterranean region for centuries. The Allied military campaign in North Africa during World War II was the beginning of the end for the Axis powers. The worldview lesson examines the faith system of Islam.

Lesson 21 - Of Jasmine and Spring: The Story of Tunisia
Lesson 22 - A Man and a Canal
Lesson 23 - The Berbers
Lesson 24 - The North Africa Campaign in World War II
Lesson 25 - Faith System: Islam

Memory Verse   Memorize Psalm 36:6 by the end of the unit.

Books Used

The Bible
Exploring World Geography Gazetteer
Patricia St. John Tells Her Own Story

Project (Choose One)

1) Write a 250-300 word essay on one of the following topics:
   - Imagine you were someone subject to political or religious oppression. How would you feel? What actions would you take to address the situation?
   - Identify three things (events, inventions, or trends) that have changed the world in your lifetime and tell why they have had an impact.
   - What do you think is the best way to reach Muslims with the gospel of Christ?
2) Research Berber carpets and draw a pattern for one.
3) Plan a civil engineering project for your area (such as a bridge, road, or canal) that would have a major geographic impact. Tell what would be involved in the project and what would need to take place. Draw a sketch of the completed project.

Literature

In an unpretentious, conversational narrative, Patricia St. John records her memories of a remarkable life. She begins with the heritage of Christian faith and missions that her parents established. She shares the adventures of a joyful childhood that led into a rich and useful life of service in the kingdom of God. Her memories include the England of her early and late life, North Africa where she served for decades in her middle years, and many travel experiences in a host of other countries. She reflects on the work of God in and through her family and friends and reveals her love and value for the individual people she met. Those who have enjoyed Patricia St. John's many works of fiction will enjoy learning the sources of her inspiration. Her story is one of service wherever she was called to be, from an English boarding school, to wartime London, to a Moroccan mission hospital, to Ethiopian refugee camps, to simple hospitality in her modest home.

Patricia St. John was born in 1919 into a missionary family. Her father traveled for ministry throughout her childhood while her mother maintained their home base in England with their five children. Patricia grew up to serve as a nurse during World War II, then joined her brother to work for many years as a missionary in Morocco. She helped to care for many young and elderly family members, eventually settling back in England. She continued to be involved in international missions into her later years. She authored many books for children, including the beloved *Treasures of the Snow* and *The Tanglewoods' Secret*. She died in 1993.

Plan to finish *Patricia St. John Tells Her Own Story* by the end of Unit 7. You have three weeks to complete this book.

Protestors in Tunis, January 20, 2011

21 Of Jasmine and Spring: The Story of Tunisia

On December 17, 2010, Mohamed Bouazizi prepared to sell fruits and vegetables from his stand in the small town of Sidi Bouzid, Tunisia, outside of the capital city, Tunis. Bouazizi was twenty-six years old and unemployed, but he was doing what he could as the breadwinner for his widowed mother and six siblings.

However, Bouazizi did not have a permit to conduct his business. The police approached him and demanded that he surrender his cart to them. Allegedly, the police demanded a bribe to let him sell, and a policewoman slapped him during the confrontation. Frustrated, humiliated, and angry, Bouazizi rushed to the front of a government building. He set himself on fire to protest the treatment the police had given him, and in doing so he took his own life.

The Jasmine Revolution

This event triggered weeks of protests and demonstrations in Sidi Bouzid and other Tunisian cities against poverty, government corruption, and political repression. Word of the demonstrations spread primarily through social media. The government reacted harshly, and in attempting to suppress the demonstrations, troops killed some 300 protesters.

Civil conflict consumed Tunisia for the next month in what the media dubbed the Jasmine Revolution (jasmine is the national flower of Tunisia). Demonstrators demanded the resignation of President Zine el-Abidine Ben Ali. Ben Ali at first reacted harshly, but as the protests grew and spread, he offered concessions to the protesters. These concessions did not satisfy their demands, and on January 17, 2011, Ben Ali resigned and fled the country to Saudi Arabia.

An interim government enabled a series of free elections beginning in October 2011. The government put a new constitution in place in 2014, and a workable and stable political system based on that constitution restored peace in Tunisia—for the most part.

The National Dialogue Quartet was founded in Tunisia in 2013. It is a coalition of businesses, human rights advocates, labor unions, and attorneys who worked to forge a stable and lasting democratic government that could lead the country out of the unrest and uncertainty of the Jasmine Revolution period. The Quartet was so successful that in 2015 it received the Nobel Peace Prize.

Representatives of the National Dialogue Quartet visited Vienna in 2016. From left to right are Houcine Abbassi of the Tunisian General Labour Union; Wided Bouchamaoui of the Tunisian Confederation of Industry, Trade and Handicrafts; Abdessattar Ben Moussa of the Tunisian Human Rights League; and Noureddhine Allègue of the Tunisian Order of Lawyers.

The Arab Spring

The Jasmine Revolution in Tunisia contributed to protest movements in several other countries across North Africa and in the Middle East in 2011. The media called these movements the Arab Spring. People living in other countries felt some of the same frustrations as many in Tunisia felt. The movements grew in part because of young people's use of social media to share information and ideas. Protesters demanded greater political freedom, economic opportunity, and other reforms. In many of the countries, protesters demanded the resignation of the head of government who had been in office for decades. Government authorities generally responded by cracking down on the demonstrations at first, but in some countries reforms did take place in varying degrees.

Demonstrations began in Egypt in January 2011, demanding the resignation of president Hosni Mubarak. Mubarak had been in office since 1981 following the assassination of Anwar Sadat. Egypt suffered from poor economic conditions, government corruption, and a lack of true political freedom. Mubarak resigned in February and handed power to the military, who killed over 800 protesters. Mubarak was later tried, convicted, and imprisoned. The conviction was eventually overturned but Mubarak remained in custody because of other issues. Over time one government replaced another, and radical Islamic groups such as the Muslim Brotherhood clashed with the military over power and control. Egypt became relatively more stable, but it is still not a free democracy as Americans understand that term.

Protests began in Yemen in January of 2011 against president Ali Abdullah Saleh, who had held office since 1978. Government forces killed hundreds of demonstrators. Saleh handed power to his successor in February of 2012 but unrest continued. Rebels killed Saleh during a battle with his supporters in late 2017.

In Libya, protests began in February of 2011. The government of dictator Muammar al-Gadaffi used extreme violence against protesters and killed thousands. The United Nations and NATO intervened for a time to protect citizens. Dissidents captured Gadaffi and executed him in October of 2011. Warring militias took control of different parts of the country. During this unrest, attacks by

Lesson 21 - Of Jasmine and Spring: The Story of Tunisia

Islamist militant forces at Benghazi in 2012 resulted in the death of the American ambassador and three other Americans there. Libya remains an unstable country, much like Iraq has been since the fall of Saddam Hussein in 2003.

In Bahrain anti-government demonstrations formed in February 2011, demanding political reforms and the removal of the prime minister, Prince Khalifa bin Salman Al Khalifa. Prince Khalifa was the king's uncle and had been in power since 1971 when Bahrain achieved independence. Clashes occurred with some fatalities. In January 2012 the king announced some reforms and gave parliament slightly more power, but some unrest continues.

Syria began experiencing violence in March 2011, with demands for greater freedoms, the release of political prisoners, and the removal of Bashar al-Assad, who had been president since 2000 following the death of his father, who had seized power in 1971. However, Bashar refused to resign. He clung to power and Syria dissolved into areas held by warring militias that are often supported by other countries or groups wanting to control the outcome. Assad's government met protests with extreme violence, including the apparent use of chemical weapons.

Generally peaceful demonstrations took place in Morocco, Algeria, Jordan, Oman, Iran, Saudi Arabia, and the United Arab Emirates, although some clashes and fatalities occurred. The monarchy ruling Saudi Arabia has instituted some social reforms there. Iran has seen continuing sporadic demonstrations against government policies.

This turmoil is primarily Muslim vs. Muslim. Some of it is simply an attempt by one faction to wrest control from another faction, but some of the early protests involved people sincerely wanting greater human and political rights and a change from what had become the status quo. The trigger or flashpoint for this turmoil was Tunisia.

A World Crossroads

Tunis is the critical geographic location where the cultures of Africa, Europe, and Islam intersect. The coastline of Tunisia and the coastline of Sicily (an island which is part of Italy) are less than 100 miles apart. Morocco is only eight miles away from Gibraltar in Spain in southwestern Europe, but Tunisia is the point in Africa that is closest to the heart of the European continent.

Tunisia is home to about 11.5 million people, ten percent of whom live in the capital of Tunis. Two-thirds of the Tunisian population are considered urban. Most of the people live near the northern and eastern coasts. Fewer people live in the semiarid southern part of the country that leads toward the Sahara Desert.

Two mountain ranges of the Atlas Mountains in northern Africa cross Tunisia west to east. Only a few are over 2,000 feet high, and the tallest is only 5,066 feet above sea level. Between these ranges lies fertile, productive farmland. About two-thirds of the country is suitable for farming. Beneath the earth, explorers have discovered oil, natural gas, phosphate, iron, and salt. Areas along the coast enjoy a Mediterranean climate, so tourism is a big industry. Several rivers flow in Tunisia, but only the Majardah does not dry up in the summer.

One example of an issue related to Tunisia's geographic location is the 2006 agreement between Tunisia and Malta to conduct joint exploration of the continental shelf that they share. The countries hope to exploit oil reserves that many experts believe to be present. The agreement resolved an issue that had festered for 35 years.

History

Tunisia's strategic location and natural resources have led to many invasions and settlements over the centuries. This in turn has created a rich and diverse culture there.

Ruins of Carthage

The Phoenicians established a settlement at Carthage, which was near the site of modern Tunis, in 814 BC. Carthage became a leading Mediterranean civilization in its own right. After a long rivalry with Rome, Rome conquered Carthage in 146 BC and ruled it for almost six hundred years. The Romans laid out the basis of much of the road system that is in use in Tunisia today. In 439 AD, the Vandals defeated the Romans in Carthage.

The Byzantine Empire (the Eastern Roman Empire based in Constantinople), ousted the Vandals from Carthage in 534 AD. Muslim invaders swept through the area in the mid-600s and took control of it. Muslim Moors who were forced out of Spain in the 1200s moved to Tunisia. The Muslim Ottoman Empire, which controlled much of the Middle East and northern Africa for centuries, gained control in 1534. It continued to claim authority until the end of World War I, although its control for the last many years was practically nonexistent. Within this Muslim-led context, Jews and Christians lived relatively peacefully in Tunisia for centuries.

France established actual, practical control over Tunisia in 1881 by declaring it a protectorate. An independence movement developed there before World War I, but it only became a viable force in 1934 as the New Constitution Party. Finally in 1956 France granted independence to Tunisia.

The new government under Habib Bourguiba sought to modernize the country quickly, but it was a one-party government that allowed no opposition and sought above all else to maintain unquestioned power. The Bourguiba government granted women more rights than in any other Muslim country, but it repressed both Islamic fundamentalists and reformers. After thirty-one years the country was suffocating politically, and in 1987 Zine el-Abidine Ben Ali led a bloodless coup that replaced Bourguiba.

Unfortunately, Ben Ali's regime was about as repressive as the former one. This was the background in 2010 when the Jasmine Revolution erupted in Tunisia.

Harvesting Dates

Tunisia Today

The population of Tunisia is almost completely Sunni Muslim. Ethnically the population is mostly Berbers, a pre-Arab people who have lived in northern Africa for millennia.

Tunisia is a modern country with a well-developed network of roads, railroads, and airports. Most people speak Arabic, but French is a second language and most newspapers are published in French. Radio and television stations broadcast in French, Arabic, and Italian. In modern urban areas most people wear western-style clothing and many buildings feature European architecture. In rural areas more traditional clothing and building styles hold sway. Life expectancy in Tunisia is among the highest in Africa.

Tunisians are generally welcoming, hospitable, and tolerant of others. Family is still central to the lives of many if not most Tunisians. Moderate political parties have done well in recent elections. This openness and progressive attitude is probably a big reason why Tunisia suffered two major terrorist attacks in 2015.

A level of unrest continues in Tunisia. Protests erupted in 2017 at an oil pipeline and pumping station in the south of the country. Protesters believed that the government and contractors were not providing as many jobs for young people as they had promised. A spokesman responded that the government had offered many jobs but that "sometimes there are more demands than the possibilities."

The Mediterranean beach at Monastir, Tunisia, is a popular tourist destination. It the background is the Ribat of Monastir, an Islamic fortification founded in 796 and expanded over the centuries.

Unit 5: North Africa

Atlas Mountains, Tunisia

Tunisia has a rich heritage and many positive aspects. It has come a long way in making the most of its geographic position, but in some respects it still has a long way to go.

The Bible compares the righteousness of God to mighty, steadfast mountains.

Your righteousness is like the mountains of God;
Your judgments are like a great deep.
O Lord, You preserve man and beast.
Psalm 36:6

Assignments for Lesson 21

Gazetteer Study the maps for Africa and North Africa and read the entry for Tunisia (pages 19, 20, and 26).

Worldview Write a paragraph telling what you know about Islam.

Project Choose your project for this unit and start working on it. Plan to finish it by the end of this unit.

Literature Begin reading *Patricia St. John Tells Her Own Story*. Plan to finish it by the end of Unit 7.

Student Review Answer the questions for Lesson 21.

Transiting Through the Suez Canal

22   A Man and a Canal

Ferdinand de Lesseps was not an Egyptian, but the work he oversaw changed the face of Egypt and the dynamics of Egyptian economics and politics from his time on. He was not a financier or much of an administrator, but his project transformed the world's travel and commerce seaways. De Lesseps was not an engineer, but people around the world praised his feat as an engineering marvel.

Ferdinand de Lesseps was a diplomat, an entrepreneur, and a promoter. In these roles, he brought about the creation of the Suez Canal.

A Brief History of Canals

The man-made waterways we call canals are a classic example of human geography: people encountering the physical geography of a place and changing it to accomplish their purpose. The two main purposes of canals are transportation (to move people and goods) and the moving of water for irrigating crops, creating or enhancing a water supply, or providing drainage.

Canals can benefit the economic activities of people, but they can negatively affect plant and animal life in a place. For example, a canal may disrupt migration routes that animals had taken.

People have built canals in many times and places around the world. Mesopotamia gives us the earliest evidence we have of canals, built thousands of years before Christ. The Grand Canal of China connects the Yellow River and the Yangtze River. At 1,104 miles, it is the longest canal route in the world, although parts of that length are natural waterways. It is also the world's oldest canal still in use. Various Chinese dynasties added to the canal's length, but the oldest parts date from the 400s BC. The Romans built numerous canals throughout their empire.

Venice, Italy, is set on many islands that have natural and man-made waterways running among them. Northern Europeans have built numerous canals that connect rivers and cities. The Erie Canal in New York state opened the western United States for more rapid settlement and greater economic development. It also helped make New York City the financial capital of the world. Canals have enabled commercial traffic to thread through the Great Lakes and to avoid Niagara Falls. The narrow Isthmus of Corinth connected the Peloponnesian Peninsula with the rest of Greece. From ancient times workers hauled ships on rollers from one side of the isthmus to the other to avoid having to go around the Peloponnese. In 1893 Greece completed a canal that opened the isthmus to water traffic.

The Suez Canal was not the first attempt to dig a canal in Egypt. Historical records indicate several attempts, but the records are unclear as to how successful they were. In the 1800s BC, Egyptians apparently dug a narrow canal between the Nile River and the Bitter Lakes, which are north of (and at that time were connected to) the Red Sea. Desert winds blew sand into the canal and prevented its continued operation. Pharaoh Necho attempted to build a new canal about 600 BC, but he abandoned the project after an estimated 100,000 workers died. Darius the Great had more success in building a canal, and Ptolemy II opened the route again about two hundred years later, but again it did not remain open.

Napoleon seized control of Egypt for France in 1798. He investigated the possibility of a canal but concluded that such a project was impossible. French control of Egypt only lasted a few years, but it raised interest among the nations of Europe in the possibilities for becoming involved in that part of the world.

The Historical Setting

Geography is a key factor in trade. People produce goods or services or minerals depending on what is possible for them in their respective places. Then they have to transport those goods to the people in other places with whom they carry on trade. Geographic factors such as mountains, rivers, deserts, and oceans, as well as human factors such as friendly or unfriendly people who live along the trade route, can make trade either easier or more difficult.

In the late Middle Ages, Europeans developed a strong market for spices and other goods from Asia. The trade routes that brought those goods to Europe involved long, expensive journeys overland through the Middle East; and Arab raiders often attacked the trade caravans. A major impetus for opening a water route from Europe to Asia around Africa and for Columbus' venture west from Europe was the desire

This painting of trading posts at Canton (now Guangzhou), China, from around 1805 shows the flags of Denmark, Spain, the United States, Sweden, the United Kingdom, and the Netherlands.

to establish safer, quicker, and more reliable trade routes between Europe and Asia. This desire for a better trade route led many in Europe to consider a canal at Suez.

A second factor was the industrial revolution taking place in Europe, especially Britain, in the mid-1800s. Europeans were building factories, and factories used the raw materials that businessmen could obtain in Asia. Factory owners and workers were becoming financially able to purchase more consumer goods. This ability increased the demand for manufactured products. Factories fed the consumer market, and raw materials fed the factories, and trade provided the raw materials.

A third factor was the growing efforts by European countries to establish overseas colonies. British interests in India and French interests in Indochina (Southeast Asia) are just two examples of this drive to establish not just trade but permanent colonies in other parts of the world. A canal at Suez would provide a quicker and cheaper means for sending settlers and soldiers to the colonies than going overland or around Africa.

Finally, a technological revolution was taking place that enabled improvements in sea travel. Steam-powered ships were replacing the wind-driven vessels that people had used since ancient times. Sailing around Africa would get Europeans to Asia—eventually. The journey took four months. A water route through Suez would cut over 4,000 miles from the trip.

Lesson 22 - A Man and a Canal

De Lesseps' Dream

Ferdinand de Lesseps was born in France in 1805 into an extended family of diplomats. As a child and young man, he spent several years with his family in various diplomatic posts, including seven years in Egypt.

While there, he became fascinated with the idea of a canal that would connect the Mediterranean Sea directly with the Red Sea instead of utilizing the Nile River. In later years he returned to France but continued to research the possibilities of a canal through the Sinai Peninsula.

In the 1850s Egypt was technically part of the Ottoman Empire. The Ottoman ruler was not interested in a canal in Suez. However, the viceroy of Egypt, Mohammed Said, was a friend of de Lesseps; and in 1854 Said gave de Lesseps permission to build a canal. The agreement called for Egypt to receive 15% of revenue from canal tolls, with 10% going to investors and 75% to the company that would build and operate the canal. De Lesseps formed the Suez Canal Company and began raising money from investors for the project. Thousands of everyday French citizens invested small amounts, but this only generated about half of the money needed. According to the agreement, Egypt had to supply the rest, which put a strain on the Egyptian government's budget. This was only the first way that the canal hurt Egypt.

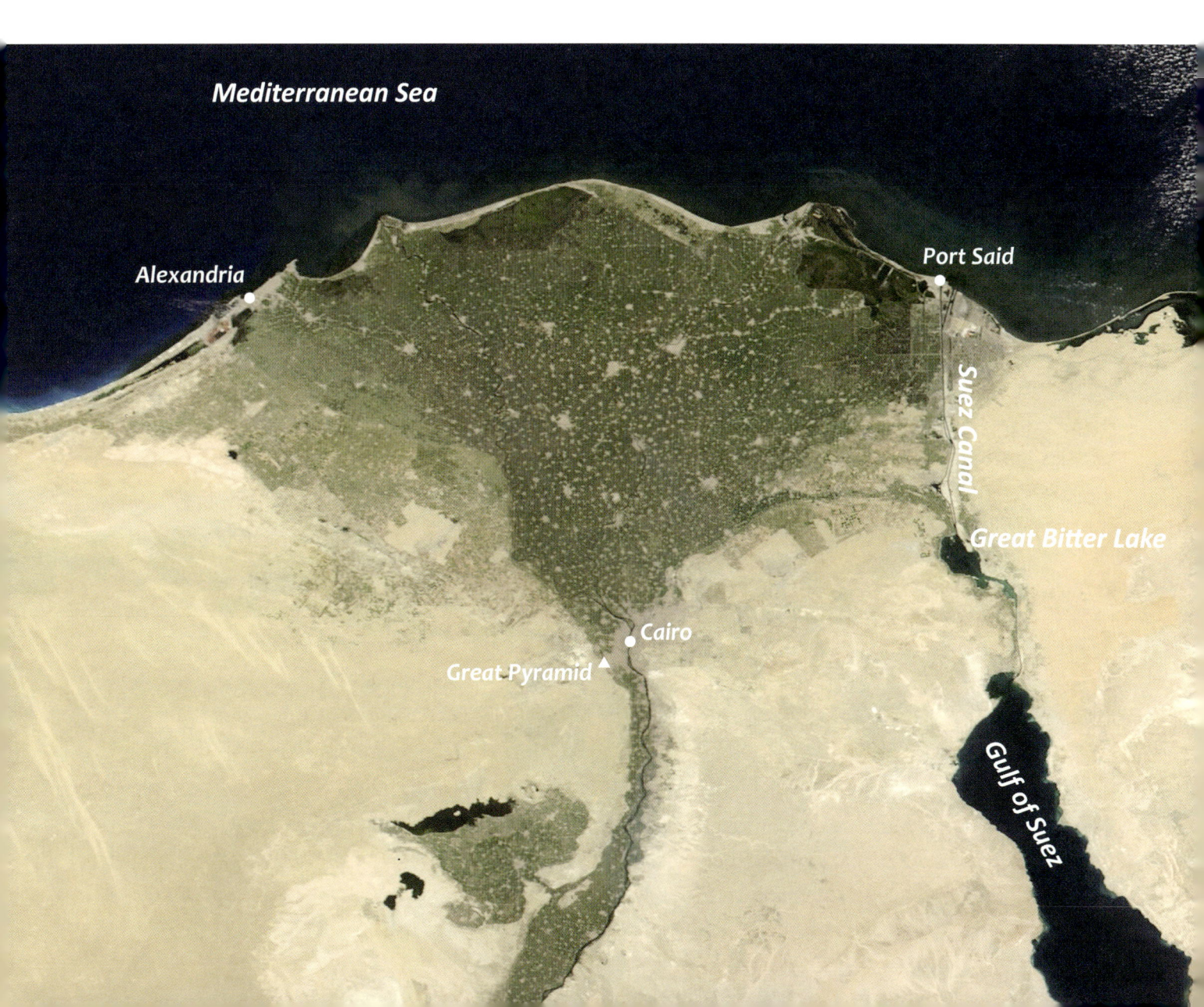

Port Said (c. 1860)

Work began on April 25, 1859. The task consisted mostly of digging a 120-mile ditch through desert sand, although workers had to remove some layers of rock also. The canal did not require the construction of locks because both ends were at the same sea level and the canal did not go through any higher elevations. The route was not a straight line but utilized the Bitter Lakes and other lakes in the area. Another hardship for Egypt was the requirement that they provide the labor for the project. The Egyptian government required an estimated 60,000 workers to move to the canal area. Many brought their families, and Egypt had to provide food and shelter for these people.

The men worked 18-hour days for low pay; some critics compared the forced work to slave labor. One group of workers started at the Mediterranean and dug south. They began at the city of Port Said, which de Lesseps built and named for the Egyptian viceroy. Another group started at Suez on the Red Sea and dug north. Reports differ as to whether the workers received simple tools such as pickaxes and shovels or whether they had to dig with their bare hands.

The workers also had to build a separate canal to bring fresh water to the construction site. They used the same path Egyptians had used in 1800 BC. One estimate says that 25,000 workers died in the effort, but no one kept official records.

After a few years of labor, it was obvious that the estimated six-year project would take much longer to complete if it were possible at all. De Lesseps had no money to buy the large dredging machinery he needed, so Said had to borrow money from European banks. This enabled the project to go forward.

The dredging machines were placed on barges. They consisted of a series of large buckets on a chain, similar to the chain on a bicycle. The buckets would dip under the water, collect mud and sand, and deposit those materials on the shore or on other barges. The dredging machines enabled the work to proceed much more quickly. The workers completed the canal in 1869, and on November 17 of that year, amid great celebration, the first ships passed through the canal. At the age of 63, De Lesseps saw his dream fulfilled.

Lesson 22 - A Man and a Canal

Politics, Politics

The canal quickly became so important that dramatic events that affected international affairs took place there for decades. The canal was a grand achievement for de Lesseps and a boon to many countries of the world, but the project left Egypt in dire financial straits. Great Britain watched the project with concern because the French, their longtime rivals, oversaw it. Taking advantage of Egypt's tenuous financial situation, in 1875 Great Britain bought all of the stock in the canal that Egypt owned. This meant that canal ownership was about half French, about half English, and none Egyptian. The Suez Canal Company became a joint French-English enterprise that operated the canal.

Following a period of political instability in Egypt, in 1882 Britain sent troops to occupy Egypt to, as the British government said, protect its investment. Britain remained the dominant force in Egypt until 1936, when it withdrew. The canal played a significant part in Britain's effort during World War I, when the country used the canal to move troops between Europe and Asia.

Egypt remained poor and poorly governed. In 1952, the charismatic Egyptian military officer Gamal Abdel Nasser led a coup against the monarchy and seized control of Egypt. Nasser believed that Egypt's economic future depended on the construction of a large dam on the Nile at Aswan, to control flooding and to generate electricity. In 1956 the United States and Great Britain agreed to finance the dam, but then withdrew their offers because of concern that Nasser was becoming an ally of the Soviet Union. A few days later, Nasser announced that Egypt was taking over the canal from the Suez Canal Company and that the tolls Egypt would collect would pay for the Aswan High Dam.

France and Britain, assisted by Israel, invaded Egypt to regain control of the canal. The United States condemned the invasion, and the United Nations arranged for a cease-fire. A UN peacekeeping force entered the area. The invasion forces withdrew, and the UN-supported settlement continued Egyptian operation of the canal. As it turned out, the Soviet Union loaned Egypt the money it needed to build the dam.

Egypt closed the canal in 1967 after Israel seized control of the Sinai Peninsula during the Six-Day War. Egypt reopened it in 1975.

The Canal Today

At first, the canal was 26 feet deep, 72 feet wide at the bottom, and 175 feet across its most narrow point on the surface of the water. Over

Dredging is a regular part of maintaining a canal, keeping it free of sand and other debris. Dredging machines have also been used to widen and deepen the Suez Canal over the decades. The images below show dredging equipment on the canal around 1865 and around 1900.

Tugboats Pulling Cargo Ships in the Suez Canal

the years, workers have widened and deepened the canal several times to accommodate larger vessels. Currently the minimum depth is 66 feet and the narrowest surface measure is 741 feet across. It is the longest canal in the world without locks. The canal is now large enough to handle 97% of the vessels used in the world's maritime traffic and is one of the world's most heavily used shipping lanes. However, its width only allows ships to travel one way or the other in the main canal. Side canals every few miles, like railroad siding tracks, allow ships to pass each other. Today an average of fifty ships per day pass through the canal. When a ship enters the canal, a trained Egyptian pilot comes onboard and guides the vessel along the route. A complete transit takes about 15 hours. After all the conflict about who would operate the canal, it is today truly an international waterway.

In 1980 Egypt opened an automobile tunnel under the canal to connect the main part of Egypt with the Sinai Peninsula. In 2015 Egypt completed the most recent expansion of the canal, which included a new 22-mile parallel channel that allows for larger ships and two-way traffic at that location.

De Lesseps' Other Dream

With the opening of the Suez Canal, Ferdinand de Lesseps became an international celebrity; but he wasn't through dreaming. He longed to create an around-the world waterway near the equator, so he set his sights on building a canal across the Isthmus of Panama in Central America.

De Lesseps obtained the rights to build a canal in Panama and once again began to raise funds. Construction got underway in 1881. After six years, however, de Lesseps abandoned the project and gave up his dream. He would not build a canal through Panama.

Several factors contributed to this outcome. First, de Lesseps stubbornly wanted to build a sea-level canal like the one at Suez; but the rugged mountain terrain in Panama and the realities of Atlantic and Pacific sea levels made that impossible. Second, the technology that would create the machinery needed for the project was not available. Another 20 years or so would pass before it came into existence. Third, accidents and tropical diseases took a heavy toll on the workforce, which was largely made up of black Caribbeans. Given the setting, the manpower

Lesson 22 - A Man and a Canal

De Lesseps Family (c. 1880)

needed for the project simply was not available. Fourth, de Lesseps and his company were guilty of mismanagement and corruption. He ran the company poorly and deceptively. In the setting of Panama, de Lesseps' shortcomings became painfully obvious.

It was only when the United States obtained the rights to build the canal across Panama that all of the factors for success came together. The American work began in 1904, and the completed Panama Canal opened for traffic in 1914.

Ferdinand de Lesseps married in 1837. He and his wife had five sons, only two of whom lived to adulthood. She died in 1853. De Lesseps married again in 1869, shortly after the Suez Canal opened. He and his young second wife had eleven children. De Lesseps died in 1894.

The prophet Ezekiel shared a message from the Lord about canals in Egypt.

Moreover, I will make the Nile canals dry
And sell the land into the hands of evil men.
And I will make the land desolate
And all that is in it,
By the hand of strangers; I the Lord have spoken.
Ezekiel 30:12

Assignments for Lesson 22

Gazetteer   Read the entries for Egypt, Libya, and Sudan (pages 22, 23, and 25).

Worldview   Copy this question in your notebook and write your answer: What are some questions you have about Islam?

Project   Continue working on your project.

Literature   Continue reading *Patricia St. John Tells Her Own Story*.

Student Review   Answer the questions for Lesson 22.

Celebrating a Berber Wedding in Morocco

23 — The Berbers

Imagine that you lived next to a busy intersection. Would you ever think that you could provide your family's income by stopping cars, pointing a gun at the drivers, and either robbing them or demanding that they pay you money to let them pass? Would you consider kidnapping their passengers and selling them as slaves? No?

The Barbary pirates did something very much like that for centuries.

The Setting

The Mediterranean. Some have called the Mediterranean Sea the incubator of Western civilization. Think about the remarkable number of civilizations that have developed on its shores, including Egypt, Phoenicia, Israel, Greece, and Rome. About 2,500 miles east to west and 500 miles north to south, this sea "between the lands" (for that is the meaning of its name) has provided avenues for trade, exploration, colonization, and warfare.

The sea's westernmost point is the eight-mile-wide Strait of Gibraltar, which funnels ships between the Mediterranean Sea and the Atlantic Ocean. The eastern limit includes the Dardanelles, the Sea of Marmara, and the Bosphorus Strait, which leads into the Black Sea. These make up the Turkish Straits, which you learned about in the lesson on the World War I Battle of Gallipoli. Between the coast of North Africa and the island of Sicily is an underwater ridge that divides the Mediterranean into distinctive western and eastern parts.

The Maghrib. The Arabic name for the region of North Africa that stretches from Libya through Tunisia and Algeria to Morocco is the Maghrib, which is Arabic for west. This was the area of the westernmost spread of Islam following the death of Muhammad. You might think that North Africa is nothing but sand from the coast through the Sahara, but its geography is much more complicated than that. Paralleling the western part of the North African coast are the Atlas Mountains, a series of ranges that stretch for more than twelve hundred miles. Some are snow-capped at times, and some of the nearby land is green with grasses and trees. These mountains separate the Mediterranean basin from the Sahara to their south.

The Berbers

The mountains and the desert are the homeland of the Berbers, who are as complex in their ethnic makeup as the lands on which they live. The Berbers see themselves as divided into several subgroups

Lesson 23 - The Berbers

Berber Goatherd in Morocco

or tribes. They speak many distinct dialects. The Atlas Mountain ranges have resulted in the Berbers dividing themselves into these independent and often isolated groups. When some of the Berbers have wanted to hide out, the mountains have provided a secure refuge.

The ancient Berbers worshiped various elements of nature, such as the sun, the mountains, the water, and the trees. Apparently they commonly sacrificed their first-born children, but the Romans outlawed this practice when they ruled the region. Some Berbers still practice magic rituals along with their Muslim faith.

They call themselves the Amazigh. The ancient Greeks and Romans considered them to be barbarians, a word they used for people who did not speak Greek or Latin. To those north Mediterraneans any other language was gibberish, which they ridiculed as bar-bar-bar sounds. This became what the Greeks and Romans called other peoples: barbarians. The term for this region of North Africa became Barbary or the Barbary Coast; Berber is simply a variation of that.

The Berbers became skilled in animal husbandry and in the making of high quality carpets that used the wool from their sheep. Berbers still produce handmade, homemade carpets; but large-scale manufacturers use the term Berber carpet for a particular style and quality of carpet that sometimes includes man-made fibers. However, the main economic activity that characterized the Barbary Coast for centuries was piracy that preyed on sea traffic.

Barbary Pirates

The Berbers (or Barbary pirates) utilized their geographic location at the western end of the Mediterranean to rule the shipping routes by means of organized crime. Barbary pirates were active at least as early as Roman times, and likely before that. The city of Algiers in Algeria was the longtime headquarters of the pirate activity. With the coming of Christianity, some Berbers became Christians. Later, the Muslim tide spread into northern Africa and the vast majority of Berbers adopted Islam. Their conversion, however, did not end their piracy.

Barbary piracy took the form of stopping ships from other countries, plundering their cargo, and demanding a payment to allow them to continue. The pirates sometimes seized crew members as hostages. If the hostages were not ransomed by their native country or wealthy relatives, they could be sold into slavery. The Barbary pirates also demanded payments (otherwise known as blackmail or protection money but which they characterized as gifts) from other countries, in return for which the pirates promised not to stop ships from these paying countries. Wealthy backers who sponsored the pirate ships received a ten-percent cut of the take.

Why did the Barbary people engage in such criminality? For at least two reasons. First, sin runs deep in the human heart. Even our modern world has people who make their living doing illegal things. They were pirates; in our day we have human traffickers and drug dealers.

We must remember that the Barbary pirates were not the only people to traffic in slaves. Just about all ancient civilizations bought, sold, and used slaves. Slave traders came from Christian nations also. We must remember that Great Britain and France pursued the policy of impressment in the late 1700s and early 1800s. This practice involved stopping and boarding ships on the high seas, supposedly to search for seamen from their countries; but French and British ship captains "impressed" or took several American seamen as well. This maritime arrogance was one cause of the War of 1812 and almost caused an American war with France in the 1790s.

Second, the Barbary piracy worked. This was crime that paid, at least for a time. Building and maintaining a navy is expensive, and risking that navy in a war is also expensive. Sending a ship loaded with cargo onto the open sea is a great risk for the owners of and investors in that ship. Trade can be profitable and can make a huge difference for individuals and in a country's economy. To buy off the Barbary pirates probably seemed worth it. That view is wrong, as is the failure to value the lives of those who were sold into slavery. The Barbary pirates took advantage of these moral failures and made a profit in doing so.

Model of Barbarossa's Galley

Lesson 23 - The Berbers

A Little History

As we indicated above, the Barbary Coast was known for piracy from ancient times. Thus, it didn't take much of a trigger to intensify the activity in the age of exploration. The Spanish expelled Muslims from their country in 1492. Europeans called Muslims Moors, from the Latin word for dark or black. Many Spanish Moors settled on the Barbary coast. For Berbers, families are all-important and an injury requires retaliation. In response to the expulsion of Moors from Spain, the Barbary pirates began attacking Spanish ships. Spain replied by attacking the Barbary Coast.

Soon thereafter arose one of the most powerful of the Barbary pirate leaders, Khayr al-Din, whom Europeans called Barbarossa, Italian for "Redbeard." A Muslim Turk from the island of Lesbos, Barbarossa and his brother became involved in the Barbary pirate activity. Spanish warriors killed Barbarossa's brother in 1518, and Barbarossa pledged allegiance to the Ottoman Sultan Suleyman in exchange for reinforcements. Barbarossa captured Algiers in 1529 and made it the headquarters of Berber piracy. Four years later the Ottoman sultan named Barbarossa admiral of the Ottoman fleet. He continued to do battle with European forces for several years. Barbarossa served in the Ottoman court in Constantinople until his death in 1546.

Barbary piracy continued for centuries. They spread their activity into the Atlantic Ocean, and evidence indicates some of the pirate ships went as far as Ireland and Iceland. One observer noticed Irish captives in the Algiers slave market in 1631. The Barbary Coast continued as part of the Ottoman Empire, but Ottoman control was weak and the Berber tribes functioned largely as autonomous nations.

The Barbary Wars

Perhaps you have heard of President Thomas Jefferson's war on the Barbary pirates in the early 1800s. As you can see, and as the following

16th-Century Ottoman Illustration of Sultan Suleyman Receiving Barbarossa

paragraphs will show, that confrontation involved much more than a sudden flare-up of piracy.

From the 1600s into the 1800s, European countries often hired privateers (ships owned by individuals who were willing to take the risk of confrontation without involving national navies) to attack ships from other countries. European involvement in the overseas slave trade was common also. As a rule, during the 1700s the Barbary pirates did not attack the superior navies of Great Britain and France. Before the American colonies gained their independence from Britain, colonial shipping enjoyed this protection. After the United States became a separate country, however, Britain let the

Barbary pirates know that they should no longer see American shipping as under its protection.

In 1785 the dey (leader) of Algiers declared war on the United States and captured several privately owned American merchant ships in the Mediterranean. The weak U.S. government under the Articles of Confederation did not have the money to build a navy or to pay tribute to protect American interests. Algiers was also at war with Portugal at the time, and Portuguese ships blocked Algerian ships from going into the Atlantic. As a result, American shipping largely avoided the threat of Barbary piracy in the Atlantic for several years.

Then in 1794 Algerian pirates again seized several American ships. In response Congress authorized the building of the first six ships of a U.S. naval force. The next year, the United States signed treaties with Algiers, Tunis, and Tripoli. After the U.S. paid the tribute they demanded, Algiers released 83 American sailors they had captured. The U.S. had already signed a treaty with Morocco in 1786 and had peaceful trading relations with that country.

However, in 1801 the pasha (a position similar to that of governor) of Tripoli demanded more tribute and declared war on the United States. The U.S. defeated Tripolitan forces with a combined sea and land attack that utilized United States Marines (thus the Marine Hymn's reference to "the shores of Tripoli"). The 1805 treaty concluding this war provided for the U.S. to pay ransom for prisoners held in Tripoli but required no ongoing tribute.

In 1812 a new dey of Algiers rejected the 1795 treaty with the U.S. and declared war. He did this upon the advice of British diplomats who hoped that the declaration would distract the U.S. from its recently-declared war against Great Britain, which we call the War of 1812. That war prevented the U.S. from attacking Algiers or ransoming American captives being held in Algiers. After the War of 1812 ended, President James Madison asked Congress for a declaration of war on Algiers. Congress responded

Bombardment of Tripoli, 3 August 1804 by Michele Felice Cornè (1806). Cornè (1756-1845) was born on the island of Elba, off the coast of Italy. He emigrated to the United States in 1800.

Lesson 23 - The Berbers

by enacting the Authorization for the Use of Force Against the Dey of Algiers on March 3, 1815.

American naval officer Stephen Decatur led a squadron of U.S. warships into the Mediterranean. They defeated two Algerian warships and captured several hundred Algerian prisoners of war. Decatur proposed, and the ruling dey (different from the one who had declared war) reluctantly accepted, a treaty of peace. The dey later tried to reject the treaty, but after American and Dutch ships bombarded Algiers, the dey decided to give the treaty his final approval.

The Barbary pirates did not attack any more American ships; but they continued their pirate ways until the French conquered Algeria in 1830. This put an end to the hundreds of years of Barbary piracy.

Geography influences how people live. Sometimes people's decisions based on geography harm others. Geography is a factor that leads to war and a factor in how armies and navies conduct war. The United States and North Africa are thousands of miles apart on the face of the globe, but what happens in North Africa and in the Mediterranean Sea directly affects the U.S., sometimes in serious ways.

In the Bible, we read about how Paul experienced dangers from robbers and dangers at sea.

I have been on frequent journeys, in dangers from rivers, dangers from robbers, dangers from my countrymen, dangers from the Gentiles, dangers in the city, dangers in the wilderness, dangers on the sea, dangers among false brethren. . . .
2 Corinthians 11:26

Assignments for Lesson 23

Gazetteer Read the entries for Algeria and Morocco (pages 21 and 24). Look at the photographs of the Sahara Desert and read the captions with them (pages 254-257).

Worldview Copy this question in your notebook and write your answer: What concerns do you have about the current practice of Islam?

Project Continue working on your project.

Literature Continue reading *Patricia St. John Tells Her Own Story*.

Student Review Answer the questions for Lesson 23.

WWII Wreckage in the Libyan Desert (2009)

24 The North Africa Campaign in World War II

On June 21, 1942, British Prime Minister Winston Churchill was visiting Washington, D.C., to discuss World War II strategy with President Franklin Roosevelt. Churchill and Roosevelt were in the White House Oval Office when U.S. Army Chief of Staff George Marshall entered. Marshall told Churchill that German forces led by General Erwin Rommel had defeated British forces at Tobruk, Libya's only natural harbor, and captured the city. Rommel had declared his intention to continue across North Africa through Egypt to the Suez Canal. Great Britain oversaw Egypt and the Canal at that time. Roosevelt turned to Churchill and asked, "What can we do to help?"

Erwin Rommel (pointing) in North Africa

Roosevelt was offering American help in Allied effort to take control of northern Africa from the Axis powers of Germany and Italy. Because of the critical events that took place there and in other parts of the world during the ensuing months, Churchill later called this period the "hinge of fate" that determined the outcome of the war.

The Geographic Significance of North Africa in World War II

Contrary to the wording you sometimes read in history books, war does not "break out" or "erupt." The events of many years can build to a decisive breaking point, and people tend to have long memories about what happened years before. Because war involves the possession of territory, geography plays a crucial role in war.

During the 19th century, European nations competed for control of different parts of Africa. These nations included Belgium, France, Germany, Great Britain, Italy, Portugal, and Spain. Conflict continued into the first half of the 20th century.

Let's consider the actions of Italy as an example. Benito Mussolini, the Fascist dictator of Italy, had seized power in 1922. He wanted to expand Italy's power and prestige by controlling more overseas

Lesson 24 - The North Africa Campaign in World War II

colonies. He wanted to quit playing second fiddle to Adolph Hitler, the German leader, in their Axis alliance. Mussolini also wanted to avenge a loss that Italy suffered in Africa in 1896.

Italy had established a colony in Eritrea, East Africa in 1890, during the European race for Africa. Italian forces tried to invade neighboring Ethiopia (once called Abyssinia) in 1896, but the Ethiopians defeated them in the Battle of Adowa (or Adwa) that year. It was the first time an African army had defeated a European colonial power. A border incident between Italian Eritrea and Ethiopia in 1934 in which Italian soldiers lost their lives heightened the tension.

Not one to forget an embarrassment, Mussolini ordered an invasion of Ethiopia in 1935. The world watched in horror as the Italians rolled over the outmanned Ethiopians and ousted Emperor Haile Selassie. Selassie made a desperate plea for help to the League of Nations, but the League was powerless to respond.

At the beginning of World War II in 1939, Britain controlled Egypt and Sudan. Italy controlled Ethiopia and Eritrea to the south and Libya to the west. France fell to Germany in June of 1940 and Germany installed a puppet government there that was loyal to Germany. Therefore loyalties in the French territories of Tunisia, Algeria, and Morocco were divided.

The strategic importance of North Africa in World War II was not so much the region itself but rather its geographic connections. Several

American planes delivered supplies to Allied forces in North Africa.

European nations had colonies in sub-Saharan Africa, so whoever controlled North Africa could defend or threaten the colonies further south. The Allies wanted ready access to Middle Eastern oil and wanted to prevent the Axis powers from having that access. Thus, we cannot overemphasize the strategic importance of North Africa.

The Battles in North Africa

Every place where a battle occurs has its own unique geography. From the islands of the Pacific, to the bitter cold of Russia, to the trackless deserts of North Africa, soldiers have to prepare and compensate for where they fight. The heat of North Africa caused difficulties for soldiers on both sides of the conflict. Fighting in North Africa involved infantry soldiers but also the widespread use of tanks, whose tracks helped them move across the sand.

On June 10, 1940, Italy declared war on France and Great Britain. Italian forces far outnumbered British forces in northern Africa at the time. Italy launched an invasion of Egypt, but the British fought back and by December the Italians were near collapse. In early 1941 Hitler ordered German tank and infantry divisions into North Africa to help the Italians.

WWII Tanks at the El Alamein War Museum in Egypt

Unit 5: North Africa

Bernard Montgomery (right) in North Africa

General Erwin Rommel led the Nazi forces. Rommel was a brilliant tank warfare general whose clever moves earned him the nickname the Desert Fox. For the next two years Axis and Allied armies pushed back and forth across the North African desert, trading advances and retreats. In August of 1942 Churchill named Field Marshal Bernard Montgomery to lead the British and Commonwealth forces. From that point Allied strength generally grew and Axis strength generally weakened. Montgomery led a decisive victory at El-Alamein, Egypt, in a battle that lasted from October 25 until November 5, 1942.

Another blow to the Axis was the Allied invasion in Morocco and Algeria on November 8, 1942. Called Operation Torch and led by General Dwight Eisenhower, the invasion had as its goals to secure French North Africa, link up with the British, and drive the Axis forces out of Africa in a giant pincer move. Another important leader who came onto the scene at this time was General George Patton. Patton was an expert in tank warfare. Shortly after the Japanese bombed Pearl Harbor, Patton set up a desert training center near Indio, California, to prepare for war in North Africa. Patton oversaw the Allied landing at Casablanca, Morocco. He was an effective counterbalance to Rommel's expertise.

The first direct confrontation between German and American forces in the war took place at Kasserine Pass, a gap in the Atlas Mountains in Tunisia, in February of 1943. Rommel won the day, but the Americans regrouped and pushed back. Axis forces went into retreat, Rommel returned to Germany, and the last Axis forces in North Africa surrendered in May of 1943.

Key Battles in the North Africa Campaign

This photo shows Allied troops landing on the beaches near Algiers on November 8, 1942. Notice the large U.S. flag at left. The Allies hoped that the French troops in North Africa, who were nominally aligned with Germany, would not fire on Americans. Minimal fighting occurred in Algiers, though the other Allied landings in Operation Torch met more resistance.

The Impact of the North Africa Campaign in World War II

The war at first did not go well for the Allies. Germany controlled Europe, Italy controlled North Africa, and Japan controlled the Pacific. The tide began to turn in late 1942 and early 1943. The Allies stopped the Japanese advance in the battle of Guadalcanal in the Solomon Islands in the Pacific. The German invasion of the Soviet Union failed. The Allies began to push the Germans and Italians out of North Africa.

Allied leaders wanted to get American forces involved in the war as early as possible and where that involvement would make the most difference. Soviet leader Joseph Stalin wanted to see an Allied landing in northern France, but Churchill and Roosevelt believed such a landing would fail because of the strength of the German hold on Europe (thus, geography is important, but timing is also important).

Churchill and Roosevelt decided to send American troops into North Africa because success was more likely and because it would prepare for an Allied invasion of Sicily and Italy, what Churchill called the "soft underbelly" of Europe. This is precisely what happened, which means that the D-Day invasion was actually the third major Allied landing of the war. Eisenhower learned a great deal in Operation Torch that helped make D-Day successful.

This message from President Franklin Roosevelt, distributed in both Arabic and French, encouraged people in North Africa to cooperate with the Allies.

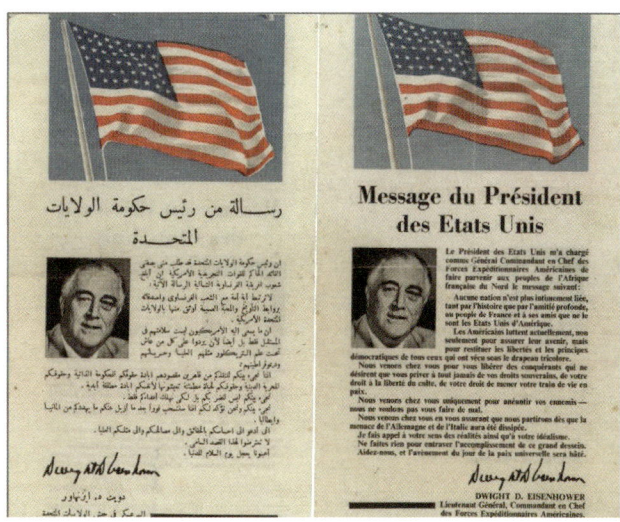

The victorious North Africa campaign not only helped secure the Mediterranean, Middle Eastern oil, and British interests; but it also drew Axis forces out of Europe, which made later fighting there somewhat easier.

It's easy to think that deserts don't matter much in terms of geography. In terms of the human geography of World War II, however, the deserts of North Africa were a matter of life and death for much of the world.

Isaiah told of the desert rejoicing to see the glory of the Lord.

The wilderness and the desert will be glad,
And the Arabah will rejoice and blossom;
Like the crocus
It will blossom profusely
And rejoice with rejoicing and shout of joy.
The glory of Lebanon will be given to it,
The majesty of Carmel and Sharon.
They will see the glory of the Lord,
The majesty of our God.
Isaiah 35:1-2

Assignments for Lesson 24

Geography Complete the map activities for Unit 5 in the *Student Review* book.

Worldview Copy this question in your notebook and write your answer: Do you think it is possible for Christians, Jews, and Muslims to get along in a civil society—not to say that everyone is right with God but to respect everyone's freedom of religion, personal property, and security? If so, how could this happen? If not, why not?

Project Continue working on your project.

Literature Continue reading *Patricia St. John Tells Her Own Story*.

Student Review Answer the questions for Lesson 24.

Page from the Qur'an

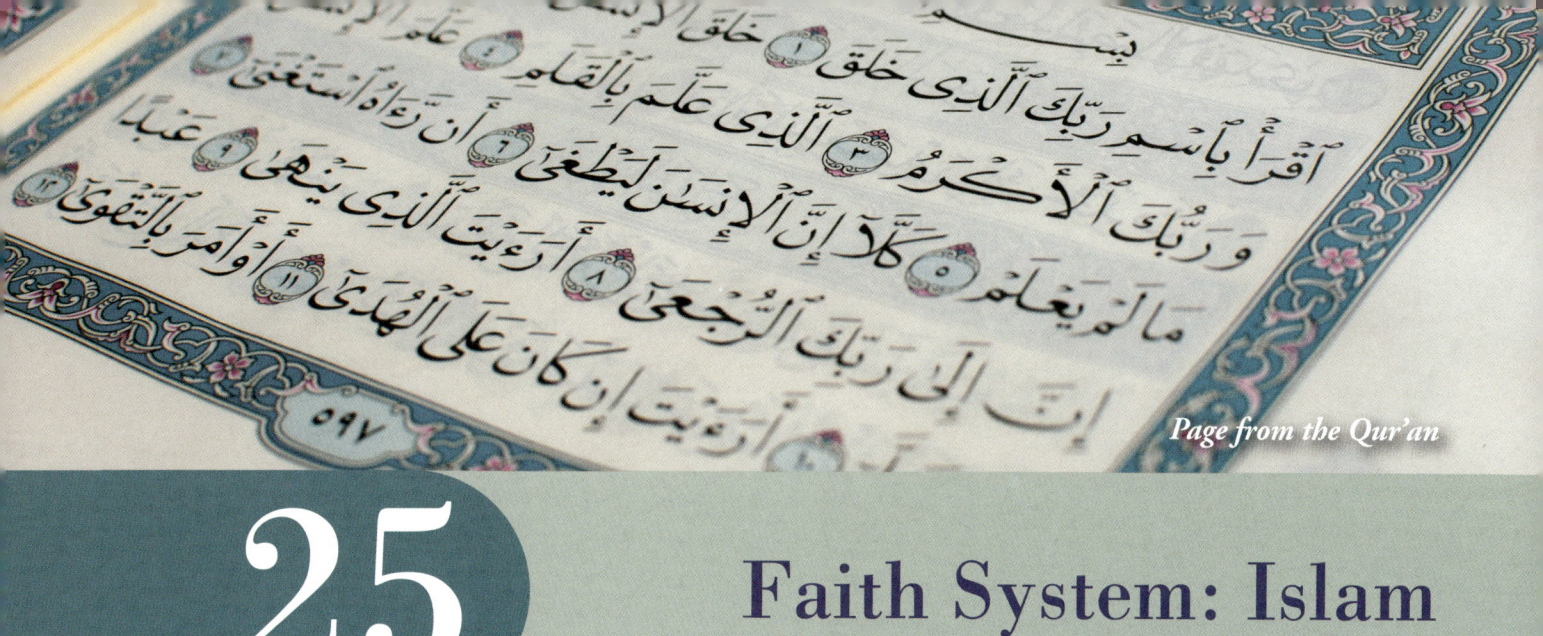

25 Faith System: Islam

In our world, ideas develop and events take place in the context of other ideas and events. This helps us understand the development and growth of the Muslim religion, also known as Islam.

In the 500s AD, most Arabs were nomads; but many lived in cities and took part in lucrative trading ventures between Europe and Asia, situated as they were on important trade routes between these two continents. At the time, most Arabs believed in many gods and spiritual beings. Some Arabic religions practiced infanticide. Generally speaking the level of morality among the Arab people was low.

In the important trading city of Mecca, in the western part of modern Saudi Arabia on the Arabian peninsula, stood a tall, black, square shrine called the Kaaba. The Kaaba housed the idols of many pagan gods as well as a black stone that most Arabs considered sacred because they believed that it had fallen from heaven. Many Arabs made pilgrimages to Mecca to worship at the Kaaba.

The Life of Muhammad

Muhammad was born in Mecca about 570 AD and grew up there. A grandfather and an uncle reared him, and Muhammad became a successful and wealthy trader. Through his work Muhammad learned about Judaism and Christianity. When he was about forty, he developed a deep interest in spiritual matters. He believed that he received revelations from a god he called Allah, which Allah wanted him to share with others.

The core message that Muhammad taught was that Allah was the one true god and that he (Muhammad) was Allah's prophet. Although Muhammad gained a few followers, in general the people of Mecca denounced him as a blasphemer. Feeling threatened, in 622 Muhammad and his small group of followers fled north to the city of

The Kaaba in Mecca

149

Dzhuma Mosque, Tashkent, Uzbekistan

Yathrib. There the people received him warmly and he began to build a large following. He changed the name of the city to Medina. Muslims count the year 622 AD as Year One in their calendar.

Muhammad wanted to unite the Arab world in one faith to eliminate paganism, to end feuds among Arab tribes, and to help people live upright lives. He wanted people to submit to Allah and to him as the prophet of Allah. The Arab word for submitted is islam, and a submitted person is a muslim.

Muhammad believed that Allah gave him permission to use force to convert unbelievers. He called this effort at persuasion jihad or holy war. Spreading out from Medina, Muhammad and his followers began intercepting trading caravans and forcing people to convert or die. Muhammad's forces captured Mecca in 630 AD. Muhammad destroyed the idols in the Kaaba but kept the black stone as a symbol of the new faith.

Muhammad traveled to Jerusalem a few years before he died. Muslims believe that he ascended to heaven from the temple mount and spoke with Allah. Muhammad died in 632, and Muslim forces captured Jerusalem in 638. Jerusalem became the third most sacred city in Islam after Mecca and Medina.

At the time of Muhammad's death, most of the Arab world had converted to Islam. The new faith had also spread to non-Arab peoples across northern Africa, into southern Europe, and east to the borders of India. In the ensuing generations this faith spread across India and into Malaysia and southeast Asia. Today the southeast Asian country of Indonesia has the largest Muslim population of any country in the world.

Islam places five basic requirements on its adherents: (1) believing that there is no god but Allah and that Muhammad is his prophet; (2) saying a set prayer at five prescribed times per day, kneeling and bowing toward Mecca; (3) giving to the poor, which in some Muslim countries has taken the form of a tax; (4) fasting during the daylight hours in the month of Ramadan when Muhammad supposedly received his revelations; (5) if possible, making a pilgrimage to Mecca at least once, with the performing of certain rituals there.

Islam was a reaction to the religions around it. Muhammad rejected polytheism as wrong. He rejected Judaism as hypocritical because Jews claimed to believe in the one God but rejected Muhammad's claim to be the prophet. He rejected Christianity as absurd and idolatrous because of its belief in the threefold nature of God.

Al-Rashid Mosque, Edmonton, Canada

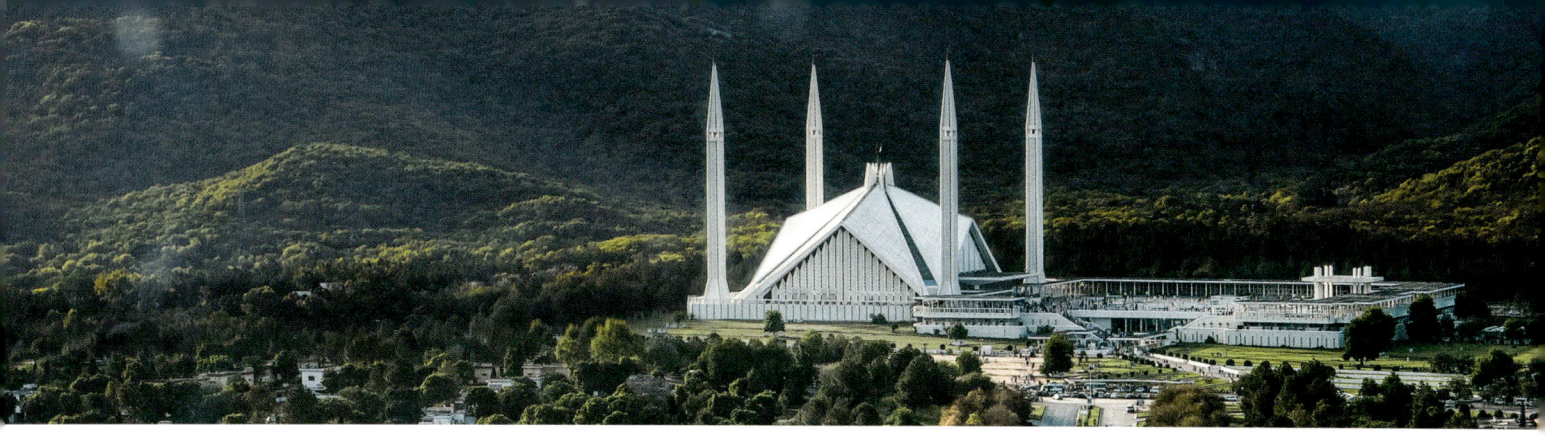

Faisal Mosque, Islamabad, Pakistan

The Muslim View of Allah

Allah is the Arabic word for deity, like *theos* in Greek and *god* in English. The central belief in Islam is that Allah, whom they consider the one true god, is sovereign above everything else. The phrase "Allahu Akbar," which Muslims use every day in their prayer times, is usually translated "Allah is great," but it really means "Allah is greatest/ultimate/absolute."

Allah is the Muslims' one and only deity. They do not accept the Christian doctrine of the Trinity. Muslim doctrine also does not include the concept of the deity indwelling humans. They believe that the idea of incarnation, deity taking on human form, is degrading to Allah. Muslim doctrine does not include the idea of Allah having a personal relationship with humans, although they believe that Allah is present everywhere and hears prayers.

In Muslim belief, Allah defines himself by his actions. What he does is just because he does it. He does not have to submit to human concepts of justice. Allah's actions are merciful and gracious because he does them.

Muslims believe that Allah created everything out of nothing and that he manages everything that happens. Nothing occurs outside of the will of Allah, although they do believe that humans have free will. Mankind is a special creation of Allah. Muslims accept a version of the Garden of Eden story, but they believe that mankind is redeemed and not fallen.

Islamic teaching places great emphasis on doing good deeds as Allah defines good, seeking to live up to Allah's standards, in order to enter paradise after death. In the Muslim worldview, life is a test; but since doing good pulls people upward it is not a burden. Islam teaches that a Muslim cannot have any assurance of his or her salvation; to believe in that assurance would be to presume to know the mind of Allah, which they would never do. Allah prevents the unbeliever from believing; the Qur'an describes Allah as the enemy of unbelievers.

Muslim teachers have offered various ideas about the end times on the earth. According to Muslim doctrine, at the last judgment, an angel will give each person a book that contains all of his or her deeds. If the angel puts it in the right hand, that person is saved; if in the left hand, that person is doomed (Surah 69:19-20).

Muslims and the Qur'an

Muslims believe that the Qur'an is a compilation of the thoughts of Allah as communicated by Muhammad and others. Islam holds that the Qur'an communicates these thoughts of Allah perfectly when rendered in Arabic. As the thoughts of Allah, they believe that the Qur'an has existed eternally in the mind of Allah. Since they believe it is Allah's ultimate message, and since they believe Muhammad was the ultimate prophet, they see no need for any further revelation or prophets.

The Hadiths are collections of sayings and writings of Muhammad as others remembered them. Various hadiths exist, and they vary in source, content, and number and quality of manuscripts.

Hassan II Mosque, Casablanca, Morocco

The authority of hadiths is secondary to that of the Qur'an.

Muhammad was familiar with the teachings of the Bible to some degree. The Qur'an cites a long list of prophets, of whom Muhammad is the last and most important. The next in importance for Muslims is Abraham, then Jesus, then other prophets and other people in Scripture as well as some not mentioned in the Bible. Muhammad accepted some things about Jesus as true: the virgin birth, His teachings, and His miracles; but Muhammad denied His divinity, crucifixion, and resurrection. The Qur'an gives a supposed quotation of Jesus in which He denies being called God (Surah 5:118-119). Islam holds that the followers of Jesus, Moses, David, and others distorted their teachings; so in the Muslim view the Bible has value but it is unreliable and not authoritative.

In addition to these changes, the Qur'an makes other alterations to the Biblical narrative. For instance, the Qur'an does not name the son that Abraham prepared to sacrifice (Surah 37:100-105). Arabs consider Ishmael to be their ancestor, and Muslims believe that Abraham began the process of sacrificing Ishmael, not Isaac. The Qur'an says that Abraham and Ishmael built the Kaaba in Mecca (Surah 2:127), but no historical evidence supports this narrative. Concerning the Israelites in Egypt, the Qur'an says that they were Pharaoh's bondsmen but that they were "a puny band" and had gardens, fountains, treasures, and sumptuous dwellings (Surah 26:55-60).

The Qur'an teaches that Allah created *jinn* (Surah 55:15). The singular of jinn is *jinni*, usually rendered in English as genie. The jinn are spirit beings below the level of angels and devils. They can assume animal or human form and can dwell in inanimate objects. According to Muslim belief, jinn take great pleasure in punishing humans. Muslims often blame them for illnesses and accidents. A jinni can even ruin a person's life. However, people can learn magical procedures that turn the jinn to their favor. Jinn are the subject of many folk stories, including some that are in *The Thousand and One Arabian Nights*.

Divisions Among Muslims

We usually speak of "Islam" as though it were a unified religion with a single set of doctrines, but in reality Islam has many divisions and subgroups. The two largest groups are the Sunnis and the Shi'ites. Sunnis believe that the leader of Islam should be elected and that traditional practices as well as the Qur'an itself are authoritative. Shi'ites believe that the leader should be related to Muhammad (who

Lesson 25 - Faith System: Islam

had no children of his own), and they accept only a literal interpretation of the Qur'an as authoritative. About 85-90% of Muslims are Sunnis, while most of the rest are Shi'ites. However, the majority of Muslims in Iran, Iraq, and Bahrain are Shi'ites. The Muslims in Lebanon are divided about equally between the two groups.

National groups of Muslims have some views that differ from each other. Palestinian Muslims do not have the same beliefs as Saudis, and the Saudis do not have the same beliefs as Iranians (in fact, Saudis and Iranians are often opposed to each other). In addition, Iranian Muslims do not believe exactly the same as Turkish Muslims. Even within a country, differing Islamic sects and ethnic groups can be at odds with each other. As with any religion, some Muslims are not as devoted to practicing their faith as others are.

The Sufi sect tends toward a mystical interpretation of Islam. The conservative Wahabi sect is found almost exclusively in Saudi Arabia. Followers of folk Islam often believe in superstitions and adopt some beliefs and practices of non-Muslims around them. The differences among Muslims often center on a particular Muslim doctrine. For instance, among Sunnis the Ash'arites believe that the Qur'an has been eternal in the mind of Allah, while the Mu'tazilites believe that the Qur'an was created. Both groups believe it to be inspired.

Corniche Mosque, Jeddah, Saudi Arabia

Ayasofya Camii Mosque, Amsterdam, Netherlands

The Practice of Islam

Islam does not have a strict, formalized religious structure, but it does have a strong informal pattern of leadership. In Shi'ite Islam, a mullah is the head of a Muslim community and leads prayers at a mosque. A mullah is a religious scholar who has received training in a madrasah or religious school. Mullah is a title of respect for an educated religious man. Ayatollah is the highest title for a Shi'ite. He also acquires training, but he works to build a following for his teachings and can become recognized as a religious and political leader, as in Iran.

The application of teachings of the Qur'an to society is called Shari'a law. Among Sunnis there are four schools of thought that differ with regard to the strictness of applying the Qur'an. One of the most strict is the Hanbalite sect, who gave rise to the Wahabis, who gave rise to the Taliban.

Islamic teaching lays out many strict rules. It forbids the consumption of alcoholic beverages and the eating of pork and of the meat of an animal that has been strangled (compare Acts 15:29). The religion has rules about what clothing men and women may wear. The practice of the religion has been male-dominated. Women have been restricted in their rights, such as the right to obtain an education; and the physical abuse of women does occur. Islam forbids the making of images of people, so Islamic art involves the use of creative designs.

Muslim Views of Scholarship and Reason

From the 700s through the 1100s, the Muslim world increased its contact with the West and with China. This led to a greater interest in geography and navigation. Muslims established universities and made advances in medicine and science. The West learned such things as Arabic numerals (which the Arabs learned from India), algebra, and advances in trigonometry. Arabic scholars translated classical Greek texts such as Plato and Aristotle into Arabic. Muslim architecture flourished. In many ways these developments paved the way for the European Renaissance as European scholars learned from Muslims.

The Abbasid dynasty in Babylon ruled the entire Muslim world from 751 to 1258 AD. Scholars from around the region traveled to Babylon and interacted with each other. Arabic became the common language for much of the Muslim world. Muslim scholars had contact with scholars from Egyptian, Greek, Indian, Chinese, and Persian civilizations. The introduction of paper from China encouraged scholarship, correspondence, literature, record keeping, banking, and literacy.

But then things changed. The Abbasid Empire broke apart, and in 1258 Mongol invaders swept it away. A movement arose among Muslims that opposed philosophical inquiry. This group believed that the Qur'an is co-equal with God and that God is the only cause in the natural world. To them, anything that happens is a discrete event caused by the will of God, not cause and effect in the natural world—end of discussion. This group said there is no need for scientific inquiry; people should just trust Allah and submit to the accepted authorities. This attitude became dominant in Islam.

While the Western Christian world encouraged research, the Muslim world rejected it. The printing press came into use in Europe about 1460, but it did not enter the Muslim world until 1727. Reversing the flow of ideas, European scholars began translating Muslim scholars. Islam did not accept free inquiry. It did not adequately handle modernity. From early on, Jews and Christians were second-class citizens in the Muslim world. Muslims practiced slavery. Muslims did not pursue self-examination and self-criticism; their attitude was more one of self-defense. Believing they had arrived, there was nothing left to do but defend themselves and attack any who disagreed.

Grand Mosque of West Sumatra, Indonesia

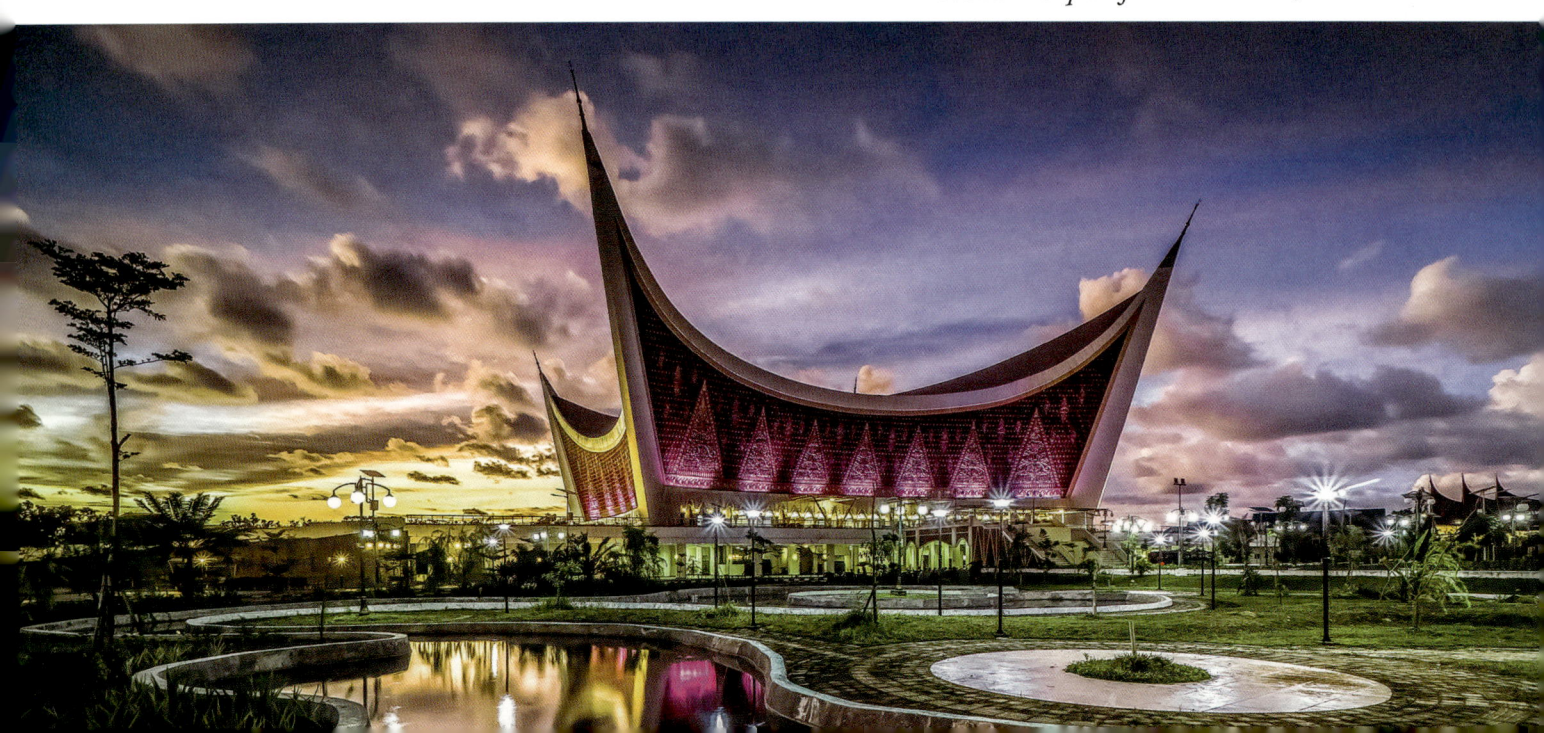

King Fahd Islamic Cultural Center, Buenos Aires, Argentina

How Muslims See Non-Muslims

For the most part, Muslims believe that their religion is the only true one, which means that other religions are not completely true and are therefore inferior. They believe that people should convert to Islam to have the truth and the possibility of entering paradise after death. Islam is not the only religion that has this outlook.

The question that commonly arises in a discussion of Islam is whether Muslims believe that non-believers should convert or be killed. As the faith spread through the leadership of Muhammad and his successors, there were times when this kind of evangelism by the sword took place. Some people welcomed the faith as a positive change from the paganism they had known. When Islam had taken control of some places, usually Jews and Christians were allowed to live there if they paid taxes and promised not to be rebellious.

There are extremists who shout "Death to the Jews" and "Death to America," and at least one Muslim leader has called the United States the "Great Satan." Clearly some terrorist sects take the lives of others in the belief that they are giving glory to Allah. However, obviously not all Muslims believe this way because most Muslims are not challenging people to convert or die. A low percentage of Muslims support violence, but that percentage is higher in some countries.

Political Worldviews

Islam sees no required separation of religion and state, but Muslims have a variety of opinions about how their faith affects their political involvement. Some Muslims are in favor of democratic government, while others want to see the creation of Islamic states that follow Shari'a law. Others are content to maintain something of a middle ground between these alternatives.

Many Muslims recoil at the immorality that Western media has spread throughout the world. They have seen Western media as a negative influence on their younger generation. However, changes are coming to the Islamic world, and many of these changes are the result of influence from Western culture.

The production and distribution of oil by Arab states has led to significant wealth for some members of society. The emirate of Dubai, one of the United Arab Emirates, has become a center for business, shopping, and other activities that we usually associate with economic wealth. Many Muslim students attend American and European universities to learn engineering, business, or other disciplines, often for the purpose of returning to their homelands to work. Muslim governments give financial support to at least some of these students.

Muslim leaders can continue to work to improve the standard of living of their people. Despite religious differences, people can work together to make a better world.

The Bible makes clear the supremacy of Christ, and it warns against other religious views. For example, Paul had this to say to Christians in Colossae who were being influenced by false teachers:

See to it that no one takes you captive through philosophy and empty deception, according to the tradition of men, according to the elementary principles of the world, rather than according to Christ. For in Him all the fullness of Deity dwells in bodily form, and in Him you have been made complete, and He is the head over all rule and authority . . . .
Colossians 2:8-10

Assignments for Lesson 25

Worldview   Write or recite the memory verse for this unit from memory.

Project   Finish your project for this unit.

Literature   Continue reading *Patricia St. John Tells Her Own Story*.

Student Review   Answer the questions for Lesson 25.
Take the quiz for Unit 5.
Take the first Geography, English, and Worldview exams.

Playing a Game on Orango Island, Guinea-Bissau

6 West Africa

Cocoa farming is a vital part of the life and economy of Côte d'Ivoire. The Dogon people of Mali live their distinctive culture around the Bandiagara Escarpment. The Akosombo Dam has had both positive and negative impact in Ghana. The music of West Africa, including Nigeria, reflects particular styles of composition and performance. The worldview lesson discusses the faith system of folk religion, which is relatively common in West Africa and which people practice on every inhabited continent.

Lesson 26 - Cocoa Growing in Côte d'Ivoire
Lesson 27 - Riches and Poverty in Mali
Lesson 28 - The Trees of Lake Volta
Lesson 29 - The Music of Nigeria
Lesson 30 - Faith System: Folk Religion

Memory Verse   Memorize Colossians 3:16 by the end of the unit.

Books Used   The Bible
Exploring World Geography Gazetteer
Patricia St. John Tells Her Own Story

Project (Choose One)

1) Write a 250-300 word essay on one of the following topics:
   - Do many farmers in your area grow a particular crop? What are ways in which this crop impacts where you live? This could include the hiring of temporary workers, processing facilities, secondary businesses that are related to the primary crop, and other ways.
   - What project has been both a blessing and a bane to the people where you live? For example, a dam might have brought electricity while flooding much farmland or a factory might have created jobs while expanding urban sprawl.

2) Draw a picture of an escarpment (see page 164).

3) Compose a call and response song and perform it with a group. See Lesson 29.

26 Cocoa Growing in Côte d'Ivoire

Cacao Trees, Côte d'Ivoire

"If you're not courageous then you cannot have a cocoa farm."

These are the words of Genevieve Yapipko, wife, mother, and cocoa grower in the West African country of Côte d'Ivoire. She and other growers need courage for many reasons.

- The physical demands of growing cocoa are significant. Cocoa production involves eighteen steps, including growing the trees, cutting the pods from the trees, breaking open the pods with a stick, scooping out the sticky beans, letting them ferment for a week, drying them, putting the beans into sacks, and taking the beans to a warehouse for sale. (Technically the cacao plant produces cacao beans. When these are processed at high temperatures, the result is cocoa. The entire process is usually described as cocoa growing.)

- Wives and mothers who are involved in growing cocoa are often the first to arise in the morning and the last to retire at night. They must care for their families before and after working their cocoa farms. Many women also care for elderly parents and tend a garden to help feed their families.

- Cocoa workers are notoriously underpaid. The world cocoa market is lucrative for companies that produce and sell cocoa products, but the people who actually harvest the beans usually make less than two dollars per day. Companies that process the cocoa and that make chocolate products receive together about eighty percent of the retail price of the finished products; those who do the harvesting work receive about seven percent.

- The price that growers receive for their crop can fluctuate wildly, and that price depends on what happens in cocoa markets in places like New York and London, far from Côte d'Ivoire.

The cocoa growing industry in Côte d'Ivoire is a prime example of the interaction between people and their geography.

Background

The country now called Côte d'Ivoire is about the size of New Mexico. For centuries, it has been the home of many tribes that speak many languages. French missionaries arrived there as early as 1637. French interest in that region of West Africa continued, and just over two centuries later, between 1843 and 1845, France signed treaties with tribal chiefs that established a French protectorate over the region.

Côte d'Ivoire was part of a larger area overseen by France and called French West Africa. Côte d'Ivoire became a French colony in 1893. It was during France's colonial rule that cocoa growing became an important part of the economy. The country gained its independence in 1960.

The country was long known as the Ivory Coast because of the trade in elephant tusks that took place there. However, gathering the tusks all but wiped out the once-abundant elephant population. An international agreement in 1989 banned the sale of ivory, although some ivory still makes its way onto the international black market. Ivory art pieces may be sold openly as long as the buyer can prove that it was made before 1947. The country officially changed its name to the Republique de Côte d'Ivoire (French for the Ivory Coast Republic) in 1986.

French is the official language, but the people also speak about sixty different native dialects. The majority of the population lives along the southern coast, and relatively few live in the arid northern region. About 45% of the people are Christian; these live mostly in the south. Forty-three percent are Muslim, and these are concentrated in the north. The rest of the people are adherents of folk religion or profess no faith.

Overall, the country has been one of the most economically successful former colonies in Africa. At the time of its independence, the Ivory Coast was one of the poorest countries in the world. Its first president after independence, Felix Houphonet-Boigny, was a physician and wealthy planter who led the country well in developing its resources and potential. It is now the world leader in exports of cocoa and cashews, and it also exports a number of other products such as coffee and palm oil. Côte d'Ivoire had many years of political stability after independence, but since 1999 the country has experienced political unrest due to government corruption and ineffective leadership.

Cocoa

Côte d'Ivoire produces over thirty percent of the world's supply of cocoa. It and Ghana grow almost two-thirds of the cocoa consumed in the world. By comparison, Europe and North America account for seventy percent of the world's consumption of

Elephant Ivory

Where you live geographically often affects the issues you face. While the international community has outlawed the selling of ivory because of the decline in the elephant population, Zimbabwe has the opposite problem. The elephant population there is on the rise and poses a hazard to crops, cattle, and people. Botswana, Namibia, and Zimbabwe contain over sixty percent of the world's population of elephants. Zimbabwe has a stock of elephant tusks and rhino horns, mostly taken from animals that have died of natural causes, that the government values at $600 million. They would like permission to sell those items to pay for conservation programs. The elephant statue above, made of ivory, is shown on display at the Sir Seretse Khama International Airport in Gaborone, Botswana.

Women Processing Cocoa Beans in Côte d'Ivoire

---

cocoa. One in six people in Côte d'Ivoire depend on cocoa for a living.

The cocoa-growing industry of Côte d'Ivoire has been a success in many ways, but it has its downside. The greatest burden of the production work falls on women. Of the eighteen steps involved in cocoa growing, women perform fifteen. They comprise about two-thirds of the labor force, even though women only own about one-fourth of the cocoa farms. Women do most of the harvesting work, while the men take the filled bags to sellers and collect the money. Unfortunately, some men spend the profits before the money gets to the family. Private and government efforts are underway to help women own more cocoa farms and receive more of the profit of their labor. Fair trade organizations encourage companies to pay a living wage to the people who grow the raw materials for the products they sell.

Another unfortunate source of cheap labor is children. Although child labor is illegal, it is a common practice in Côte d'Ivoire and other countries. The children receive little or nothing for their long hours of work, they often come in contact with harmful chemicals, and they are denied opportunities to further their education.

Several hundred thousand people live illegally in the country's national parks and protected forests. This allows them to avoid paying rent to landlords. Many of them try to make a living by growing cocoa on those lands as well. Authorities find it difficult to enforce the law and have little incentive to do so.

The price for cocoa is set by the government in negotiations with buyers. World consumption and the size of a year's crop exert a strong influence on the market price for cocoa. Generally speaking, because of the economic principle of supply and demand, the more successful the growers are in producing a large crop, the lower the price that buyers will pay. In 2019 Côte d'Ivoire and Ghana reached an agreement with cocoa buyers to establish a price floor, below which producers will not sell. Companies that produce and market cocoa products will have to share some of their profits with growers or raise their prices to meet the new arrangement.

Another negative in the geography of Côte d'Ivoire has been the loss of rainforest lands. In the last fifty years the country has lost an estimated eighty percent of its rainforests, as people have cut hardwood trees for sale to increase their income and to create more areas for growing cocoa. These deforested areas have lost a large part of their primate population as well.

Despite the difficulties, the people of Côte d'Ivoire continue to depend on growing cocoa. One recent positive development has been the start of a chocolate-making industry. Côte d'Ivoire has produced cocoa for many years, but chocolate candy has generally only been available as a luxury treat for the wealthy. This is starting to change as small businesses have begun crafting chocolate for retail sale.

As one rural department (local area) governor put it, "It doesn't make sense to ask an Ivorian what cocoa means to him. It means everything! It's his first source of income. . . . Our houses are built with cocoa! The foundations of our roads, our schools, our hospitals is cocoa. Our government runs on cocoa! All our policy focuses on sustaining cocoa!"

The human interaction with geography in Côte d'Ivoire includes the decimation of elephant herds, the right conditions for growing cocoa, and the deforestation of rainforests. As with so much about human interaction with the physical world, some is good but some is not.

Are you hungry yet? Chocolate is fun to eat, but Proverbs warns against eating too much that is sweet.

It is not good to eat much honey,
Nor is it glory to search out one's own glory.
Proverbs 25:27

Assignments for Lesson 26

Gazetteer Study the map of West Africa and read the entries for Benin, Burkina Faso, Cabo Verde, and Côte d'Ivoire (pages 27-31).

Worldview Copy this question in your notebook and write your answer: What are some sayings and practices that you have heard from people that reflect superstitions? (Examples include "Good luck!", "Knock on wood," and fear of black cats or the number 13.)

Project Choose your project for this unit and start working on it. Plan to finish it by the end of this unit.

Literature Continue reading *Patricia St. John Tells Her Own Story*. Plan to finish it by the end of Unit 7.

Student Review Answer the questions for Lesson 26.

Working in a Field Near Banani, Mali

27 Riches and Poverty in Mali

Otume is no longer a child, but he does not yet see himself as an adult—and neither do the people of his village. He has not yet participated in the special rite of passage that marks a young male's entry into adulthood.

What is in store for Otume?

His land holds great riches, but the people suffer great poverty.

The artistry of his people reflects rich traditions, but terrorism and theft threaten that artistry.

They are a people who see meaning in every human and natural activity, in everything that happens and everything that doesn't happen.

What is in store for his people and his country?

A Country of Contrasts

Mali is a large, landlocked country in West Africa, about twice the size of Texas. It is a diverse land. In the north is the Sahara desert. In the central region is the semi-arid Sahel. To the south is the subtropical Sudan, where the Niger River runs. The great majority of people live in the south, near the river.

A former French colony, Mali is the third largest producer of gold in Africa; however, all that gold does not make the country wealthy. Mali is one of the poorest countries in the world. About half the people live below the accepted world poverty income line of $1.25 per day. Some 80 percent of the people work in farming or fishing. Only about a third of the population is literate. About 90-95 percent of the people are Muslim. Some 3-5 percent are animists. Animism is a worldview which teaches that spirits dwell in many physical objects around us.

For centuries, the people of Mali have used mud bricks to build houses and other buildings. Mali has hundreds of mosques built in this style. Every year workers apply fresh mud plaster to maintain the structures.

Dogon Community Built Along the Bandiagara Escarpment

The Dogon of the Bandiagara

Mali has many ethnic groups, but this lesson concerns the Dogon people, who number about 250,000. Otume is a Dogon. He lives in one of the approximately three hundred small Dogon villages on the brow of and around the foot of the Bandiagara Escarpment. This formation and the people who live there constitute one of the most striking geographic, architectural, and cultural settings in Africa, if not the world. The escarpment is a natural wall about ninety-five miles long and about 1,600 feet high. It forms the edge of a plateau and towers over a rocky, sandy plain.

Many centuries ago, the Tellem people settled there and carved dwellings out of the side of the cliff. The Dogon invaded later, took over those dwellings, and built other distinctive structures at the base of the escarpment. The Dogon society has existed largely unchanged for centuries.

A Dogon Village

As we mentioned earlier, everything in Dogon life has meaning. The people built their villages where they did for the purpose of strategic defense. On top of the escarpment, in the cliff-side dwellings, and at the base of the cliff are all positions that the Dogon can defend easily and that give them a clear view of any attackers.

The structures in a typical Dogon village at the base of the escarpment are arranged to symbolize a human body lying down. The first building constructed in a village is the *toguna*, which means "house of words." This structure represents the head of the body and is the place where the men of the village gather and talk. The straw roof on the toguna has three layers that symbolize the cliff, the plain, and the plateau. Eight posts, carved with figures that represent ancestors, hold up the roof. The low, flat roof is suspended not far above the men's heads when they are seated. This arrangement

Lesson 27 - Riches and Poverty in Mali

prevents a man from standing up quickly to fight if a discussion becomes heated. Near the toguna is the *tei* or public square. In this area on special days the people hold ceremonies; on regular days it is a children's playground.

The walls of houses and other structures are made of straw with mud or dung. A typical family home has two stories, with a door but no windows. They are decorated with sculptures of male and female ancestors. The most prominent home is the *ginna*, where the senior man of the village lives with his family. The ginna is at the center of the village, and the families of men with comparatively less prestige live farther and farther away from the center. Each family will likely move several times over the years as the man's status changes in the village. The Dogon do not have electricity or indoor plumbing.

Each village has several storage structures called granaries. These are square with thatched roofs and stand somewhat taller than typical family homes. They are built on platforms of rocks that serve as

Toguna

Granaries

barriers to insects and rodents. Each granary has a square door with carved images that represent family history. A male granary is a place where men store millet or sorghum, although one particular kind of male granary is a shelter for older men. Female granaries are places where women store clothing and jewelry or foods for cooking. Men may not enter female granaries. Farm land is assigned to village residents, with the best plots closest to the village given for use to the oldest men.

Animism and the Dama

Most Dogon people are animists. They believe that everything has a spirit and that the living and the dead coexist. Their belief system identifies three main gods: Amma, the sky god; Lebe, the earth god; and Nommo, the water god. Each deity has a role to play in the life of the person and the village. The spiritual leader of a village is the *hogon*, who is the oldest man in the village. Each village has a *binu* or shrine that serves to keep peace with the spirit world. An individual or a family might have an individual shrine as well. The Dogon do not have a written language, but they do make limited use of hieroglyphics painted on rocks that convey spiritual realities as the Dogon see them.

Dogon Hieroglyphs

The Dogon hold three events to mark death. The *nyu yana* is the initial period after a death that includes the mourning and the burial. The *dama* marks the end of the period of mourning when the soul departs the world of the living. The third event is the *sigui* and is only held every sixty-five years. It marks the replacing of one generation with the next.

The dama is an elaborate dance ceremony that requires extensive preparation and practice. Men in their late teens or early twenties who have been circumcised may take part. Some villages hold one dama a year on behalf of all who have died that year. The costumes are elaborate with colors and decorations that bear specific meanings. The most significant part of the costume is the wooden mask that is tied to the dancer's face with cloth bands. The masks represent various things: animals, people, or spiritual beings. One kind of mask features a wooden blade that extends up over fifteen feet tall. This symbolizes the living world's connection with the spirit world. The dancing is carefully planned and is performed in groups and as solos. The music that accompanies the dance includes drums, a stringed instrument, individual vocals, and group chants.

When Otume is able to take part in a dama, it will mark his full passage into adulthood. He will finally be seen as a man.

Threats to the Dogon Way of Life

The Dogon culture has existed largely unchanged for centuries, but changes are taking place. One change that presents a relatively mild threat is the increasing number of tourists who come to the Bandiagara. As information about our world spreads and as more people can afford to travel to exotic places, more tourists visit the Bandiagara region. The presence of outsiders changes their way of life the way tourists would if they walked around your house and looked in your windows day after day. They introduce new ideas and new realities into Dogon life. A dama mask is used only one time, for one dance. Sometimes the villagers sell used masks to tourists.

Another threat is the rising demand for Dogon antiquities. As the Dogon people and their artistry have become more widely known, a market has developed for Dogon artifacts. Some people are willing to pay thousands of dollars for a mask, a statue, or other piece of Dogon handiwork. Unfortunately, some are willing to steal Dogon works of art to take advantage of this market. Dogon antiquities that are reputed to be several hundred years old have brought high prices from collectors.

A third threat to the Dogon way of life comes from the fighting among groups that want to control Mali. These groups fight each other as well as the Malian military. Several of the groups are factions of Islamist militants. The groups see value

Dogon Dancers

Lesson 27 - Riches and Poverty in Mali

in controlling Mali for what they can gain from it, but an additional perceived benefit is being able to use Mali as a staging area for attacks elsewhere. One tactic that some of the groups use is indiscriminate killing that they hope will intimidate their opponents into ending their opposition. An attack on a Dogon village of three hundred in June of 2019 left ninety-five people dead.

What Will Happen?

Geography shapes everyone's life. The Dogon people have developed a distinctive lifestyle because of their worldview and because of their geographic setting. They have strong traditions that they have maintained for centuries. But Dogon life is changing as the world around them changes. Sometimes new ideas, such as the gospel, bring progress. Sometimes new ideas bring destruction. What will become of the Dogons' home and lifestyle? What will become of Otume?

Paul lived through great changes in his life and in the world around him. He could not see the future, but he could see the goal to which he was striving.

Therefore, being always of good courage, and knowing that while we are at home in the body we are absent from the Lord—for we walk by faith, not by sight—we are of good courage, I say, and prefer rather to be absent from the body and to be at home with the Lord.
2 Corinthians 5:6-8

Assignments for Lesson 27

Gazetteer Read the entries for Guinea-Bissau, Liberia, Mali, Mauritania, and Niger (pages 36-40).

Worldview Copy this question in your notebook and write your answer: What do you believe about the spiritual realm? Does it have angels, demons, and/or any other beings? What powers do such beings have?

Project Continue working on your project.

Literature Continue reading *Patricia St. John Tells Her Own Story*.

Student Review Answer the questions for Lesson 27.

Plates at a Market in Ghana

28 The Trees of Lake Volta

The dam that created the lake provides the electricity that brought the country into the modern world. On the other hand, sometimes the dam doesn't generate all of the electricity that the country needs. In such cases, large industry keeps going, but the citizens sit in darkness.

The lake pushed tens of thousands of residents out of their homes, but it has led to a lucrative underwater tree harvesting endeavor.

The lake provides work for thousands of fishermen. It also serves as the place where children are forced into low-paid or unpaid labor and where parents sometimes sell their children younger than ten years old into slavery in the fishing industry. Sometimes the selling price is about $250 or less.

The lake is a mixed blessing.

Ghana

Like many countries in sub-Saharan Africa, Ghana used to be the colonial possession of a European power. It was the British Gold Coast colony, which merged with Togoland and became the independent country of Ghana in 1957.

Ghana was the first former colony south of the Sahara to become independent. Its leader was the dynamic Kwame Nkrumah. After a difficult period of instability, Ghana developed into a fairly stable republic with about twenty political parties. Some problems of corruption and poor administration persisted, but many observers considered it one of the most successful countries in Africa south of the Sahara. It led the way in independence for thirty other countries in the region that emerged from colonial rule and became independent in the decade after Ghana took that step.

Ghana is a small country, slightly smaller than Oregon, but it has a market economy, great wealth in mineral resources, and thriving agriculture (cocoa is a major crop, grown on over half of the arable land). Over fifty percent of the people are under twenty-five years of age.

Ghana carries on significant international trade. In fact, local traders dealt with Europeans from the 1400s. Unfortunately, one of the aspects of trade there for many years involved enslaved persons.

Akosombo Dam and Lake Volta

Nkrumah had a vision for how his country could emerge from the status of a colony to become a player in the modern world. That vision involved Ghana's geography. Nkrumah wanted the country to build a dam across the Volta River, which runs from

168

Lesson 28 - The Trees of Lake Volta

north to south in the eastern part of the country. This hydroelectric dam could provide much needed electricity. The resulting lake could encourage commercial fishing. The reservoir would enable the irrigation of farmland. It could help navigation (and thus commerce, including tourism) in a large area of the country.

A few people had considered such a dam as early as 1915. Plans were drawn up in the early 1950s. Construction began in 1961 and was completed in 1965. Ghana provided half of the financing, with the rest coming from the United States, the United Kingdom, and the World Bank (an international lending institution). Nkrumah was so closely associated with the project that it became known as Nkrumah's Baby.

Akosombo Dam created Lake Volta, which covers 3,283 square miles. It is the largest manmade lake in the world. The dam is the main source of electricity for all of Ghana. Besides powering homes and businesses, the electricity generated enabled Kaiser Aluminum to build a smelting plant that capitalizes on the aluminum deposits in the country. The lake displaced about 80,000 Ghanians from some 15,000 homes in about 740 villages and forced them to move to newly created towns. About two million people now live along the lake's shores.

President Nkrumah visited President John F. Kennedy in Washington, D.C., in 1961.

At times, the dam generates enough electricity to meet Ghana's needs and to sell surplus power to other countries. During times of drought, however, the power supply runs short, leaving hospitals, homes, and businesses in the dark while Kaiser Aluminum continues to operate. The sometimes-inconsistent power supply has earned the Ghanian president the nickname of Mr. Dumsor (*dum* means off and *sor* means on in the local Asante language).

Akosombo Dam and Power Station

Fishing on Lake Volta

The lake and its over sixty species of fish have enabled the development of a large fishing industry. Unfortunately, unscrupulous labor traffickers sometimes obtain the services of children by buying them from their impoverished parents. An estimated twenty thousand boys work long days for many hours per day in the fishing industry and receive little or no pay.

Underwater Timber Harvesting

About half of the surface of Lake Volta has partially submerged trees poking out of the water. When the dam was being built, Ghana did not have the equipment to harvest the trees that the lake would cover, and harvesting the trees would take longer than the construction timeline for the dam. So they remained. The trees complicate fishing, as nets get tangled in the submerged branches. They make navigation on the lake difficult because they can puncture the bottom of boats. Over three hundred people have drowned in such accidents. They also leave a valuable natural resource untapped.

This situation sounds unusual, but in fact it has occurred in several places around the world where dams have been built. These places include British Columbia in Canada, Suriname and Brazil in South America, and Malaysia in southeast Asia. The trees in Lake Volta include valuable rot-resistant hardwood species. The lack of oxygen underwater means that the trees do not deteriorate.

A Ghanian company harvests the timber. The cutting equipment is stationed on a floating barge. During cutting, propellers directed by GPS hold the barge in place. Workers guide the remote-controlled cutting equipment underwater. Once the workers cut a tree, the equipment lifts it out of the water, trims it, and places it on a floating log trailer. The company processes the wood for lumber and for charcoal on the local and world markets.

One benefit of this underwater harvesting is less destruction of Ghana's surface forests, which cutting has reduced by 90% since 1900. One concern about the harvesting involves whether the cutting might alter fish habitats, but the cutters take care to harm the environment as little as possible. Overall, experts see the benefits as outweighing possible detriments.

Lesson 28 - The Trees of Lake Volta

As in many places in the world, the interaction between geography and human activity is complicated. People altered the flow of the Volta River to produce electricity. The resulting lake increased fishing and boating activity but gave human traffickers a place to harm the lives of others. Given the presence of the manmade lake, the underwater timber harvesting provided work and a source for wood products that spared a portion of the remaining surface forests. This kind of interaction is what human geography is about.

Submerged Tree in Lake Volta

The trees of the Lord drink their fill,
The cedars of Lebanon which He planted.
Psalm 104:16

Assignments for Lesson 28

Gazetteer Read the entries for Equatorial Guinea, The Gambia, Ghana, and Guinea (pages 32-35).

Geography Complete the map skills assignment for Unit 6 in the *Student Review Book*.

Worldview Copy this question in your notebook and write your answer: In your mind, what is the difference between honoring your ancestors and worshiping your ancestors?

Project Continue working on your project.

Literature Continue reading *Patricia St. John Tells Her Own Story*.

Student Review Answer the questions for Lesson 28.

Drummers in an Olojo Festival Parade, Nigeria (2018)

29 The Music of Nigeria

In 1796 British physician Mungo Park was on an expedition through the interior of Africa when he came to West Africa. He traveled with a group that was marching enslaved Africans to the Atlantic coast for sale. According to his account, the group included thirty-five slaves, fourteen free black African men, each accompanied by one or two wives and some domestic slaves of his own, and six *jillikeas* or singing men. These men sang frequently, either to relieve the fatigue that others felt or to obtain a welcome from people the group did not know.

Park told another story about a delegation from one West African town that traveled to another town to buy grain. Accompanying the delegation were four drummers and four male singers. Park understood that this was a common part of the transaction that was taking place.

You likely wouldn't think about singers accompanying captives on their way to a slave market or a vocal and percussion group as part of a trade delegation, but these incidents demonstrate how music has been an integral part of life in West Africa. Music continues to play this vital role today.

Southern West Africa is a region of many ethnic and tribal groups who now live in various nation states. Much of what we say about music in this lesson is true for West Africa as a whole, but our emphasis is on the country of Nigeria.

A Land of Diversity

Nigeria is diverse geographically. Just over twice as large as California, Nigeria has arid plains in the north where perhaps twenty inches of rain fall per year. It has hills and plateaus in the central region, coastal plains in the south, and mountains in the equatorial southeast, which receive as much as 120 inches of rainfall annually.

Nigeria has a diverse population. With about 214 million people, it has the largest population of any country in Africa, although it is only the fourteenth largest country in Africa in terms of land area. This means that Nigeria has a high population density compared to the rest of Africa. That population includes over 250 ethnic groups, although the three largest (Hausa, Igbo, and Yoruba) make up about fifty-five percent of the population. The Hausa language, since it is the language of the largest ethnic group, is the most widely spoken language; but the official national language is English, reflecting its history as a one-time British colony. Nigeria gained its independence in 1960 but is still a member of the British Commonwealth.

Lesson 29 - The Music of Nigeria

Nigeria is diverse religiously. Just over half of the people are Muslim, and the majority of these live in the northern part. About forty-five percent are Christian. However, a significant number of those who profess Islam or Christianity also practice folk religions. Ethnic and religious tensions have overshadowed Nigerian life for many years.

Nigeria is diverse economically. It has the largest and most productive national economy in Sub-Saharan Africa. The southern region is the most economically developed. The poor soil and unfavorable climate of the central area make for sparse economic activity, while the north has a wide range of economic endeavors. For many years the largest growth factor in the Nigerian economy overall has been the development of its significant petroleum reserves.

Nigeria is diverse culturally. The many ethnic groups have developed their own distinctive cultural practices and beliefs. However, some ethnic groups are similar to others in some practices. One factor for these similarities has been the migrations of some of the ethnic groups, who have spread their practices to various regions. Another factor has been trading activities, through which people from one ethnic group have brought not only goods to sell but information about their own cultural practices, including music. In addition, the influences of western and Arabic cultures have made for another source of diversity.

One of the richest aspects of Nigerian culture is the music.

The Functions of West African Music

West Africans in pre-modern times generally did not keep written records about their music and did not use musical notation. They did have strong oral traditions that they passed down from one generation to the next. European travelers in Africa often commented on how important music and dancing were to the peoples they met.

Nigerian Musician Performing at a Festival in China (2007)

"A Griot of the King of Boundou" (1890). Boundou was a kingdom in what is now Guinea.

Historically, most West African music in general and Nigerian music in particular has been functional. In other words, music has served particular purposes and has been played at particular events as opposed to being purely for entertainment.

For instance, people have played music to celebrate a baby's birth, as part of the rite of passage when the tribe recognizes that a child has become an adult, to celebrate a couple's betrothal and later at their marriage, when a hunt begins and at its successful conclusion, at the beginning and end of warfare with another tribe, at religious festivals, to pay homage to a king, when a king made a state visit to the ruler of another tribe, at gatherings of secret societies, and at funerals—in other words, from birth to death and at significant times in between.

Much music has had a spiritual motive, either in worshiping a deity, making an appeal to a deity, or recognizing a deity's influence in their lives. Music has also been a way to teach history through the lyrics of songs, reminding the current generation and teaching the younger generation the stories and beliefs that a tribe holds dear.

In recent years, with the growing worldwide interest in popular music because of recorded music, radio, and the Internet, Nigerians have developed their own pop music industry. These songs are largely for entertainment, but they often retain some elements of traditional Nigerian music.

Kinds of West African Music

One of the most common forms of traditional music in West Africa is the call-and-response style. In this form, one person or one subgroup will call out a line, and the rest of the group will reply, sometimes repeating the line and sometimes responding with a follow-up idea. A common application of this style is the field holler, when a leader calls out a line and the other workers respond, all in keeping with the rhythm in which they complete their field work.

Another form is the music of the *griot* or *djeli*. A griot is an individual who wanders from village to village and sings the news or sometimes repeats stories of history to the people.

The incidents at the first of this lesson tell of other functions that music has traditionally played in West Africa. As you can see, music has permeated West African cultures such that it plays a part in just about every area of life.

Attributes of West African Music

Typical attributes of West African and Nigerian music include:

- *heterophony*, which is the playing or singing of a melodic line with variations played or sung at the same time.

- *pentatonic scale*, which consists of five notes as opposed to the eight-note scale common in Western music. Pentatonic music is also

- *polyrhythm*, which involves multiple different beats played at the same time, usually on percussion instruments.

- *melisma*, the practice of sliding from note to note or holding a note to a higher or lower pitch. This is done especially in singing.

Instruments

Anthropologists have discovered clay pots in Nigeria from the 1100s and 1200s AD which show drums and other instruments that are still used today. West Africans have fashioned instruments from materials they have had available to them, such as animal skins that they would stretch over hollow logs, sometimes with rattles attached to the rims.

Instrumentalists play many different kinds and sizes of drums. One popular style is the talking drum, which is usually hourglass-shaped and which has strings on the side which the player can tighten or loosen by hand. Nigerians regulate the sound of these drums to mimic the human voice.

Other widely-used instruments include trumpets of various lengths and wooden clappers. Musicians pluck the strings of some instruments such as forms of the lute and the zither, while they play other stringed instruments with a bow. Flutes are also common. The nose flute is played by holding the flute to one's nose and breathing out through the nose.

Instrumentalists often attempt to imitate the sounds they hear around them, such as birds or thunder or the wind blowing through the trees.

Drummers at the Lisabi Festival in Nigeria (2020)

Singing and Dancing

In traditional Nigerian songs, some parts and styles were reserved for men, some for women, and some for children. Some singers sang solo, while at other times a group would sing.

Ancient West African rock paintings portrayed dancers. Musicologists believe that the movements of dance help shape the rhythms of West African music, which one expert described as sophisticated and intricate. One common style of dance was the ring shout, which was a slow, shuffling, circular dance, often performed in worship and accompanied by singing.

Traditionally, every Nigerian could and was expected to participate in musical performances in some way. However, music was held in such high esteem that some royal courts maintained professional singers, dancers, and musicians, especially in places that were influenced by Islamic practices.

The Impact of Slavery

It is important that we understand and appreciate the role of music in West African cultures without comparing their practices to those of Europeans and Americans. However, the practice of slavery

Musical Group in Brazil (1846)

This painting called The Old Plantation *is from around 1790. Notice the enslaved persons enjoying the music of a drum and a banjo-like instrument.*

impacted music both in Africa and in the New World in ways we should recognize.

The forced removal of millions of Africans from their homelands to the New World took them from the scenes of their cultural heritage, which included their music. They went from music being a huge part of their lives to the horrendous middle passage across the Atlantic in overcrowded slave ships where musical practices were difficult or impossible. Part of robbing these people of their humanity and making them objects for sale and use involved robbing them of their culture, language, and musical settings.

Enslaved persons brought to the New World were not just "Africans." They were members of a wide variety of ethnic groups with widely different beliefs and practices. Being thrown together with other slaves to whom they could not speak and with whom they shared no ties of background made the horrors of slavery even worse.

At work in the lives of these slaves were two conflicting forces. On one hand, slave owners usually did not want them to retain memories and practices from their homelands. Owners thought that their slaves would be more productive if they forced slaves to turn from their past and focus on the work before them. Thus many owners forbade slaves from engaging in any musical practices.

On the other hand, despite the hardships and the restrictions, Africans in the New World found identity, solace, and hope in the music they retained and adapted to their new situation. We are indebted to slaves for several elements of modern American music, such as gospel music. An instrument used in Africa and reproduced in the Americas was the "bonjoe" (people spelled it various ways), which was the forerunner of the modern banjo. What Africans maintained, developed, and accomplished is a testimony to their resilience of spirit and to the power of music.

Paul spoke of the functions of music in the fellowship of the church.

Let the word of Christ richly dwell within you, with all wisdom teaching and admonishing one another with psalms and hymns and spiritual songs, singing with thankfulness in your hearts to God.
Colossians 3:16

Assignments for Lesson 29

Gazetteer Read the entries for Nigeria, Senegal, Sierra Leone, and Togo (pages 41-44). Read the excerpt from *Travels in the Interior of Africa* by Mungo Park (pages 258-260). Answer the questions on the excerpt in the *Student Review Book*.

Worldview Copy this question in your notebook and write your answer: What do you believe is the proper place of fear in relation to God? Should Christians live in abject terror of God?

Project Continue working on your project.

Literature Continue reading *Patricia St. John Tells Her Own Story*.

Student Review Answer the questions for Lesson 29.

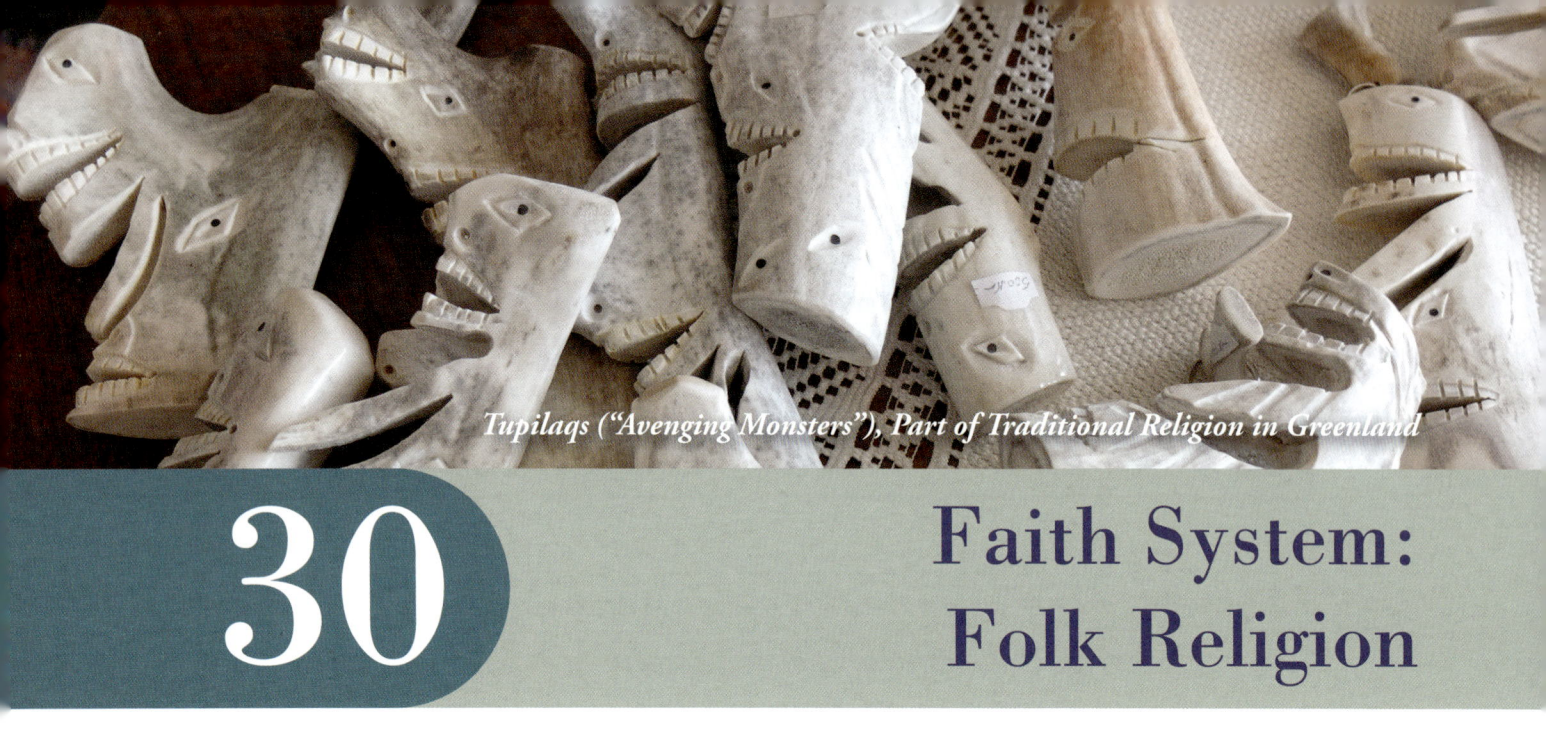

Tupilaqs ("Avenging Monsters"), Part of Traditional Religion in Greenland

30 Faith System: Folk Religion

The range of worldviews is not so simple that it can be divided into those who believe in God and a spiritual realm and those who deny God and any spiritual realm. Some religious or worldview systems believe in a spiritual realm, but what they teach looks very different from the ideas of theism with which we are most familiar.

An estimated 400 million people practice what is known as folk or traditional religion. People on every inhabited continent participate in folk religions. Animists or practitioners of folk religions believe in a spiritual realm; but the various groups in these religions have many different ideas about the nature of that realm, such as the existence and identity of a supreme being and the presence and power of lesser spiritual beings and spirits. Various ethnic groups have developed a number of different expressions of folk religion, but enough similarities exist for us to give a broad overview of these practices.

Folk religion arises from a group's worldview but at the same time it contributes to its worldview. The older generation passes down its beliefs and practices to the younger generation, and in this way the religion continues. Folk religion usually predates the coming of other faiths, such as Christianity, Islam, and Buddhism, into a group.

Worldview

Those who practice folk religions believe that a spirit world exists that is filled with spiritual beings and forces. They also believe strongly in an impersonal or mystical power, sometimes called a life force or life essence, which can do both good and evil. Practitioners believe that these beings and forces are present in and have a strong influence on the world in which people live.

At the highest level of the spirit world is the supreme being. Most folk religionists believe that this being created the world but that ever since this supreme being has had little to do with the world. Most folk religion practices address the spirits and beings on lower levels of the spirit world.

Traditional religions believe in a secondary level of lesser deities and then a third, even lower level of non-human spirits and the spirits of ancestors. The lesser deities and non-human spirits live everywhere, in trees, rocks, rivers, mountains, the soil, the wind, and so forth. Each has a specific area of activity, such as hunting, the weather, digestion, pregnancy, birth, marriage, burials, building a house, and business activity. Thus these beings deal with every aspect of human life. Folk religionists believe that the spirits of ancestors can do good or ill for their descendants who live on the earth. The worldview known as animism holds that spirits dwell in many physical objects, ready to do good or to do harm to humans.

Many folk religionists believe in the evil eye, the ability of some people to cast harmful spells on others. They also believe in the reality of witchcraft, magic, sorcery, spirit possession and exorcism, reincarnation, and life after death that includes a significant degree of reward or punishment.

Practices

Followers of traditional religion usually believe in seeking the ideal of unity or harmony between the spirit world and the human world. The desire to maintain this harmony and the tendency they see for the spirits and deities to be easily offended result in folk cultures identifying taboos (forbidden behaviors) and superstitions as well as the practice of rituals. The practices of folk religion are most often attempts by its followers to regain control of their world when spiritual forces appear to them to have upset their world. Some folk religion practices are attempts to prevent problems from arising in major life events. Maintaining the moral order of the community is a high priority.

Adherents of traditional religions engage in many practices that they hope will pacify the spirits and help them have a successful life. These include trusting in juju (amulets or charms), maintaining in their homes shrines dedicated to the spirits of their ancestors or other spirits, offering sacrifices to the spirits for protection, and keeping other objects they consider to be sacred, such as skulls, carved figures, spears, and animal horns. Folk religion practitioners engage in ceremonies to honor their ancestors. Many consult traditional healers when they or a family member is ill. In their minds, illness is a spiritual problem and not just a physical problem.

Shinto Shrine and Tree in Osaka, Japan

This spiritual leader is conducting a ceremony among the Pataxós people in Brazil.

The Diviner

Almost every group that practices folk religion recognizes a person whom they believe has special powers and a special relationship with the spirit world. This person has different titles depending on the group, such as diviner (for the ability to divine or obtain the will of the spirit in question) or sorcerer or priest. The term *shaman* is most closely associated with the Shenjiao folk religion in Northern Asia, but people sometimes use the term for any such person anywhere. A medicine man usually deals primarily with physical illnesses because he supposedly knows the potencies of various plants and other substances as well as the rituals and incantations appropriate for different illnesses. Westerners sometimes called this person a witch doctor, but most people now see that term as condescending.

A person does not go to school to become a diviner. Usually a community recognizes someone they believe has the ability to fill this role, or a person feels a "call" to this role. Depending on the ethnic group, both men and women can fill this position. A young diviner will usually learn oral traditions from an older diviner; but there is no creed, set of doctrines, or scripture that a diviner must first learn.

Unit 6: West Africa

Folk religion practices reveal the doctrines that the diviner and the people of the group believe.

A diviner usually wears special clothing and uses special accessories to engage in his activities. Masks that represent ancestors, heroes, gods, or dynamics of the cosmic order are common. The diviner learns the chants, incantations, and dances that tradition says a given situation warrants. Sometimes a diviner faints, engages in hysterics, or falls into a trance while conducting a ritual. People believe that a diviner can communicate with the spirit world to make his appeal to the appropriate deity or spirit. Sometimes people think that the diviner takes on the form of an animal. The diviner often uses a drum while engaging in a ritual. Sometimes a diviner will "sell" a spirit to a neighboring village for a time to help that village try to deal with a problem.

Practitioners of traditional religion believe that a diviner possesses many powers, including revealing the future, locating a lost animal, or identifying a thief. They believe that a medicine man can extract an offending spirit or substance from a sufferer's body.

Altar in Mopti, Mali

Carved statues called hogangs mark paths and boundaries among the Ifugao people of the Philippines.

A Christian Response

What is a proper and effective response for Christians to take to folk religions? What can followers of Jesus do to help those who believe in traditional religion come to a life-changing faith in God and in Jesus Christ? Westerners sometimes see adherents of folk religion as being members of a "primitive" culture or engaging in "primitive" beliefs. This is a condescending view.

Christians can receive a couple of reminders from folk religionists. For one, they remind us that the spiritual realm really does exist and affects our everyday lives. The modern, Western tendency is to separate the spiritual from the material; as a result, many people have come to see the spiritual realm as optional and unimportant. Christians need to demonstrate our faith in God and our loyalty to Christ in real-life situations.

Folk religion also reminds us that family and community are important and that our actions have an impact in those contexts. We need to show respect for the status and integrity of our families, and we need to live as salt and light and to redeem the opportunities we have with others (Matthew 5:13-16, Colossians 4:5-6).

As we have said before, the most important need in seeking the conversion of folk religionists is for such practitioners genuinely and deeply to change their worldview. Becoming a Christian is more than merely intellectually affirming the truth of certain Christian doctrines. An evangelist referring to certain passages of Scripture might not be enough to convince a follower of traditional religion.

It is very common for a practitioner of folk religion to agree to the existence of God and even to confess Jesus as Lord, but still to maintain largely unchanged his or her belief in the spirit world. Such people who adopt the Christian or the Muslim faith often still keep what they consider sacred objects, perform the same rituals, and consult diviners and medicine men or women as before when they believe the need arises. Vodou (also spelled voodoo) is a folk religion in Haiti brought there by African slaves which blends African, Roman Catholic, and Caribbean beliefs and practices.

There is "one God and Father of all, who is over all and through all and in all" (Ephesians 4:6). There is one Lord and one Spirit (Ephesians 4:4-5). Yet Paul also refers to "the spiritual forces of wickedness in the heavenly places" which Christ has conquered and against whom we do battle (Ephesians 6:12), and the "rulers and authorities" over which He triumphed (Colossians 2:15). Paul's point is that whatever forces and realities exist in the spiritual realm (what he calls the "heavenly places"), Christ has triumphed over them in the cross and the resurrection. Satan exists and works against us (Ephesians 6:11-12, 1 Peter 5:8-9), but we have the answer to his wiles in Christ.

God's truth is that the true Supreme Being is not detached from us. Instead He loves us, works in the world for His purposes, and wants what is good for us. He desires an ongoing relationship with us as His children. We pray and obey His will not to keep the evil spiritual forces off our backs but to live in the assurance of His presence with us. God's spiritual power does not exist just to put out fires and solve problems but to help us grow in Christ. We do not have to live in constant fear in a world that is out of control as a result of spiritual forces that we do not understand. Instead, we live in this world victorious and confident with the faith that God is in control. The greatest truth, the central idea and central reality of the universe, the greatest force in the world, is love; and God has demonstrated that love toward us in the cross.

As a result, we do not have to live in fear of what lesser spiritual forces might do to us. These spiritual forces are headed for eternal destruction. Life does not have to be a battle; the battle has been won on our behalf. We can put aside the old ways and take on the new ways of Christ—not by performing the proper incantations but by trusting what Christ did on our behalf that makes us overwhelming conquerors (Romans 8:35-39).

For our struggle is not against flesh and blood, but against the rulers, against the powers, against the world forces of this darkness, against the spiritual forces of wickedness in the heavenly places. Therefore, take up the full armor of God, so that you will be able to resist in the evil day, and having done everything, to stand firm.
Ephesians 6:12-13

Assignments for Lesson 30

Worldview Write or recite the memory verse for this unit.

Project Finish your project for this unit.

Literature Continue reading *Patricia St. John Tells Her Own Story*.

Student Review Answer the questions for Lesson 30. Take the quiz for Unit 6.

Tundavala Gap, Angola

7 Central Africa

Millions of people live as displaced persons around the world, including people from the Central African Republic. The Mbuti or Pygmy people are struggling to maintain their way of life. Like many nations of Africa, Cameroon lives with the legacy of having been a colony of European nations. The Democratic Republic of the Congo has experienced numerous outbreaks of the Ebola Virus Disease. The worldview lesson tells the story of Dr. Kent Brantly, who survived the case of Ebola he contracted while serving as a medical missionary in Liberia.

Lesson 31 - Life in a Refugee Camp
Lesson 32 - The Mbuti: A Threatened Way of Life
Lesson 33 - Cameroon: Legacy of Colonialism
Lesson 34 - Beautiful and Deadly: Ebola in the DRC
Lesson 35 - Personal Application: Kent Brantly, Physician and Missionary

Memory Verse   Memorize John 15:13 by the end of the unit.

Books Used   The Bible
Exploring World Geography Gazetteer
Patricia St. John Tells Her Own Story

Project (Choose One)

1) Write a 250-300 word essay on one of the following topics:
   - Tell what you think about racial prejudice and what people can do to oppose it.
   - Tell what you admire about Kent and Amber Brantly and Nancy Writebol regarding their service to others. (See Lesson 35.)
2) Write a five-minute drama portraying life in a refugee camp, using dialogue among several actors, with the play ending in repatriation.
3) American missionaries in some African countries often communicate with the people who live there by using Pidgin English. Research Pidgin and make a list of 20 words and phrases in Pidgin and their English equivalents.

UN Peacekeepers in Bangui, CAR (2014)

31 Life in a Refugee Camp

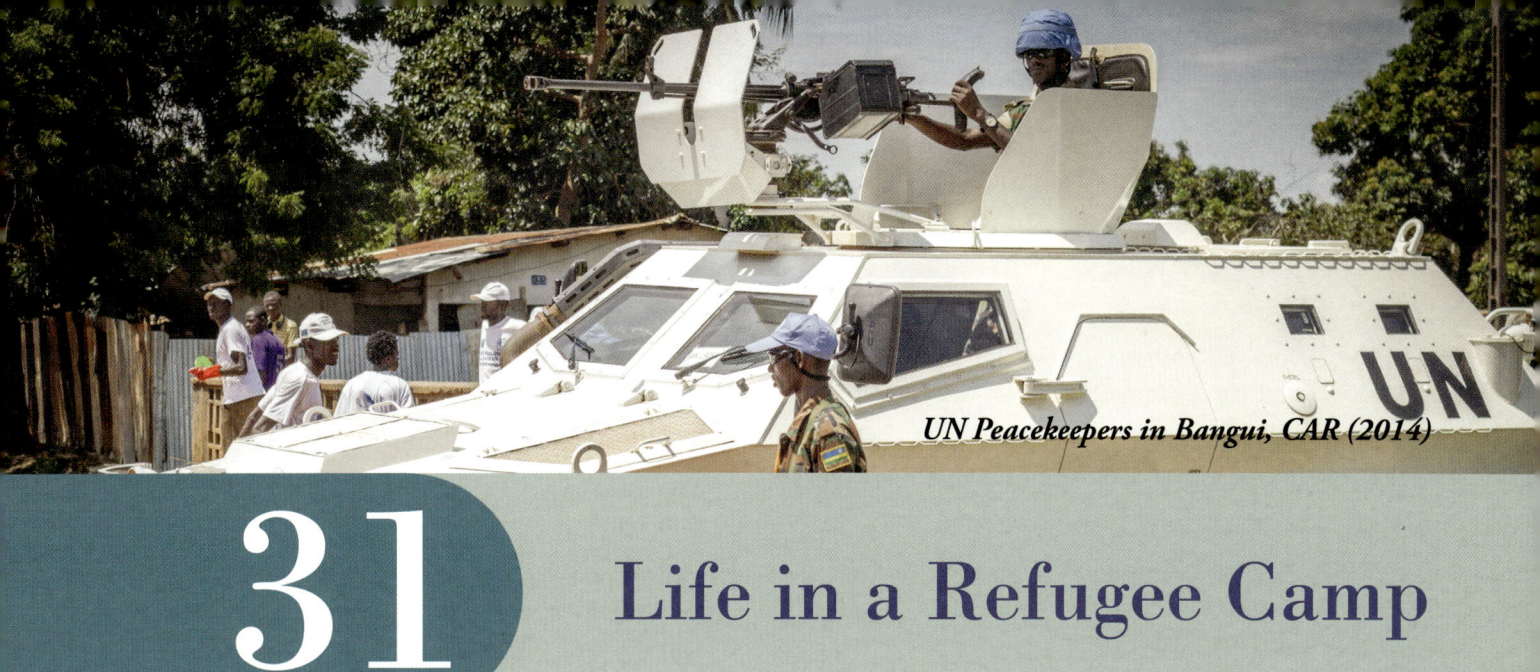

Hundreds of displaced citizens of the Central African Republic (CAR) sing to a group of thirty of their fellow citizens as the new refugees arrive at the Gado Badjere camp near Garoua Boulai, Cameroon. The camp already houses thousands of refugees. The displaced are welcoming the displaced. Desperate as they feel, the new arrivals are able to smile as they hear a familiar song in a familiar language. Even through war, poverty, and their lives being turned upside down, the refugees can sing a song of hope.

The Reason for the Crisis

The French colony of Ubangi-Shari gained independence in 1960 and became the CAR. Its government, usually led by military officers, was never stable and never functioned well. Civilians gained control of the government in 1993, but even then opposing groups bickered and sometimes fought with each other.

A military coup led by General Francois Bozize ousted President Ange-Felix Patasse in 2003. Bozize was re-elected president in 2005 and 2011, but many believe the elections were not fair and free. In late 2012 a coalition of Muslim rebel groups began attacking towns and villages. The government dissolved in 2013, Bozize fled the country, and rebels took over the capital of Bangui. A coalition of Christian militia groups began fighting the Muslim rebels. After a transition period, an election in 2016 elevated Faustin-Archange Touadera to the presidency; but he was not able to restore order to the country.

As a result of the fighting and lack of security, over one million Central Africans have fled their homes to find a safer place to live. They sometimes have to walk for weeks, often hiding in forests without adequate food. Over 600,000 moved to other parts of the CAR, officially becoming internally displaced persons (IDPs), while an estimated 575,000 left the country altogether and became refugees. About a quarter million of the latter number escaped to Cameroon, while the rest have scattered among the Democratic Republic of the Congo, the Republic of the Congo, Chad, Sudan, and South Sudan.

Life for the Refugees

About 27,000 people live in the Gado Badjere camp, one of several camps in Cameroon. The people came with the clothes on their backs and what they could carry. The makeshift housing, built with materials that relief agencies could obtain, has been home for many people for several years. Some 60-70 babies are born in the Gado Badjere camp each month. More refugees arrive nearly every month.

The office of the United Nations High Commissioner for Refugees (UNHCR) operates the camps and attempts to maintain order. Several charitable relief agencies provide what they can. The camps usually do not have adequate food, clothing, or medical or hygiene supplies. Rarely do adults have opportunities for work to earn money, and children are rarely able to attend school.

The UNHCR and relief agencies send out appeals for money to give to the refugees and to buy food for them, but the helpers cannot meet all the urgent needs fast enough. The UNHCR provides some training for jobs. In the camps, the refugees gather wood for fires to cook food when they have it. They protect what they have from getting stolen, and they wait. Helping agencies are not always able to get to IDPs because of security issues within an unstable country.

Cameroon and other host countries cannot afford to absorb the increased population and provide services, supplies, and jobs for them. Thus the UN and other relief agencies make constant appeals for hundreds of millions of dollars to ease the situation.

Cathédrale Notre-Dame, Bangui

Because of decreased violence in some areas, about a quarter million IDPs and refugees have been able to return home in the CAR. Most of the hundreds of thousands of displaced persons want to return to their homes, but not until the violence stops. They have seen too many family members and friends killed or disappear to want to go back any sooner.

The Life They Left Behind

The CAR is at the center of the African continent. It is mostly a rolling plateau, with mountain ranges in the northeast and northwest. The north is generally treeless while the south is mostly covered with rainforests. Many rivers flow through the country and connect with rivers in neighboring countries. The CAR has a rich diversity of animal life.

French and Sango are the official languages, but people speak many other languages as well. The people think of themselves as clans or language groups as opposed to ethnic groups. About eighty percent of the population is Christian (either Catholic, Protestant, or Independent) and ten percent is Muslim. Another ten percent follow traditional or folk religion, although many of the other ninety percent hold to some traditional beliefs as well.

Unfortunately, the refugees do not have bright prospects for their lives if and when they get home. The CAR has just about every problem you can name.

The CAR does not have a large population base. The country is about the size of Texas or France. Texas has 28.3 million people, while France has 67.1 million. The CAR has only about 5.5 million people, and many of them have left the country.

The CAR is one of the poorest countries in the world. Because of poor health care, malnutrition, and violence, average life expectancy is less than fifty years. The country's one hospital, located in Bangui, is below standard, while rural clinics are scarce and

Village Along the Ubangi River in the CAR

poorly supplied. Most Central Africans live on subsistence farming. Traditionally the men clear the fields and the women cultivate the crops.

Central Africans export some crops, including cotton, coffee, and timber. Diamond mines in the western part of the country produce nearly half of the country's export earnings, but government mismanagement and the civil violence have made the mining unstable and less profitable than it could be.

Transportation is a major problem. The country is landlocked, so traders have to use river transport to take products to neighboring countries to export them. The country has no railroads and only about four hundred miles of paved roads. The education system has been decimated because of the violence. Teachers are scarce and many students are not able to attend school. Only about one-half of men and one-fourth of women over age fifteen are literate.

Historically, the area was involved in the slave trade. Today, the CAR is a nexus for human trafficking.

The World Situation

The CAR has the most severe refugee problem in the world, but it is not the only place of such suffering. The UNHCR estimates that 68.5 million people live as forcibly displaced persons worldwide. Forty million people are IDPs, twenty million are refugees living in a country other than their home country, and 3.1 million are seeking political asylum. The UN separately classifies another 5.4 million Palestinians as displaced persons.

Is There Hope for Peace?

In February 2019 the CAR government signed a peace agreement with fourteen armed groups. The signees expressed hope that the arrangement would work, although similar agreements signed in 2014, 2015, and 2017 all broke down.

At the beginning of this lesson we described one group of CAR refugees expressing welcome in song. Another group expresses their hope for peace in the same way. Five CAR refugees formed the group Peace Crusaders that sings every weekend in the northern Cameroon town of Garoua. They sing in front of churches and mosques and as they walk along the streets. They plead with people to talk with each other instead of resorting to violence to resolve their conflicts. They call on the warring factions in their home country as well as the world community to end the violence so they and all other displaced Central Africans can go home. The Peace Crusaders say they will keep singing until the CAR has peace.

David spoke of finding strength in the Lord and giving thanks in song.

The Lord is my strength and my shield;
My heart trusts in Him, and I am helped;
Therefore my heart exults,
And with my song I shall thank Him.
Psalm 28:7

Assignments for Lesson 31

Gazetteer Study the map of Central Africa and read the entries for Angola and the Central African Republic (pages 46, 47, and 49).

Worldview Copy this question in your notebook and write your answer: What does this verse mean to you: "For to me, to live is Christ and to die is gain" (Philippians 1:21)?

Project Choose your project for this unit and start working on it. Plan to finish it by the end of this unit.

Literature Continue reading *Patricia St. John Tells Her Own Story*. Plan to finish it by the end of this unit.

Student Review Answer the questions for Lesson 31.

Mbuti Village in the Ituri Forest (2010)

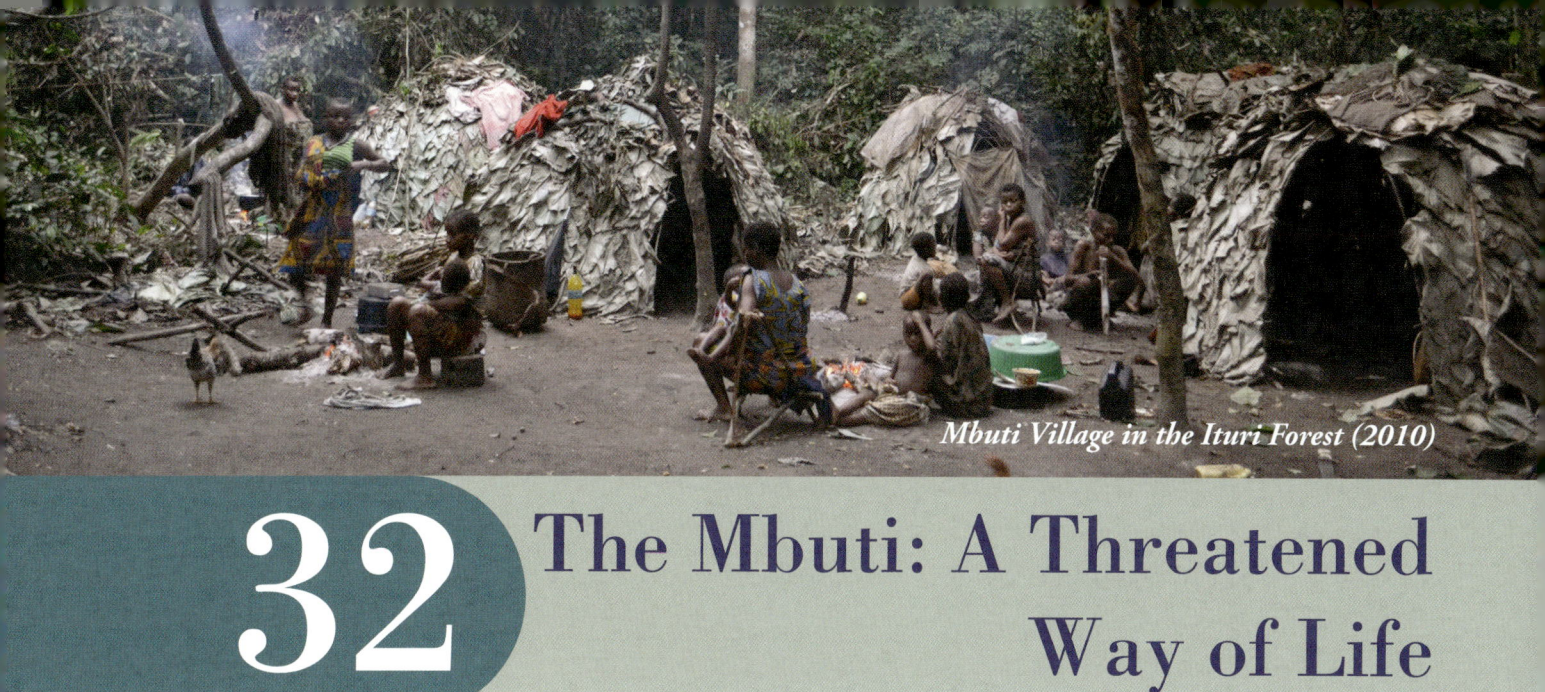

32 The Mbuti: A Threatened Way of Life

They are losing their battle for survival.

About 250,000 of the ethnic group commonly known as Pygmies live in small groups in a relatively narrow band on either side of the equator across Central Africa, from Cameroon and Gabon in the west to the region of Lake Victoria in the east. The largest tribe are the Mbuti (sometimes called the Bambuti), who live primarily in the Ituri Forest in the eastern part of the Democratic Republic of the Congo. Other Central African Pygmy tribes include the Twa, the Tswa, and the Babinga. The San live in the Kalahari Desert further south in Africa.

The term Pygmy is from a Greek word that means dwarf or dwarfish. Ancient Greek and Egyptian writings from two thousand years before Christ contain references to dwarfish people. Probably the best known physical characteristic of Pygmies is their relatively short stature, usually under five feet. The women are on average shorter than men. Their limbs are appropriately proportioned, so they are not really dwarfs. Scientists are not sure why Pygmies are shorter than average humans, although intermarriage is probably a major factor as to why the trait continues. Non-Pygmies rarely marry a Pygmy. The Negritos of Asia are another group with short stature.

One characteristic of many Pygmies is their front teeth being chipped down to a point. This is usually done with a machete. The Pygmies believe that this makes them more attractive.

The Pygmies are the largest known group of hunter-gatherers in the world today. They do not generally own land. They are at home living a nomadic life in the tropical rainforest, capturing game with nets or using bows and arrows. The Mbuti erect simple, round huts in a place, stay there for perhaps a month, and then move on.

The Mbuti generally have no chiefs or council of elders. They make decisions by general discussion. They do not practice fine arts except for simple

Mbuti Man with a Hunting Net

Members of the Mbuti Community of Bahaha (2020)

pottery, but singing and dancing are important elements of their community life. The Mbuti have a simple belief system, focusing on the spirits they believe inhabit the rainforest. They apparently take little thought of the afterlife.

The Mbuti Struggle for Survival

Today the Mbuti are struggling to stay alive and maintain their way of life. Their primary conflict is with the Bantu. The Bantu make up the majority of black Africans in the region and hold positions of power. Most Bantu harbor strong prejudice against the Mbuti; some even question whether Pygmies are really human.

In many places the Bantu have forced the Mbuti to leave the rainforest and live in permanent settlements near towns. Torn from their familiar habitat, the Mbuti pick up odd jobs as they can in order to survive. Deforestation projects have taken away the most common habitats of the Mbuti. The Mbuti usually obtain little or no schooling, and few can read or write. Accustomed to living off the land, when they lose the ability to do that, they have few options. They often do not find it easy to adjust to life in modern communities.

The Man Displayed in a Zoo

Prejudice against the Pygmies is longstanding. One of the saddest demonstrations of this fact is the story of Ota Benga. This story shows the terrible effect of racism, but it is also an illustration of the consequences of the theory of evolution as a worldview.

In the early 1900s, the commonly accepted worldview of scientists as well as the general public was the belief in evolution. This included a belief in the "survival of the fittest" as the way the world operated. Those who held this view usually believed that white Europeans and Americans were the fittest and most advanced humans and that black Africans, especially Pygmies, were the least evolved and the least fit. A 1904 *Scientific American* article described Pygmies as "small, ape-like elfish creatures, furtive and mischievous." Many evolutionists believed that the black race would eventually become extinct.

Lesson 32 - The Mbuti: A Threatened Way of Life

The organizers of the 1904 World's Fair in St. Louis wanted the fair to demonstrate human progress "from the dark prime to the highest enlightenment." Their plan was to put representatives from various people groups around the world on display so that visitors could see the evidence of the various levels of human evolution for themselves.

William McGee, the president of the American Anthropological Association and chairman of the ethnology department of the fair, sought out people who would be willing to go to various parts of the world. McGee wanted these people to recruit representatives of various ethnic groups who would be willing to come to St. Louis as volunteers to be put on display at the fair. A University of Chicago professor brought nine Ainu people from Japan. Another group came from the region of Patagonia in South America. Representatives of various American native nations agreed to come.

Samuel Verner, a South Carolinian who described himself as a white supremacist, agreed to bring a number of Pygmies from Africa. Verner said that he was an expert on scientific matters in Africa, especially Pygmies. In a 1902 article in *The Atlantic*, Verner had written, "Are they men or the highest apes? . . . . Have they degenerated from larger men, or are the larger men a development of Pygmy forefathers?" He described Pygmies as, "The most primitive race of mankind, almost as much at home in the trees as the monkeys." Verner obtained a young man named Ota Benga from a village where Benga was being held captive. Verner's accounts of how he got Benga varied in later years, but the area Verner visited was a well-known slave market.

Benga and the other Pygmies were displayed at the fair, along with the representatives of other ethnic groups. After the fair, Verner returned the Pygmies to Africa, as he had agreed to do, but he convinced Benga to return to the United States with him. In 1906 Verner arranged to deliver Benga to the Bronx Zoo in New York City.

At first, Benga simply roamed the grounds of the zoo. Then zoo officials encouraged him to spend time in the primate house. Later Benga was locked in the primate display. The display garnered an enormous response. Thousands of people came to the zoo. They followed him around as he walked the grounds, and then they observed him in the cage.

The display made headlines in New York newspapers. "Bushman Shares a Cage with Bronx Park Apes" read one headline in the *New York Times*. A subhead on the story read, "Some Laugh Over His Antics, but Many Are Not Pleased." Some claimed that Benga could communicate with monkeys and orangutans. A week after Benga's debut, zoo officials again allowed him to roam the zoo's grounds. Visitors chased him and poked at him, and Benga struck back. Three zoo workers returned Benga to the monkey house.

Black ministers in New York strongly protested the display of Benga. The group expressed their opposition to the theory of evolution. They saw the display of Ota Benga as an attempt to show that evolution was true. The prevailing views about black people mentioned above help us understand why they would think this. However, the New York Zoological Society, which operated the zoo, defended the exhibit as educational. The mayor of New York refused to become involved.

Ota Benga (second from left) was one of many indigenous peoples from around the world put on display at the World's Fair in St. Louis.

Ota Benga at the Bronx Zoo (c. 1915)

A *New York Times* editorial wondered what all of the fuss was about: "Not feeling particularly vehement excitement ourselves over the exhibition of an African 'pigmy' in the Primate House of the Zoological Park, we do not quite understand all the emotion which others are expressing in the matter." Another *Times* editorial said, "They are very low in the human scale" and expressed the opinion that a school would be a place of torture to Benga.

Twenty days after Benga went on display at the zoo, a black minister arranged to take Benga to the Howard Colored Orphan Asylum in Brooklyn. He lived there for about four years. In 1910 Benga was moved to the Lynchburg (Virginia) Theological Seminary and College, an all-black institution. Benga lived with the widow of a former seminary president and her children. In Lynchburg Benga taught local children outdoor skills. He improved his English vocabulary, learned how to read, occasionally attended classes at the seminary, and eventually became a Christian.

Benga expressed the desire to return to his homeland, but the onset of World War I postponed that possibility. By 1916 Benga began to show signs that he had become depressed. One evening, Ota Benga took his own life.

From the viewpoint of evolution, the treatment of Ota Benga made perfect sense. Humans evolved, the common evolutionary view held, and some humans had evolved further than others. A comparison of Western civilization to the "primitive" lifestyle of Pygmies, as well as the Pygmies' physical characteristics, seemed to prove this. After all, the subtitle of Darwin's *On the Origin of Species* was *The Preservation of Favoured Races in the Struggle for Life*. The controversy was not just about racism. It was about racism fueled by the worldview of evolution. The theory of evolution played into the hands of the Nazis of Germany, who applied the idea of inferior (and thus unfit) groups to Jews, the Roma people (often called gypsies), and other groups of whom they disapproved and who they attempted to destroy.

From God's point of view, the treatment of Ota Benga was a cruel wrong. The Bible teaches that all people are valuable because they are created in the image of God. Jesus demonstrated how valuable every person is when He died for the most educated white man and for the Pygmy who has never learned to read and write. The Bible speaks of nations and ethnic groups, but it never lifts up one race above others. The whole perspective of racial identification is a worldview that people have developed.

The apostle John wrote about Christ's love for the whole world, not making any distinctions or value judgments and not putting one group of people above another.

[A]nd He Himself is the propitiation for our sins; and not for ours only, but also for those of the whole world.
1 John 2:2

Assignments for Lesson 32

Gazetteer Read the entries for Chad, the Republic of the Congo, and Gabon (pages 50, 52, and 53).

Worldview Copy this question in your notebook and write your answer: When have you let fear stop you from doing something you ought to do? Or, when have you overcome fear and done something you thought you ought to do?

Project Continue working on your project.

Literature Continue reading *Patricia St. John Tells Her Own Story*.

Student Review Answer the questions for Lesson 32.

Yaounde, Cameroon

33 Cameroon: Legacy of Colonialism

In October of 2018, 85-year-old Paul Biya was elected to serve a seven-year-term as president of the West African country of Cameroon. It would be his seventh term as president. Biya received 71% of the vote.

Biya had become president in 1982 after the resignation of Ahmadou Ahidjo, who had been president since the country achieved independence in 1960. Biya had been prime minister under Ahidjo and had served in the cabinet for twenty years.

If it sounds as though Biya's political party has a lock on power in Cameroon, that's because it does. With firm control of media outlets in the country, it also has a reputation for corruption, cronyism, and failing to protect civil rights and freedoms.

In the first two decades of independence, Cameroon was an economic success story. It had a strong agricultural sector, industry was growing, cocoa was developing as a major crop and export, and petroleum exporting became a large part of the economy. Then falling prices on the world market, coupled with internal instability, upset the picture. Cameroon is recovering economically, but all is not well. One major issue dividing the country is the legacy of unrest from its era spent under colonial rule, not by one or two foreign powers, but by three.

Prime Geography

Cameroon is blessed in many ways with its geography. Slightly larger than California, it is located at the curve of the West African coast. The Carthaginians in North Africa knew about the area before the time of Christ. Ethnic groups lived in their various localities, and sometimes their kings tried to rule others. In 1472 the Portuguese explorer Fernao do Po became the first European known to land there. The Portuguese saw the estuary of a large river at the coast. They named what they saw Rio dos Camaroes, or River of the Prawns, because of the large number of prawns they saw there. A prawn is a kind of shellfish. Thus, Cameroon is the only country to have the honor of being named for a crustacean.

Trade developed between Europeans and the people of Cameroon. Unfortunately, much of that trade involved slaves that other Africans captured and brought to the coast.

The port city of Douala formed at the mouth of the Rio dos Camaroes, now called the Wouri River. Today it is one of the best equipped ports in West Africa. It receives goods going to and coming from the inland countries of Chad and the Central African

Lesson 33 - Cameroon: Legacy of Colonialism

Republic. Railroads carry the goods between Douala and those countries. Cameroon has other ports as well, which is significant because of the overall lack of good ports on the West African coast.

The highest point in the country is Mt. Cameroon, an active volcano 13,354 feet high and not far from the coast. The volcano last erupted in 2000. This mountainous, heavily forested area receives over 400 inches of rain per year, one of the highest totals of anywhere in the world. This contrasts to the arid northern part of the country, which usually receives about fifteen inches per year. The British called these mountains the Camerouns.

A coastal plain rises to a high plateau in the south. A central plateau descends to Lake Chad in the north. The plains are marked by inselbergs, which are mounds of erosion-resistant rock.

The first significant Muslim activity came in the early 1800s. English and American Christian missionaries brought the gospel in the 1800s and 1900s. Today the country is about 38% Roman Catholic, about 32% other Christian churches, 24% Muslim, and 2% animist. Muslims predominate in the north, while Christians and animists live mostly in the south.

Some 150 to 200 different ethnic groups live in Cameroon. Their languages reflect about 24 major African language groups. In addition, French and English are the official languages—again, a legacy of colonialism. The country has a population of about 25 million; life expectancy is about sixty years.

The Scramble for Africa

Nations have established overseas colonies for centuries. Carthage, for instance, began as a colony of Phoenicia. Rome achieved conquests in many regions. The English began planting colonies in North America in the early 1600s. In the last third of the 1800s, several European countries competed to establish colonies. These countries wanted to expand their markets, find a source for natural resources, have a place where some of their growing populations could live, and improve their standing among what many people saw as the major powers of the world.

A German manager oversees loading of bananas for export from Kamerun in 1912.

A prime target of their colonial ambitions was Africa. Seven European countries—Britain, France, Germany, Belgium, Spain, Portugal, and Italy—divided up pretty much all of sub-Saharan Africa in what was called the Scramble for Africa. Germany established a protectorate over Cameroon (or Kamerun, as the Germans spelled it). Germans created many large estates in the southwestern part of the country.

Then came World War I. Forces from the Allied nations of Britain, France, and Belgium seized control of Cameroon. Germany lost the war, so it lost its overseas colonies. The League of Nations, which came into being after the war, gave Britain and France mandates to oversee Cameroon. The French controlled about 80% of the area, which it called Cameroun. The British ruled two much smaller areas in the west of the country, which it called the Cameroons after the mountains. The French exerted more direct and active oversight of its region than the British did of theirs, and as a result the French area saw greater economic development.

Independence and After

As part of the general move in Africa toward national independence and an end to colonial rule, French Cameroun became independent in 1960. The British decided to give residents of the areas it controlled the opportunity to vote on what they wanted to do. The referendum took place in 1961. The anglophone (English-speaking) Cameroonians in the northwest of the country voted to merge into neighboring Nigeria, and most of that area did so. The anglophones in the southwest voted to join with the francophone (French-speaking) majority of the country to create a federation form of government. This allowed the English-speakers to retain a degree of autonomy. A new constitution that Cameroon adopted in 1972 established a unified republic.

In the new Cameroon, the francophones were the majority and held the political power. It became clear that they would not keep their promises of true unification and equality. English-speaking Cameroonians were effectively shut out of political power. The only legal political party, led by Ahidjo,

Gedeon Mpondo and Engelbert Mveng designed the Reunification Monument, built in the 1970s in Yaounde.

Lesson 33 - Cameroon: Legacy of Colonialism

> *Lake Nyos (pictured below) is in the Oku Volcanic Field in western Cameroon. On the evening of August 26, 1986, a huge cloud arose from the surface of the lake and spread across the ground down the valleys around it. Over 1,700 people who came in contact with the cloud died, as well as 3,500 livestock and innumerable birds and insects.*
>
> *Scientists later determined that the cloud was concentrated carbon dioxide (CO2). A pocket of magma, molten material from which igneous rock forms, lies underneath the lake. Carbon dioxide emitted from the magma enters the groundwater that accumulates in the lake. The gas settles to the bottom of the lake. Apparently some event triggered its release, causing the gas to emerge from the surface of the lake, then roll away because CO2 is more dense than air. Any human or animal that breathes more than a 15% concentration of CO2 will die instantly.*
>
> *In 2001 French scientists installed a pipe and fountain in the lake that allow the gas to escape at a constant rate to avoid buildup.*

held that power firmly. As is often the case when a country has only a single political party, corruption became common. When Ahidjo resigned in 1982, his prime minister and right-hand man Biya took over and maintained control.

Anglophones chafed at the discrimination they faced and at being shut out of the political process. Calls for change began emerging from this section of the country, with proposed changes ranging from a power-sharing agreement to complete separation as the independent country of Ambazonia. Cameroon did apply for and was granted membership in the British Commonwealth of Nations in 1995.

Biya eventually allowed other political parties to function, but his control by then was nearly complete. A failed coup attempt in 1984 gave Biya the excuse to clamp down even further. In 2008 constitutional amendments removed term limits for the president and granted him immunity from prosecution for anything he did in his official capacity during his term in office. In 2017 the Cameroon government cut Internet service to the anglophone region for a time. In 2019 the government of Switzerland agreed to mediate peace talks between the Cameroon government and opposition leaders.

Complicating matters have been occasional incursions involving kidnapping and atrocities committed by the Islamic militant group Boko Haram, which has primarily been active in Nigeria.

What Might Have Happened?

It's hard to know how things might have been different had European countries not colonized Cameroon. Would the people of Cameroon have been better off economically and politically, or worse off? Should Europeans have trained and equipped the people of Cameroon to run their own country? We cannot say. We cannot prove what did not happen. What we do know is that at least some of the political and social instability that Cameroon has suffered is a legacy of the policy of colonialism by which European nations governed it for decades.

The world has devised many ways to draw lines and to keep people separate. A repeated theme in the New Testament is the reality that unity in Christ breaks down those artificial barriers.

There is neither Jew nor Greek, there is neither slave nor free man, there is neither male nor female; for you are all one in Christ Jesus.
Galatians 3:28

Assignments for Lesson 33

Gazetteer Read the entries for Cameroon, Sao Tome and Principe, St. Helena Island, Ascension Island, and Tristan da Cunha (pages 48, 54, and 55).
Read the excerpt from *The Cruise of the Cachalot*, pages 261-264, and answer the questions in the *Student Review Book*.

Worldview Copy this question in your notebook and write your answer: What would be hard about being very ill and not able to have any visitors? What would be hard about having a loved one who was that sick?

Project Continue working on your project.

Literature Continue reading *Patricia St. John Tells Her Own Story*.

Student Review Answer the questions for Lesson 33.

Mount Nyiragongo, Virunga National Park

34 Beautiful and Deadly: Ebola in the DRC

"My Congo is full of people who are capable of doing amazing things."

These were the words of a nurse whose first name is Patient—Patient Muhindo Kamavu. The amazing things he was talking about were the courageous actions of Congolese volunteers and medical professionals in clinics in the Democratic Republic of the Congo (DRC) in 2018 and 2019. They were treating victims of an outbreak of Ebola Virus Disease (EVD).

Deadly

EVD, commonly called Ebola, is caused by a virus. It spreads by direct contact from a carrier or the carrier's bodily fluids to another person or animal. Typical carriers include bats, primates, rodents, insects, and humans. Initial symptoms include sudden fever, significant bodily weakness, muscle pain, and a sore throat. The disease progresses to cause diarrhea, vomiting, and internal and external bleeding. Treatment has primarily involved treating the symptoms because victims generally die from the effects of the virus, such as dehydration and organ failure.

A high percentage—often 50% to 90%—of those who contract the disease in a given outbreak die. Some who recover from the disease develop immunity to it, although survivors can develop post-Ebola syndrome. This syndrome can include joint and muscle pain, vision problems to the point of blindness, neurological problems, and other physical issues that can lead to disability. Researchers are making progress in developing drugs to treat and prevent the disease.

The worst outbreak of the disease to date occurred from 2013 to 2016. The outbreak was centered in Liberia but encompassed an area that included several West African countries. In that outbreak some 28,000 cases were reported and about 11,300 victims died. This was the outbreak during which Dr. Kent Brantly, an American physician, and Nancy Writebol, an American missionary, contracted the disease but recovered. (You will learn their story in the next lesson.)

The second-worst outbreak emerged in 2018 in the eastern part of the DRC. During the first year of the outbreak, about two thousand people came down with the disease and over fourteen hundred died. This was the tenth occurrence of the disease in the DRC since it first appeared in 1976.

Congo River

Beautiful

The DRC is an amazing, complex country. Situated astride the equator, it is the second-largest country in Africa after Algeria and is about one-fourth the size of the United States. The DRC is home to the second-largest rainforest in the world, after the Amazon rainforest in South America.

The country experiences heavy rainfall, high temperatures, and high humidity. It has abundant plant life, including trees one hundred fifty feet tall. Its animal life includes elephants, hippopotami, crocodiles, manatees, hundreds of species of butterflies, over 265 species of birds (including some migratory species from Europe) and many species of primates.

Mountains in the east give way to the central geographic feature of the country, the huge Congo River basin. Extending about 1,200 miles from north to south and almost the same distance from east to west, the basin covers 1.3 million square miles and includes many tributaries of the Congo River. As the tributaries feed into the Congo, the river widens from 3.5 miles to about 8 miles, even wider in times of flooding. Understandably, not many bridges span the river.

The Congo River is 2,900 miles long, the second longest in Africa after the Nile. It rises in the highlands of Zambia over a mile above sea level and about 430 miles from the Indian Ocean. It then flows in a wide, counterclockwise arc, mostly through the DRC, to its mouth on the Atlantic coast. The river provides the largest system of navigable waterways in Africa. However, the thirty-two waterfalls over the lower course of the river as it approaches the Atlantic make this last part impossible to navigate. The river forms the Malebo Pool at the head of the lower river. A railroad completed in 1898 provides transportation from the Pool to the coast. The DRC has a tiny, 25-mile-long Atlantic coastline.

The DRC has fertile soil, and the Congo River offers tremendous potential for hydroelectric power; but the unstable political situation and the difficulty of getting construction crews and materials up the river mean that much of the country's economic potential lies untapped. Thus, the DRC is an impoverished country, with high infant and maternal mortality rates, significant malnutrition, and a serious lack of access to pure water.

The country we know today as the DRC became a colony of Belgium in 1908. Known then as the Belgian Congo, it gained its independence in 1960 but has been politically unstable almost continually ever since. Multiple armed groups literally battle for control of the country to this day. From 1971 until 1997 the country was called Zaire, which means river.

About half of the DRC is Roman Catholic, about one-fifth Protestant, and about one-tenth Muslim. Another ten percent are Kimbanguists. This is a cult founded by Simon Kimbangu (1887-1951), a charismatic religious leader who claimed the ability to perform miracles. The rest of the population adhere to folk religions.

The Democratic Republic of the Congo is not to be confused with the Republic of the Congo, a much smaller country (slightly smaller than Montana, with about five million people) to the west of the DRC and on the other side of the Congo River. On the DRC side of the Malebo Pool is Kinshasa, the capital of the DRC; on the other side is Brazzaville, the capital of the Republic of the Congo. The two countries are sometimes referred to as Congo (Kinshasa) and Congo (Brazzaville).

The Republic of the Congo was a French colony that also gained its independence in 1960 and has had a somewhat more stable political life. The Republic of the Congo has a longer (one hundred mile) coastline on the Atlantic. The official language of both countries is French, but the people speak many tribal languages in their everyday lives. The region and the river are named for the Kongo people, who live near the mouth of the Congo River.

These people are disinfecting their hands during response to an Ebola outbreak in the North Kivu province of the DRC in 2019.

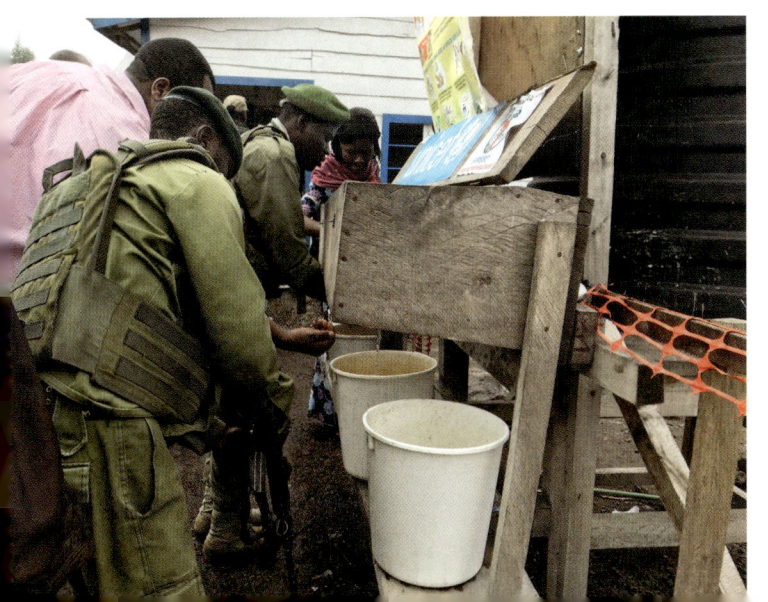

Ebola

The Congo River's tributary system contributed the name of the health crisis that the country faces. The Ebola virus carries the name of the Ebola River, a river in the northern part of the DRC which empties into the Mongala River, a tributary of the Congo. The first outbreak of the virus took place in the region of the Ebola River.

Many factors hampered the treatment of those infected with the Ebola virus during the 2018-2019 outbreak. The disease spread so quickly that clinics, which did not always maintain essential hygiene procedures, were overwhelmed; but medical professionals and trained helpers eventually put better techniques in place. Training the personnel needed to work in the treatment centers took some time.

In addition to the inadequate facilities, the political unrest was a significant factor in the spread of the disease. Militant groups attacked several treatment centers, whether to stir up fear or out of the belief that opposition parties operated the clinics. A significant number of victims of the disease feared going to the clinics, and the victims' continued presence in their communities increased the spread of the disease. This fear had many sources. Some victims feared being attacked by militant groups while traveling, while others believed that the clinics were attempts by foreign groups to profit from the outbreak. Some had the idea that former government leaders were trying to kill off their opponents, while other victims believed that the disease was simply a myth.

Amazing

In the face of this deadly disease and the dangers of militant activity, how do people live? Some live by performing the amazing actions Patient Muhindo Kamavu referred to.

Kamavu himself had worked in previous Ebola outbreaks. He and other healthcare professionals associated with Doctors Without Borders worked quickly to improve practices at treatment centers and to train additional workers.

People who survive Ebola generally have an immunity to the disease. As a result, many survivors give of their time to help in the treatment of others. Faustin Kalivanda survived, but his wife and daughter did not. After testing free of the disease, Faustin became a nurse assistant at the same hospital where he was a patient. He did this to show how survivors can overcome the fears of others and contribute to the treatment of the disease.

Some survivors of the 1976 outbreak volunteered to be examined in a study to help doctors understand what might await survivors of the current outbreak. Survival not only has physical consequences but also social and psychological consequences.

Children are especially vulnerable to Ebola because their little bodies suffer so greatly from fluid loss as a result of the disease. Child patients with Ebola have a much higher death rate. Jeanine Masika, herself a survivor who lost ten family members to the disease, spent her days cradling, rocking, and feeding infants with Ebola and assisting with their medical care. Health care workers had to wear a surgical mask, goggles, hooded coveralls, an apron, rubber boots, and two pairs of gloves. Because of the high temperatures, they could only wear the equipment for one hour at a time; but they did so to help the children. Survivors could sometimes wear lighter protection because of their immunity. For a time, when Masika finished her shift at the treatment center, she headed for the hospital where her own daughter was ill with malaria.

Children are often victims of the disease in another way. When their parents are in treatment or when they succumb to the disease, the children need care from willing volunteers. Claudine Kitsa had to place a child in care while she suffered from Ebola. Having survived, she worked at a childcare facility to help the children of other victims.

Amazing, indeed.

The Lord is my strength and song,
And He has become my salvation.
Psalm 118:14

Assignments for Lesson 34

Gazetteer   Read the entry for the Democratic Republic of the Congo (page 51).

Geography   Complete the map skills assignment for Unit 7 in the *Student Review Book*.

Worldview   Copy this question in your notebook and write your answer: When has your faith in God motivated you to do something that was really difficult?

Project   Continue working on your project.

Literature   Continue reading *Patricia St. John Tells Her Own Story*.

Student Review   Answer the questions for Lesson 34.

West Point, Monrovia, Liberia

35   Kent Brantly, Physician and Missionary

"For to me, to live is Christ and to die is gain" (Philippians 1:21).

This is the worldview by which Dr. Kent Brantly lived as he looked death squarely in the face.

This lesson is the first of several that show how a person's worldview influences how he or she lives and the decisions that person makes.

In Africa

In 2014 Dr. Brantly was on the medical staff of a mission hospital in Monrovia, Liberia. A graduate of Abilene Christian University and the Indiana University School of Medicine, Brantly had been in the country with his wife Amber (a nurse) and their two small children for less than a year.

In the summer of 2014 Amber and the children returned to the United States to attend a family wedding. Dr. Brantly planned to fly back to the U.S. closer to the time of the wedding. However, on July 23, a few days after his family left, Brantly became seriously ill. Tests soon determined that Brantly had contracted Ebola Virus Disease, which had been spreading rapidly throughout West Africa.

Brantly was in charge of the Ebola unit at the hospital and had treated several Ebola patients. Only one had survived. The mortality rate in what became the largest Ebola outbreak in history was over 70%. Where the Brantlys lived, the death rate was even higher. About the same time that Dr. Brantly became ill, another American worker at the hospital, Nancy Writebol, a nurse's assistant, also contracted Ebola. This occurred even though, throughout the Ebola outbreak, all medical workers at the hospital took extreme precautions to avoid contracting the disease.

The Illness and Treatment

Brantly had a fever for sixteen consecutive days as well as other symptoms that greatly weakened his body. During Brantly's illness, on one day he felt so bad that he thought he was going to die. As part of his treatment, Brantly received a blood transfusion from the one young man who had been to the hospital and had survived Ebola.

The U.S. State Department and other agencies worked on a plan to evacuate Brantly and Writebol to medical facilities in the U.S. while minimizing the risk of spreading the highly contagious disease. Brantly and Writebol had the opportunity to take an experimental Ebola drug, ZMapp, that had been tested on monkeys but that no humans had ever taken before. However, only one course of three doses was available in Liberia.

Sign Warning About Ebola in Liberia (2014)

The medical experts handling their cases decided to give Brantly and Writebol one dose of ZMapp each, then transport them to Emory University Hospital in Atlanta in time for them both to complete their courses of medication. To make the trip, Brantly and Writebol would have to wear special protective suits and fly in a special protective chamber inside a small plane.

Meanwhile, Amber Brantly was six thousand miles away in the United States with their children, wondering if she would ever see her husband again.

The plight of the two medical missionaries suffering from the disease they had been trying to fight made international news. Many people had great concern about bringing Ebola patients to the United States for treatment. At the same time, millions of people around the world were praying fervently for Kent and Nancy.

The staff at the mission hospital in Liberia accomplished the evacuation successfully on August 1, and the staff at Emory Hospital received Brantly and Writebol and provided the treatment they needed to make a full recovery. They were released about three weeks later within a couple of days of each other. Kent and Amber and their children, as well as Nancy and her husband, were reunited.

Continuing Their Ministry

After Dr. Brantly recovered, he and Amber met President Barack Obama in the Oval Office at the White House. *Time* magazine's Persons of the Year for 2014 were "The Ebola Fighters." Dr. Brantly's picture was on the cover of the magazine.

The Brantlys moved to Ft. Worth, Texas, where they had lived previously. Kent and Amber used their fame to focus world attention on fighting the Ebola outbreak in West Africa. He later taught in the residency program of John Peter Smith Hospital in Ft. Worth, which serves the city's poorest residents. He cared for low income patients at a clinic in Ft. Worth. Amber volunteered with a Christian refugee

Lesson 35 - Kent Brantly, Physician and Missionary

settlement organization and with a group that promotes racial reconciliation among Christians. They lived in an apartment complex where many refugees and immigrants lived.

Going Back to Africa

In 2019 the Brantlys announced that they planned to return to Africa. They intended to work at a mission hospital in Zambia in southern Africa. They planned to work with Brantly's cousin, Peter Snell, who is also a physician. The Snell family is also part of the mission team. The two physicians made a two-year commitment to work at the hospital, which is not uncommon for medical missionaries.

Dr. Brantly's Worldview

In his statement at the press conference following his release from Emory, Dr. Brantly said:

> I prayed that God would help me be faithful even in my illness, and I prayed that in my life or in my death that he would be glorified. I did not know then, but have learned since, that there were thousands, maybe even millions, of people around the world praying for me. . . .
>
> I cannot thank you enough for your prayers and your support. But what I can tell you is that I serve a faithful God who answers prayers. Through the care of the Samaritan's Purse and SIM missionary team in Liberia, the use of an experimental drug, and the expertise and resources of the health care team at Emory University Hospital, God saved my life—a direct answer to thousands and thousands of prayers.

Kent and Amber Brantly believe that their lives are a calling. They are deeply grateful that God spared Kent's life. That gratitude has led Kent to "live a life that is faithful to the calling He's given me. Right now, I think that means moving my family to Zambia to serve at a Christian mission hospital—to serve the poor and have compassion for the people in need and to participate in God's work of making all things new and fixing the broken things in this world."

Before Kent and Amber went to Liberia, they wanted to serve people in a way that would bring others to know God. Kent said that, after all he went through, some days he believed that God was saying, "Here is a press conference with eighteen million viewers. Here, testify to world leaders."

President Barack Obama met with Kent and Amber Brantly in the Oval Office in September of 2014, as seen at left. In October President Obama spoke about the fight against Ebola at an event with American health care workers, including Dr. Brantly. In the photo at right, Amber Brantly listens at this event with their young children. To her left is Dr. Anthony Fauci, director of the National Institute of Allergy and Infectious Diseases.

Thinking about confronting another outbreak of Ebola in the future, Brantly also said:

It's not a matter of not fearing. It's a matter of choosing to have compassion despite fear . . . . We're trusting that God has opened the doors and He'll pave the way. . . . Our calling is to be faithful wherever we are, to be good stewards of opportunities, to be responsible with what we have been given, to try to do good, and to serve those whose paths we cross.

---

*Greater love has no one than this,
that one lay down his life for his friends.*
John 15:13

Assignments for Lesson 35

Worldview — Write or recite the memory verse for this unit.

Project — Finish your project for this unit.

Literature — Finish reading *Patricia St. John Tells Her Own Story*. Read the literary analysis and answer the questions in the *Student Review Book*.

Student Review — Answer the questions for Lesson 35. Take the quiz for Unit 7.

Murchison Falls, Uganda

8 East Africa

Creative entrepreneurs are giving their fellow Rwandans new opportunities through clothing. William Kamkwamba had to drop out of school, but that didn't stop him from building his own windmill in Malawi. Christians in Kenya are working to provide physical and spiritual nourishment to their neighbors. Long distance runners from Ethiopia (and nearby Kenya) have excelled in international competitions. The worldview lesson helps you identify common influences on a person's worldview and how you can think clearly about your own worldview.

Lesson 36 - Hope Instead of Hate in Rwanda
Lesson 37 - William and His Windmill
Lesson 38 - Give Water, Give Hope, Give Life in Kenya
Lesson 39 - Long Distance Runners from Ethiopia
Lesson 40 - Where Did You Get That Worldview of Yours?

Memory Verse   Memorize 1 John 3:17-18 by the end of this unit.

Books Used

The Bible
Exploring World Geography Gazetteer
A Long Walk to Water

Project (Choose One)

1) Write a 250-300 word essay on one of the following topics:
   - Discuss the history of the marathon race, from its ancient origin to modern times.
   - Have you ever been involved in a project to help others, such as building a Habitat for Humanity house or collecting donations for tornado victims? Write about your experience and what you learned from it.
2) Plan a shop in which you will sell goods that you make. Draw the floorplan, make a (model) sign, and plan what you will make, whether you will involve others, and how you will advertise your shop.
3) Interview someone who has made a difference in your home town or county. Contact this person ahead of time to schedule an appointment. Write down ten questions you would like to ask him or her. Be prompt and respectful for the interview. Listen attentively to the person's answers to your questions. Be sure to express appreciation for the person's time when you are done.

Literature

A Long Walk to Water is based on the true story of Salva Dut of Sudan. During the long civil war in Sudan, large numbers of people lost their lives or were displaced for many years. Many of these were young men, known as the Lost Boys of Sudan. Salva's story begins in 1985 and covers several decades. The book's second, parallel narrative set in 2008 tells the story of Nya, an eleven-year-old girl. The stories of Salva and Nya include the horrors of civil war, life in a refugee camp, ethnic conflict, and the desperate search for life's most basic needs. Salva and Nya's stories eventually converge in this hopeful and redemptive book.

Linda Sue Park was born in Illinois in 1960. Her parents were Korean immigrants who became United States citizens. She is an accomplished writer for children. She won the Newbery Medal in 2002 for *A Single Shard*. *A Long Walk to Water* was a *New York Times* bestseller. Park lives in Rochester, New York, where she continues to write. She also travels widely to speak and support good work in literature and writing.

Plan to finish *A Long Walk to Water* by the end of this unit. You will not have a new literature assignment for the next unit, so you can extend your reading into the next unit if you would like.

Lake Kivu, Rwanda

36 Hope Instead of Hate in Rwanda

Priscilla attended a small Christian college in Oklahoma. She graduated in 2015. The next year she started a company that designs children's clothes. As of 2018, her company had four employees.

This could simply be the story of a young woman seeking to fulfill her dream of designing clothes and being her own boss, but the story goes much deeper. Priscilla Ruzibuka's shop is in Kigali, the capital of her native Rwanda. In Rwanda, the clothes industry is part of a movement of national renewal and economic revival.

Beautiful and Impoverished

Rwanda is a small, landlocked country in East Africa. It is one of the most densely populated countries in Africa. Rwanda is slightly smaller than Maryland, but its population of twelve million is twice that of Maryland. About three-quarters of the population is rural. Kigali's population is about 1.1 million.

The country boasts abundant natural beauty, including volcanic mountains, river valleys, lakes, and grassy plains. Much of the country lies on a high plateau, so even though it is just south of the equator most of the country enjoys a cool, pleasant climate. Volcanoes National Park in the northwestern mountains is a refuge for the threatened mountain gorilla and is a popular tourist destination.

Rwanda is one of the poorest countries in the world. Its lack of seaports and railways makes trade difficult and expensive, although coffee and tea are major exports. It imports much more than it exports. Most people are subsistence farmers.

During the 1990s, Rwanda became known around the world not for its scenery but for a terrible incidence of ethnic genocide.

A Tragic History

A major aspect of life in Rwanda has been ethnic conflict. About 85% of the people are Hutu, while around 15% are Tutsi. Less than one percent of Rwandans are Twa, a Pygmy people. Most of the Hutu are farmers, while many Tutsis have become business owners. These two groups are similar in language and physical characteristics, but their distrust of each other goes back for many years.

The area was part of the colony of German East Africa from the 1890s through World War I. After Germany lost the war, Belgium began overseeing the region as a League of Nations mandate. The Belgians favored the Tutsis and enabled them to control most of the political and economic power.

In 1959 a group of extremist Hutus rebelled against the Tutsi government and overthrew the Tutsi king. The rebels killed some 20,000 Tutsis in that revolt. Rwanda gained its independence from Belgium in 1962, and at the time the Hutus controlled the government. Tens of thousands of Tutsis fled in fear to nearby countries as exiles. The two groups continued to have conflict which often resulted in violence. In addition, once the Hutus were in power, rival Hutu groups began fighting each other for control.

A rebel group formed among the next generation of Tutsis in exile, and in 1990 the Tutsis began a civil war. In response, in 1994 the Hutu-led government initiated a genocide on Tutsis. An estimated 200,000 Hutus were involved in the attacks that took the lives of about 800,000 Tutsis and moderate Hutus over a period of about one hundred days. The genocide wiped out about three-fourths of the Tutsi population of Rwanda. Another two million people, both Hutus and Tutsis, left the country for neighboring countries; most of these returned over the next two or three years as calm returned. The genocide ended when Tutsi forces defeated the Hutu army and local militias.

The Cathedral of Butare, pictured below, was built in the 1930s and is the largest in Rwanda. The Roman Catholic Church gained prominence in the country during the time of Belgian control. Some Catholic leaders supported the Hutu majority and contributed to distrust between Hutus and Tutsis. During the genocide, some Catholics participated in the atrocities, while others resisted the violence to protect victims. In 2016 the Conference of Catholic Bishops in Rwanda admitted to and apologized for the role their Church had played in contributing to the genocide.

Lesson 36 - Hope Instead of Hate in Rwanda

The Healing Process

After the genocide ended, Rwandans moved toward healing and reconciliation. The Tutsis regained political power. A Hutu became president and a Tutsi, Paul Kagame, became vice president, although Kagame was the more influential leader. The Hutu president resigned in 2000, and Kagame became president. He has been president ever since, maintaining firm control and getting repeatedly re-elected with over 90% of the vote.

Trials of those accused of leading and participating in the genocide took place in various courts, including courts created under international law, courts in the Rwandan legal system, and local community courts.

Emergence of the Clothing Industry

In more recent years, even with some small degree of political unrest continuing among extremists in both groups, the country as a whole has tried to put ethnic differences aside and work together. One area of economic and cultural growth has come in the development of a domestic clothing industry.

The Kagame government has supported a Made in Rwanda initiative for producing clothes and fashion accessories. The government does not charge taxes on imported fabric and has sent designers to international trade shows. Franklin Hub Kigali is a non-governmental organization (NGO) that supports over one hundred designers. The NGO provides training and tools such as sewing machines and pattern cutters. The leader of Franklin Hub Kigali lost his father and brother in the genocide.

Through her business, Ki-Pepeo Kids, Priscilla Ruzibuka wants to help women who have been affected by the genocide and who do not have many opportunities open to them. She wants to offer employment that involves something other than their usual work as maids. Her clothes use traditional Rwandan designs and prints. She received a grant from the United States Agency for International Development, with which she purchased electric sewing machines. In 2017 the government paid for her to attend a children's clothing trade show in New York City. Ruzibuka is now receiving orders from other countries.

This 2010 photo of Kigali, Rwanda's capital, features a Paul Kagame re-election campaign billboard.

An organization of seven designers in Kigali hosts an annual fashion show. The 2017 event had over 800 people in attendance. Gift shops and other retailers in the U.S. now carry the products of Rwandan artists and designers.

The expansion of the domestic fashion industry has its critics. For many years a large part of the Rwandan economy has involved importing secondhand clothes from the United States and other western countries and selling them to the public in small shops. The government has followed a plan to reduce and eventually eliminate these imports in order to help domestic clothes production and sales. Retailers who sell secondhand clothes complain that this plan is taking their livelihood from them. Some Rwandans do not believe that they can afford designer clothes made in Rwanda. The designers want not only to build their industry but also to

Above: Used clothes market in Kigali, Rwanda *Below: Seamstresses working in Tyazo, Rwanda*

restore a sense of pride about Rwanda and about things made in Rwanda.

Another player in the clothing industry in Rwanda is a Chinese-owned factory that employs Rwandans and produces uniforms, 80% of which are exported to other countries. For some time businesses have built factories in China to reduce their labor costs. Now this Chinese company has built a factory in Rwanda to reduce its labor costs.

Many Rwandans get a sense of hope from the domestic clothes designing industry. Designer Teta Isibo said that her generation of Rwandans has "the opportunity to create our own reality and not feel held back because of the genocide. I see a future for us that's bright, one where fashion thrives."

Paul spoke of the need to put on the spiritual garments of those who have been chosen by God:

So, as those who have been chosen of God, holy and beloved, put on a heart of compassion, kindness, humility, gentleness and patience; bearing with one another, and forgiving each other, whoever has a complaint against anyone; just as the Lord forgave you, so also should you. Beyond all these things put on love, which is the perfect bond of unity.
Colossians 3:12-14

Assignments for Lesson 36

Gazetteer Study the map of East Africa and read the entries for Burundi, Comoros, Mauritius, Rwanda, and Uganda (pages 56-58, 65, 67, and 72).

Worldview Copy this question in your notebook and write your answer: What have been the biggest influences in your life (people, books, whatever)?

Project Choose your project for this unit and start working on it. Plan to finish it by the end of this unit.

Literature Begin reading *A Long Walk to Water*. Plan to finish it by the end of this unit.

Student Review Answer the questions for Lesson 36.
Read "Who, What, How, Why, and Why Not: A Primer for Literary Analysis of Fiction" on pages 5-10.

Sunset Over Lake Malawi

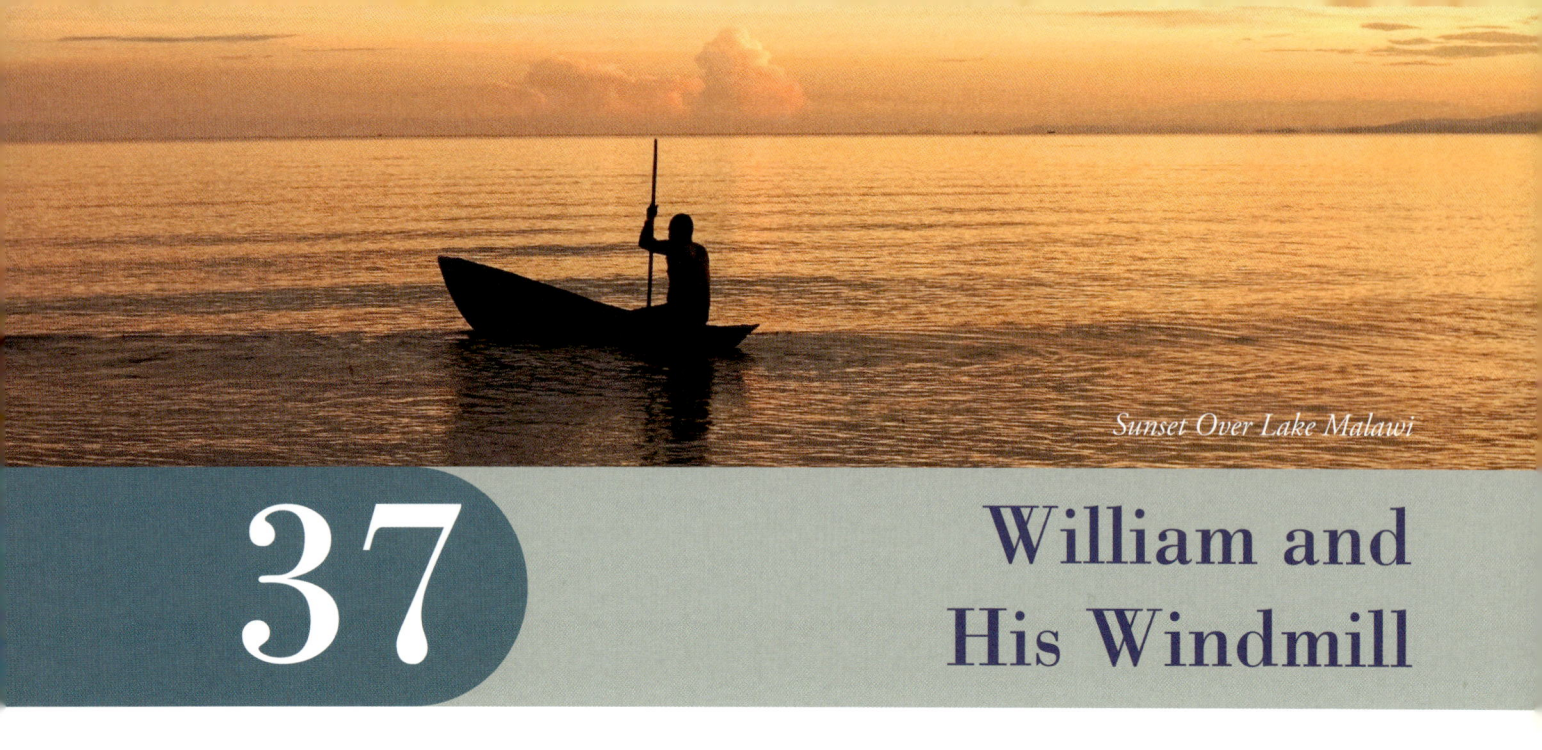

37 William and His Windmill

Geography can create a dramatic and challenging environment in which people live. The study of human geography involves how individuals meet and overcome the challenges of geography to create a better life for themselves and others.

Malawi

Malawi is a small, impoverished, landlocked country in southeastern Africa. It lies alongside Lake Nyasa, which Malawians call Lake Malawi. The lake lies in the Great Rift Valley, which runs the length of the country north to south. About 80 percent of Malawians are Christians and 15 percent are Muslims, although many people in the country also believe in magic, wizards, and other elements of folk religion.

The country has areas of great natural beauty, including mountains, savannas, and forests; but its citizens are only able to farm about 40 percent of the land. Most of the 21 million people in Malawi live by subsistence farming. This means they barely get by. The standard crop is maize (corn). The main element of the standard diet is *nsima*, a thick porridge made of maize. Some people grow tobacco to sell for income. A common style of house in Malawi has two rooms with mud walls and a thatched roof.

In a typical year in central Malawi, after the rains come in December through February, farmers plant in March and harvest in May. The people live off the harvest until September, when food starts to become scarce. The seed and fertilizer for the next year's crop take almost all of their remaining money. They get by again until the harvest in May.

Home and Corn Crop in Malawi

Clearing trees from a field in Malawi

In the 1980s, only about two percent of Malawians had electricity, and even that was subject to frequent outages because of the unreliable national power system. One reason these outages occurred and continue to occur is that people have cut down many trees for cooking and to cure tobacco. Because of this deforestation, rains more often produce floods that wash away soil and minerals. This clogs the river and the hydroelectric dam, which shuts down the turbine that produces electricity. In the 1980s most people used kerosene lamps to light their homes or simply went to bed when it got dark.

William Kamkwamba

William Kamkwamba was born in 1987 in the village of Wimbe, near the town of Kasungu. His father was a typical farmer eking out a living. William's father had become a Christian after living a rough life earlier. One day William's father told him, "Respect the wizards, my son, but always remember, with God on your side, they have no power."

William was inquisitive and resourceful. As a child he made his own toy trucks with materials he had at hand. He and a cousin enjoyed listening to the radio. They wanted to find out how a radio worked, so they took one apart and learned the functions of the various parts. They began a small business repairing radios, but they needed a reliable source of electricity to see if the radios that they worked on actually played. Batteries were expensive, so the business didn't last long. One day William saw a dynamo powered by a bicycle that generated electricity, and this intrigued him. He recalled, "I'd become very interested in how things worked, yet never thought of this as science."

Turning a Need into an Idea

In December 2000 the rains came late. When they did come, they were heavy and caused flooding.

Fertilizer was expensive for spring planting. Then Malawi experienced its worst drought in fifty years. The drought pushed many of the people who lived on the edge over the edge. Many people suffered hunger, and many died of starvation and disease. William's family could afford only one meal a day, a small supper of nsima.

William had attended primary school, but he had to drop out of secondary school because his family could not afford the eighty dollars annual tuition. William was still eager to learn, however, so he began to visit the small village library housed in the primary school. It had three shelves of books, mostly used American textbooks.

On the cover of an eighth grade science book, *Using Energy*, was a picture of windmills. William was enthralled. "Someone built those," he said

William and His Windmill (2007)

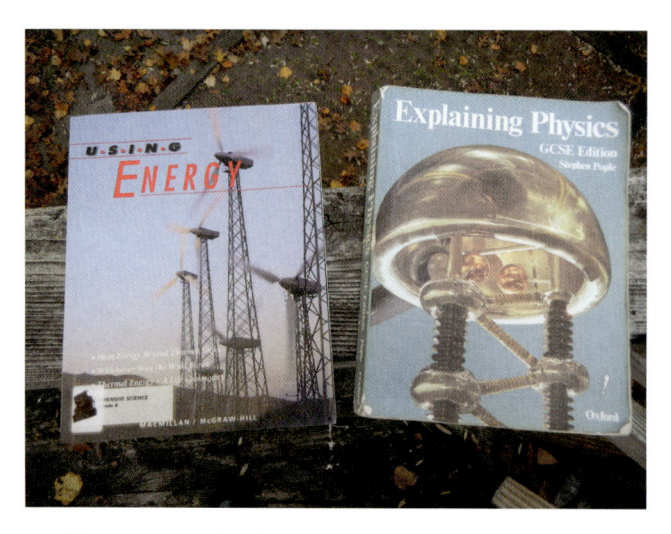

William gleaned information from copies of these two textbooks to build his windmills.

to himself, "so I can build one, too." One thing that Malawi has is a lot of is wind. William didn't know English well, even though English is one of the official languages of Malawi (the country is the former British colony of Nyasaland); but he studied diagrams in the book and let them teach him what the English text said. He realized that a windmill could generate electricity like the bicycle did with the dynamo. A windmill could also operate a well pump to provide irrigation for crops.

William was able to obtain a dynamo, and he gathered the other parts from the village junkyard. Many people in his village thought he was crazy. Some people believed that William caused the drought. His windmill stood about sixteen feet high and generated enough electricity to power one light bulb in his family's house. William was fourteen years old.

William improved his original windmill to make it able to generate more electricity. People started showing up to charge their cell phones rather than go to the nearby marketplace and pay the fees that the vendors demanded—and power outages occurred there frequently, anyway. He also built a radio transmitter that broadcast a short distance. He figured out how to charge a storage cell battery for times when the wind wasn't blowing. Later he built a water pump for irrigating the fields in his village.

Lesson 37 - William and His Windmill

Opportunities

An official of the Malawi Teacher Training Activity, a non-governmental organization that helped with education in the country, learned about the windmill and arranged for Malawian national radio to interview William. Then newspaper reporters starting coming. William was invited to speak at the TEDGlobal Conference in Tanzania in 2007. His trip there was his first plane ride. Later he gave another TEDTalk.

Word about William and his windmill spread around the world. People offered to be his financial sponsors so that he could obtain further education and continue to help his village. He made a trip to the United States, where he was amazed at New York City. He visited a wind farm in California that has 6,000 huge wind turbines. The base of each turbine was larger than his house. The wind farm could provide power for all of Malawi. William thought, "I had to be dreaming this."

When William was nineteen, he was able to go back to school. He received financial assistance to attend two other schools in Africa: African Bible College Christian Academy in Lilongwe, the capital of Malawi, and the African Leadership Academy in Johannesburg, South Africa. Before going to Johannesburg, William received assistance to attend an English intensive course at Cambridge University in England. He then enrolled in Dartmouth College in Hanover, New Hampshire, and graduated in 2014.

William Kamkwamba speaking with Chris Anderson at TEDGlobal 2007 in Tanzania

William Kamkwamba has been invited to speak in several countries and has worked on projects around the world. *The Wall Street Journal* published a profile of him. His inventions have been on display at the Museum of Science and Industry in Chicago. He and Bryan Mealer published a book, *The Boy Who Harnessed the Wind*, in 2009.

William and his wife Olivia now help lead the nonprofit organization Moving Windmills. They work with local leaders and individual farmers in Malawi to secure a viable future for that country. The organization builds low-cost water wells, installs solar-powered pumps and energy systems, provides schools with new facilities and learning materials, and supports community development programs. Moving Windmills is also creating an Innovation Center to provide young people with tools and mentorship to help them create solutions to other problems. This organization has provided every home in his village with a solar panel for the roof.

William Kamkwamba's story is about one person using his God-given creativity and willingness to learn in order to overcome the obstacles he faced, many of which involved the geography where he lived. In his first TEDTalk, when he described creating the windmill, he made a statement in his imperfect English that became a theme at the conference. It can be an encouragement for anyone who wants to accomplish something in his or her life. William said:

"I try, and I made it."

The psalmist used wind to describe the majesty of God.

He makes the winds His messengers,
Flaming fire His ministers.
Psalm 104:4

Assignments for Lesson 37

Gazetteer Read the entries for Madagascar, Malawi, Mozambique, Tanzania, Zambia, and Zimbabwe (pages 63, 64, 66, 71, 73 and 74).

Worldview Copy this question in your notebook and write your answer: How do you know who is trustworthy and who is not?

Project Continue working on your project.

Literature Continue reading *A Long Walk to Water*.

Student Review Answer the questions for Lesson 37.

Naivasha, Kenya

38 Give Water, Give Hope, Give Life in Kenya

Erastus Kavuti was a member of the Kenyan Air Force. In 1989 he came to the United States for training. While he was at Lackland Air Force Base at San Antonio, Texas, he became a Christian. Later he went to Chanute Air Force Base at Rantoul, Illinois, for a nine-month stint. While there, he studied the Bible once a week with a member of a church there.

Kavuti expressed a desire to return to his village of Tulia in Kenya and teach the gospel to his family and friends. When he completed his six-year military commitment, the church in Rantoul and another one in Arkansas paid his way to attend a two-year preacher training program at the Great Commission School in Nairobi, the capital of Kenya. The Christians who provided this support formed Caring for Kenya (CfK) as a non-profit organization. From its beginning, the purpose of Caring for Kenya was not to impose Americans and American ways on the people of Kenya, but to equip and train Kenyans to help and teach other Kenyans. In 1996 Erastus planted a church in Tulia and began teaching others.

"The Church That Cares"

Kavuti said, "We want to be known as the church that cares. We want to meet physical needs and develop relationships so that we will have opportunities to meet spiritual needs." One of the first people that Kavuti brought to Christ was James. James had three years of medical training and was licensed to operate a medical clinic in Kenya. Christians in the United States paid for James to attend the Great Commission School in Nairobi also.

By 1997 the church in Tulia had seventy-five members. The church had been meeting in a storefront, but Kavuti believed that having a building would send the message that the church was there to stay. Caring for Kenya purchased a small tract of land and materials for a building. The Christians in Tulia made the bricks and built the church building themselves. They began talking about operating a medical clinic and beginning a Christian school for children. They also confronted a major issue in the geography of their country by developing a ministry to dig water wells.

Background: Kenya

The country of Kenya is in East Africa on the coast of the Indian Ocean. It is just over twice as large as Nevada and has 54 million people. The equator runs through it.

Kenya is blessed with great natural beauty. The narrow coastal region is tropical. The land gradually rises to the western part, where the Great Rift Valley runs through the country north to south. The Rift actually runs from the Jordan River in southwest Asia through Mozambique, a distance of about 4,000 miles. It averages two to three thousand feet deep, with parts of it much deeper. The Rift is thirty to forty miles wide.

At its southwest corner, Kenya borders Lake Victoria. Africa's second highest mountain, Mount Kenya, 17,058 feet in altitude, lies in Kenya. The country has varied and abundant wildlife that attracts many tourists.

Kenya has a diverse population. It includes about forty different ethnic groups, most with their own language. The two official languages are English (Kenya was a British colony for many years) and Swahili (what Kenyans call Kiswahili). About three-fourths of the population is Christian, eleven percent Muslim, and the rest are followers of folk religion.

A Need: Water

One thing which Kenya does not have in abundance is one of the essentials of life: clean water. Much of the inland is arid. About forty percent of the population does not have easy access to safe, clean water. This includes people in urban slums and in rural villages like Tulia. As a result, Kenyans often use water from unsafe ponds, rivers, and shallow wells. The alternative is to walk several hours each day to a safe water source or spend time every day boiling water to make it safe, as well as acquiring wood for fires on which to boil water.

Because of the effort involved in obtaining safe water, the people who must do this cannot be productive in helping to support their families. Many Kenyans lack clean water to irrigate their gardens and farm crops, so they do not have the food they need for a healthy diet. Many people become sick from drinking contaminated water, which also keeps them from working. A few thousand Kenyans die from water-related illnesses each year.

Planting Onions in a Field with Drip Irrigation

Lesson 38 - Give Water, Give Hope, Give Life in Kenya

The Ministries of Caring for Kenya

Caring for Kenya is helping to fulfill Erastus Kavuti's dream of enabling churches to demonstrate that they care. CfK has provided Bibles and Bible study materials, clothes, soil testing kits, medical supplies, and school supplies. A free mobile medical clinic is now in operation. During a famine in 2006, CfK provided thousands of dollars to help with food, hospital bills, and school fees.

Here are some other ongoing CfK projects that are making a difference for the people of Kenya.

Digging Wells

One of the mottoes of the Caring for Kenya ministry is, "Give water, give hope, give life." In response to the great need for clean water, Caring for Kenya has dug many wells with motorized pumps and is moving to solar-powered pumps. These wells provide clean water for thousands of people. The ministry has also purchased equipment for drip irrigation. This method of irrigation guides water to the root systems of growing plants more efficiently than broadcast irrigation. It enables the maximum production with the minimum amount of water.

The director of Caring for Kenya, Carl Burkybile, is a retired Illinois high school agriculture teacher who makes annual trips to Kenya and teaches improved agricultural techniques. Several other Christian ministries and non-governmental organizations (NGOs) have made digging wells in Kenya part of their mission.

Education

Obtaining an education helps people acquire skills that can lead to more productive jobs. Christians in Tulia have built several elementary school classrooms on the church's property. The members of the church gathered sand from the nearby riverbed, made the bricks, and built the classrooms themselves. CfK provided funds for metal roofing sheets. The church hopes to build a secondary school as well, but until then CfK has provided tuition for older students to attend public secondary school. The government provides free elementary school, but students must pay tuition to attend higher levels. Sponsorship of a day student is $200.00 per year and $550.00 for a boarding student.

CfK has the goal of digging a well on school property that will enable a fenced-in community garden area that families can use to grow food. Once this is in place, the ministry will offer a gardening workshop to teach interested families composting, raised planting bed construction, and drip irrigation.

Here are two thank-you notes from secondary school students whom CfK sponsored:

Martin Kasina is one student sponsored by a CfK program. As other students graduate and get jobs, he has encouraged them to pay it forward by sponsoring another child in the program.

In my family I am the first born. My father died last year. My mother sells vegetables and used clothes. This year I had a B-. I love mathematics, business, and geography. I would like to be an accountant. I would like to thank you for paying my school fees. May God bless you as you continue to pay my fees. I will not let you down in academics.

—*Ruth*

My parents are farmers. I would like to thank you for the great support that you have given in ensuring the success of my secondary school. I am 6th out of 162 students. I have great interest in mathematics and physics and am working hard to become an electrical engineer. It's my prayer to almighty God to greatly bless you for the sacrifice that you make for me. After my studies I will have the heart to help the entire community as they pursue their goals.

— *Matthew*

Evangelism

Projects such as these allow local Christians to connect with other members of their community and develop friendships. Carl Burkybile has assisted with a similar project in Zimbabwe, and the church near that community garden has grown from six to eighty people in four years.

So far Kenyan Christians associated with Caring for Kenya have planted nine churches in the area around Tulia. Men from the area continue to attend the Great Commission School in Nairobi to prepare to preach and to plant more congregations.

Overcoming a Mindset of Poverty

Caring for Kenya seeks to overcome what one Kenyan Christian called poverty mental sickness. This is the worldview that says, "I am poor. My family has always been poor, and there is nothing I can do about it." CfK wants to change the Kenyans'

view of themselves and their world and help them see how they can live differently. Here are some examples of how Caring for Kenya is making a real difference in the lives of Kenyans.

On one trip that American CfK personnel made to Kenya, the team planned two agriculture workshops to teach improved farming methods. They expected about forty people to attend, but over one hundred farmers actually came.

Eunice, a young woman whose parents died of AIDS, did not have good enough grades in elementary school to be able to enroll in an academic-oriented secondary school. CfK paid her fees to attend the Tulia Vocational Institute, where she learned how to sew. When she graduated, three Christian women in the U.S. went together to buy her a sewing machine. A visiting CfK team presented the sewing machine to her, and Eunice was so excited she couldn't stand still. The education Eunice received and a $150 sewing machine will enable her to earn an income and make a difference for good all her life.

Celestine returned to her village of Nzawa in eastern Kenya after her sister died. Celestine had thirty-two years of experience working in a hospital and a mobile medical clinic. In Nzawa she took over

Eunice and Her Sewing Machine

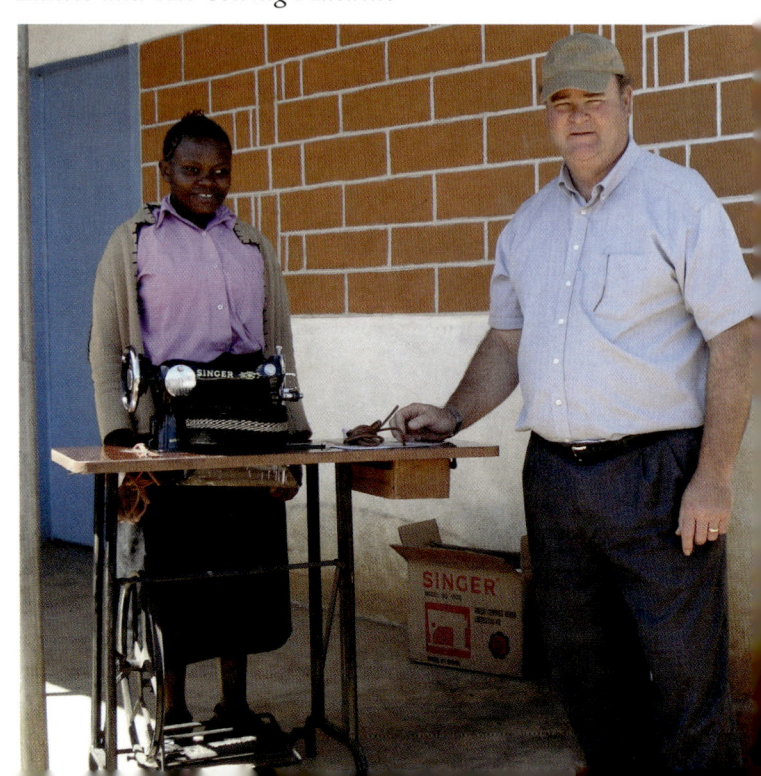

Lesson 38 - Give Water, Give Hope, Give Life in Kenya

the clinic and the secondhand clothing shop that her sister had started. An evangelist associated with CfK taught her the gospel, and she was baptized into Christ.

The mother of four adult sons, Celestine still wants to learn. She enrolled in a food preservation workshop that Caring for Kenya and another ministry sponsored. The next week, CfK workers delivered medical supplies, water filters, blankets, and mattresses to her clinic, which mostly serves the very poor. The team also enabled her to buy a new stethoscope and blood pressure machine.

Celestine planned and presented a food preservation workshop for twenty-one women in her village. She operates the medical clinic, teaches food preparation, sells used clothes, and has started a sewing co-op. Christians in America donated three sewing machines, and now women in Nzawa are learning to sew.

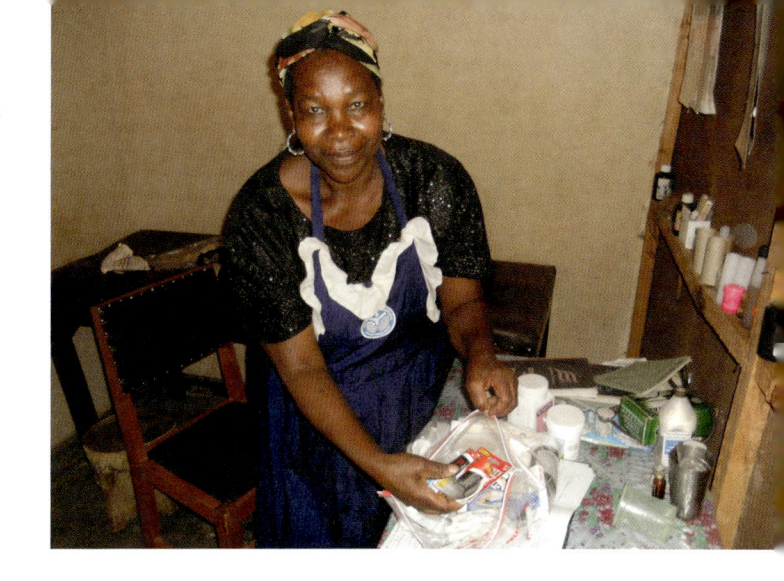

Celestine Preparing Medical Supplies

"The Most Important Gift"

When a Caring for Kenya team was preparing to depart from Kenya after one trip, the Kenyan Christians were saying goodbye and giving the Americans small handmade gifts. One Kenyan said, "The most important gift we can give is our love."

But whoever has the world's goods, and sees his brother in need and closes his heart against him, how does the love of God abide in him? Little children, let us not love with word or with tongue, but in deed and truth.
1 John 3:17-18

Assignments for Lesson 38

Gazetteer Read the entries for Djibouti, Eritrea, Kenya, Seychelles, Somalia, and South Sudan (pages 59, 60, 62, and 68-70). Read the South Sudan Independence Day Speech (pages 265-269), and answer the questions in the *Student Review Book*.

Worldview Copy this question in your notebook and write your answer: When have you seen people be inconsistent regarding what they say and what they do?

Project Continue working on your project.

Literature Continue reading *A Long Walk to Water*.

Student Review Answer the questions for Lesson 38.

2018 Cross Country Championships in Addis Ababa, Ethiopia

39 Long Distance Runners from Ethiopia

September 10, 1960, saw the running of the marathon in that year's Olympic Games being held in Rome, Italy. The winner was Abebe Bikila of Ethiopia in record time.

Bikila, who was 28, had been a shepherd before he became a bodyguard for Ethiopian emperor Haile Selassie. Bikila had taken up competitive running only four years earlier.

The course for the marathon was laid out through the streets of Rome. The runners passed the ancient Colosseum. Part of the course was on the Appian Way, paved with cobblestones and lighted with torches as dusk descended on the city. The race had begun in late afternoon to avoid the daytime heat in Rome. The finish line was at the Arch of Constantine. Bikila ran barefoot because he could not find shoes that were comfortable.

At one point, the racers ran past the Obelisk of Aksum. In 1935 Italian armed forces invaded and conquered Ethiopia as part of Benito Mussolini's desire for a world empire. Two years later, the Italians stole the 79-foot obelisk, an object of great Ethiopian pride, from the city of Aksum and took it to Rome. As Bikila ran past the obelisk in 1960, he was an Ethiopian conquering in Rome in a different way. Italy finally returned the obelisk to Ethiopia in 2005, and it was re-erected in Aksum in 2008.

In Ethiopia, millions of people listened to coverage of the 1960 race on the radio and celebrated Bikila's victory. His gold medal was the first that a black African had ever won in Olympic competition. It would not be the last. Bikila won the marathon again at the Tokyo Olympics in 1964, again setting a new record. It was the first time someone had won two consecutive Olympic marathons. Bikila had just had surgery for appendicitis forty days earlier. This time, he wore shoes when he ran.

Bikila ran again in 1968 in Mexico City, but he was unable to finish because of an injury. Another Ethiopian runner, Mamo Wolde, did win. Bikila became a paraplegic as the result of an automobile accident in 1969 and was in a wheelchair for the rest of his life. He died in 1973.

Abebe Bikila crosses the finish line in Rome.

Lesson 39 - Long Distance Runners from Ethiopia

Bikila's win in 1960 started a tradition of long-distance running champions from Ethiopia and a passion for running that thousands of Ethiopians share even today. Another long-distance champion from Ethiopia was Miruts Yifter. He acquired the nickname Yifter the Shifter because of his ability to shift into high gear late in a race with a burst of speed that usually brought him victory. Yifter won the 10,000-meter race at the Moscow Olympics in 1980. Listening to the race on the radio back in Ethiopia was a seven-year-old boy who dreamed of running and winning races himself, perhaps even in the Olympics. The boy's name was Haile Gebrselassie. He would grow up to become the man whom many consider to be the greatest long-distance runner ever.

Haile Gebrselassie

Growing up in the rural village of Asela, Ethiopia, Haile was part of a family that had ten children. He ran everywhere, such as when he was doing his chores and going to school. Haile's older brother encouraged him to run organized track at school. When Haile was eight years old and the youngest competitor, he won a 1500-meter race at school. He won junior level races and then began competing in adult competitions. As he continued winning, he participated in international events and won there also.

During his professional career, Gebrselassie compiled an impressive list of victories. He won the 10,000-meter race at the 1996 and 2000 Olympics, as well as the world championship in that distance in 1993, 1995, 1997, and 1999. He set 27 world records, sometimes beating his own record. When he was older he turned to running marathons. He won four consecutive Berlin Marathons (2006 through 2009).

Gebrselassie achieved these successes when track and field athletes were becoming professionals. Until this period, many people believed that the amateur athlete, competing just for the sake of competition and victory, was the ideal. For many years, Olympic athletes had to be amateurs. Any money they earned had to come from nonathletic jobs. For instance, the great American athlete Jim Thorpe won the decathlon and the pentathlon at the 1912 Olympic Games. For this achievement, many considered him the greatest athlete in the world. However, Thorpe had played minor league baseball in 1909 and 1910, earning just a few dollars per game. Because of this, the International Olympic Committee (IOC) stripped Thorpe of his gold medals. The IOC restored his victories and returned his medals to Thorpe's family in 1982 (Thorpe had died in 1953).

In Gebrselassie's day, as the Olympics and other athletic events were gaining popularity thanks in large part to television coverage, and as television networks and athletic equipment companies were earning millions of dollars annually, professional athletes became accepted. In 1986 the IOC decided to allow professionals to compete in the Olympics. Athletes signed endorsement contracts with equipment manufacturers. Competitions began offering cash prizes. This change enabled athletes to attract skilled trainers and to devote their full time to training (when they were not in school).

Haile Gebrselassie at the 2006 Berlin Marathon

Coffee Plantation in Ethiopia

Gebrselassie earned millions of dollars through his running. He has invested his wealth in businesses in order to help Ethiopia. The companies Gebrselassie founded or funded include construction, real estate, coffee plantations, and car importing. Those companies employ over one thousand people. He retired from competitive running in 2015 and expressed an interest in entering politics to continue helping Ethiopia.

The Passion and the Results

Thousands of Ethiopians have a passion for running. Every morning a large crowd gathers at the stadium at Meskel Square in Addis Ababa to run up through the successive levels of seats. Aspiring athletes run on rural roads for hours a day. Villages with good coaches become legendary for the number of champions they produce. The town of Bekoji, for instance, has a population of about 17,000. Runners from there have won 16 Olympic medals, ten of them gold. Like hockey in Canada, soccer in Germany and Brazil, basketball in Kentucky and Indiana, and football in Texas and Alabama, long-distance running has become the favored sport of Ethiopia. Coaches notice runners with exceptional talent at young ages and nurture and develop those talents to groom future champions.

The passion extends beyond Ethiopia. Neighboring Kenya has caught the passion as well, and it has also produced multiple Olympic and world champions in distance races. As of May 2019, Kenyan Eliud Kipchoge had won eleven of the twelve marathons he entered. He also held the world record of 2:01:39. His only loss was to another Kenyan.

In October 2019, Kipchoge ran the marathon distance in just under two hours, 1:59:40. This broke the two-hour barrier that many thought impossible. The run was not a competition, however, and so did not count as an official record. Kipchoge ran on a carefully planned six-mile circuit in Vienna, Austria. No runners competed against him, but different groups of pacesetters ran with him for different parts of the distance. He even wore a special pair of shoes for the event.

Lesson 39 - Long Distance Runners from Ethiopia

Long-Distance Olympic Medalists from Ethiopia and Kenya

● GOLD ● SILVER ● BRONZE

Men's Marathon
- 1960: Abebe Bikila, Ethiopia ●
- 1964: Abebe Bikila, Ethiopia ●
- 1968: Mamo Wolde, Ethiopia ●
- 1972: Mamo Wolde, Ethiopia ●
- 1988: Douglas Wakiihuri, Kenya ●
- 1996: Erick Wainaina, Kenya ●
- 2000: Gezehegne Abera, Ethiopia ●
    Erick Wainaina, Kenya ●
- 2008: Samuel Wanjiru, Kenya ●
    Tsegay Kebede, Ethiopia ●
- 2012: Abel Kirui, Kenya ●
    Wilson Kipsang Kiprotich, Kenya ●
- 2016: Eliud Kipchoge, Kenya ●
    Feyisa Lelisa, Ethiopia ●

Men's 10,000 Meters
- 1968: Naftali Temu, Kenya ●
    Mamo Wolde, Ethiopia ●
- 1972: Miruts Yifter, Ethiopia ●
- 1980: Miruts Yifter, Ethiopia ●
    Mohamed Kedir, Ethiopia ●
- 1984: Michael Musyoki, Ethiopia ●

- 1988: Kipkemboi Kimeli, Kenya ●
- 1992: Richard Chelimo, Kenya ●
    Addis Abebe, Ethiopia ●
- 1996: Haile Gebrselassie, Ethiopia ●
    Paul Tergat, Kenya ●
- 2000: Haile Gebrselassie, Ethiopia ●
    Paul Tergat, Kenya ●
    Assefa Mezgebu, Ethiopia ●
- 2004: Kenenisa Beleke, Ethiopia ●
    Sileshi Sihine, Ethiopia ●
- 2008: Kenenisa Bekele, Ethiopia ●
    Sileshi Sihine, Ethiopia ●
    Micah Kogo, Kenya ●
- 2012: Tanku Bekele, Kenya ●
- 2016: Paul Tanui, Kenya ●
    Tamirat Tola, Ethiopia ●

Women's Marathon
- 1996: Fatuma Roba, Ethiopia ●
- 2000: Joyce Chepchumba, Kenya ●
- 2004: Catherine Ndereba, Kenya ●
- 2008: Catherine Ndereba, Kenya ●
- 2012: Tiki Gelana, Ethiopia ●
    Priscah Jeptoo, Kenya ●
- 2016: Jemima Sumgong, Kenya ●
    Mare Dibaba, Ethiopia ●

Women's 10,000 Meters
- 1992: Derartu Tulu, Ethiopia ●
- 1996: Gete Wami, Ethiopia ●
- 2000: Derartu Tulu, Ethiopia ●
    Gete Wami, Ethiopia ●
- 2004: Ejagayehu Dibaba, Ethiopia ●
    Derartu Tulu, Ethiopia ●
- 2008: Tirunesh Dibaba, Ethiopia ●
- 2012: Tirunesh Dibaba, Ethiopia ●
    Sally Kipyego, Kenya ●
    Vivian Cheruiyot, Kenya ●
- 2016: Almaz Ayana, Ethiopia ●
    Vivian Cheruiyot, Kenya ●
    Tirunesh Dibaba, Ethiopia ●

Ethiopians Tirunesh Dibaba (left) and Almaz Ayana won medals at the 2016 Olympics. Almaz set a new world record in the 10,000 meters.

Kenyan runners Dennis Kimetto, Eliud Kipchoge, Stanley Biwott, and Wilson Kipsang compete at the 2015 London Marathon. Kipsang won this race in 2012 and 2014. Kipchoge won in 2015, 2016, 2018, and 2019. Race organizers often employ pacemakers or pacesetters to push the other runners. The pacemakers usually drop behind the leaders as the race goes on, but sometimes they end up winning!

The Why Question

The question that arises when considering this amazing record is "Why?" Why do East Africans have this record of success? Observers and experts have suggested many reasons. Perhaps it is the typical diet of Ethiopians and Kenyans. Perhaps their metabolism is especially good at turning food into energy. Perhaps it is because Ethiopians learn to run and work hard from childhood. Perhaps a factor is training at Addis Ababa, whose elevation is about 7,700 feet above sea level. This altitude tends to help people create larger red blood cells, which provide more oxygen to the body. Training at this altitude makes running at lower altitudes easier. No doubt the financial incentives encourage hard work because young athletes see running as a way out of poverty to a better life.

All of these factors probably play a part, but attitude is crucial also. Ethiopian and Kenyan athletes have developed a culture of success, a mindset that says they are champions. Haile Gebrselassie expressed his beliefs in this way:

> When you believe in something, you believe in yourself as well. I believe in God . . . . I am a religious person. I am an orthodox Christian. My family taught me how to pray.

Paul used athletic imagery to make his points in his letters, such as:

Lesson 39 - Long Distance Runners from Ethiopia

Do you not know that those who run in a race all run, but only one receives the prize? Run in such a way that you may win. Everyone who competes in the games exercises self-control in all things. They then do it to receive a perishable wreath, but we an imperishable. Therefore I run in such a way, as not without aim; I box in such a way, as not beating the air; but I discipline my body and make it my slave, so that, after I have preached to others, I myself will not be disqualified.
1 Corinthians 9:24-27

Assignments for Lesson 39

Gazetteer   Read the entry for Ethiopia (page 61).

Geography   Complete the map skills assignment for Unit 8 in the *Student Review Book*.

Project   Continue working on your project.

Literature   Continue reading *A Long Walk to Water*.

Student Review   Answer the questions for Lesson 39.

Bioko Island, Equatorial Guinea

40 Where Did You Get That Worldview of Yours?

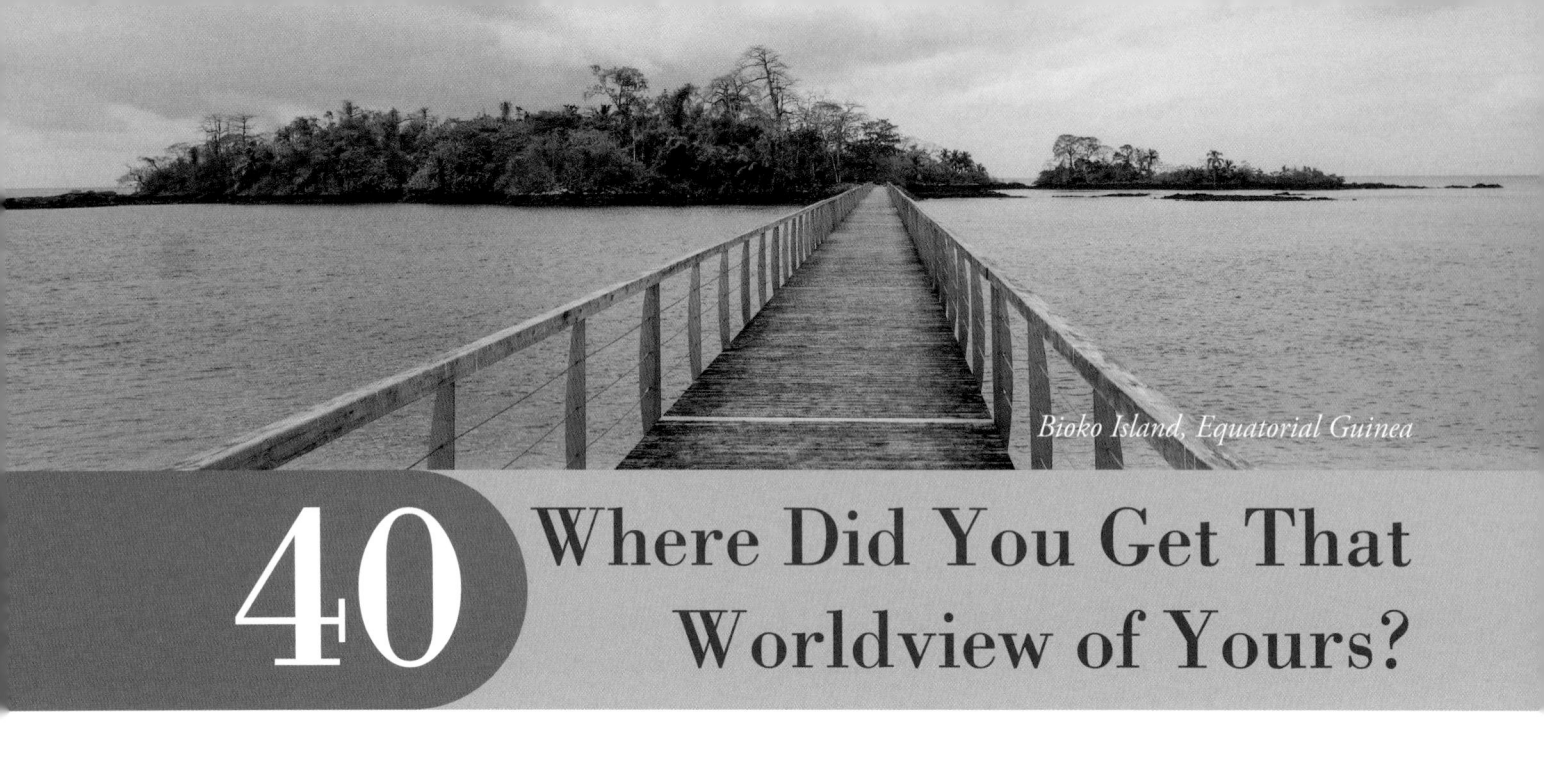

How does your mind fill in the blanks as you read these statements:

"Everyone knows that the best way to _____ (Pick the activity: avoid getting sick, prepare for retirement, find a job, etc.) is to _____." (Supply your own conventional wisdom.)

"This is the way we've always done things _____." (Pick a phrase: in this church, in this town, in our family.)

"He's a _____, and those people never _____."

"My daddy always said, 'Never trust a _____.'"

"She always _____."

Where did you get the perspectives that led you to these conclusions? Why do you see people the way you do? Why do you see yourself the way you do? Why do you see particular groups the way you do?

Digging more deeply, where did you get your ideas about God, sin, love, and truth?

A person develops a worldview as a result of many influences. These influences can include parents, religious teachings, reading, media, experiences, friends, the society and culture in which he lives, and the evidence that a person perceives in the world around him. You might like to think that you base your worldview on a careful examination of different schools of thought, but you might actually have developed a significant element of your worldview from a movie you saw or a book you read or a passing remark by someone you respect. The way that you analyze, evaluate, and adopt or reject (consciously or unconsciously) these influences results in your worldview.

As you consider the origins and content of your worldview, here are some things to keep in mind.

Lesson 40 - Where Did You Get That Worldview of Yours?

1. Be sure of your sources.

An intelligent, well-meaning person can base his or her worldview on ideas that are not true. For example, many people believe that the universe came into existence by a random, undirected Big Bang. Some people believe that no God exists. Followers of folk religions believe that humans frequently have to appease evil spirits that inhabit the world. None of these belief systems is true, but many people believe that they are true. Their beliefs, even though they are incorrect, influence their ideas and actions.

This can also be the case regarding how you see yourself and what you believe other people think about you. You might have an incorrect view of yourself and other people. Even though it is wrong, that viewpoint still influences your actions.

People around you, whom you have good reason to respect, might have developed attitudes and practices that are not best or even right. In fact, you and I might have developed some of these attitudes and practices ourselves. It is easy to develop prejudiced ideas about individuals and groups that have no basis in fact. Long-standing and even widely-held beliefs are not necessarily godly or right.

In analyzing your worldview and in seeking to develop a Biblical worldview, you should be sure why you believe as you do. You need to be sure of the reliability of the sources that have helped you form your worldview. The best source for understanding the world is the Bible. Your goal should be to base your worldview as much as possible on an accurate understanding of God's Word.

2. Be consistent.

It is important to be consistent in your worldview, but inconsistencies are easy to have. For instance, a person might believe in an all-powerful, sovereign God who rules the universe and answers prayer; but he or she might also believe in luck and say things like "Good luck" and "Knock on wood" and "You were lucky to recover from the flu so quickly." What would that person say ultimately controls the world: luck, or God? If God rules the universe, does luck have anything to do with it?

A person might say he is a Christian but read and watch material that dishonors Christ and other people. He might manipulate or use others for his own purposes and pleasure. What is his worldview? Is he a disciple of Jesus, or does he live for his own pleasure—or does he try to do both at different times? Remember that Jesus said, "You cannot serve two masters" (Matthew 6:24). Christ and self cannot both be your lord.

If someone's actions are not consistent with her expressed worldview, this indicates that her real worldview is somehow different from her stated worldview. A person's real worldview, not her stated worldview, is what guides her actions.

Maneki Neko are cat statues associated with good luck in Japanese culture.

3. Conversion to Christ should involve changing one's foundational, underlying worldview.

In Acts chapter 8, Simon practiced magic in Samaria. When he heard the gospel, he believed and was baptized. However, he later wanted to buy the ability to bestow the Holy Spirit through the laying on of hands the way he had seen Peter and John do it. Peter rebuked him for his attitude, and Simon begged the apostles to pray for him (Acts 8:9-24). Simon had come to believe in Jesus, but his worldview hadn't changed about how things worked in the world.

In the early years of the church in Jerusalem, some Pharisees came to believe in Jesus as Savior and Lord. However, in Acts 15:5, "some of the sect of the Pharisees who had believed" insisted that Gentiles who believed in Christ had first to become Jews in order to become Christians. In their minds Gentiles needed to be circumcised and keep the law in order to be faithful Christians. They were attempting to fit Christ into their basic Jewish worldview, which had not changed.

Modern missionaries who teach the gospel to adherents of folk religions (see Lesson 30) often find later that, although the people they have taught confess faith in Christ, those people still believe and practice aspects of their folk religion. Such people accept what the missionaries say as true, but they also believe that their folk beliefs are true.

Conversion means changing your worldview. Paul says the Christians in Thessalonica "turned to God from idols to serve a living and true God" (1 Thessalonians 1:9). Idolatrous practices and faith in Christ are inconsistent and incompatible. In 2 Corinthians 5:16, Paul says, "From now on we recognize no one according to the flesh." He had once known Christ according to the flesh but now he no longer looked on the Lord in that way. Paul speaks of "the renewing of your mind" (Romans 12:2) and calls on Christians to "set your minds on the things above, not on the things that are on earth" (Colossians 3:2). This radical reorientation of one's thinking is essential for Christian discipleship and for avoiding the messy business of trying to live by two conflicting worldviews. This is a difficult change to make, and it won't happen completely the moment you are born again; but it is necessary for putting off the old person and putting on the new person in Christ.

Carving of Simon at the Basilica Saint-Sernin, Toulouse, France

Lesson 40 - Where Did You Get That Worldview of Yours?

The New Testament is all about worldview, specifically, changing people's worldviews about God, Jesus, themselves, and the world. If someone comes to Christ but does not change his worldview, he can obey what he learns in church, from other Christians, and in the Bible, but still see himself, his life, other people, and the world in the same way he did before he became a Christian. He will have the same motives and desires, the same fears and failings.

Jesus wanted the Jews to think differently from the typical law-keeping Jewish mindset: not just keeping a set of rules and expectations, but turning the other cheek, going the second mile, hungering and thirsting for righteousness, and loving their enemies. He wanted them to understand the Messiah differently from how the Jews had come to think of Him as they chafed under Roman rule.

Paul constantly taught the consequences of the gospel in how Christians were to live. Whenever he got down to the core motivation for what he was saying, it was always Christ: Romans 6:1-7, 1 Corinthians 2:2, and Philippians 2::5-11 are examples of this.

To be truly converted is to accept and embrace a new worldview and to think that way (Romans 12:2). Obeying the gospel is responding to Jesus, not checking off certain requirements. The change is on-going and not a one-time thing. Developing a Christian worldview and acting on the basis of it is something in which a Christian should constantly seek to grow.

4. The majority of people in the world do not share your worldview completely.

If you grow up in a Christian home and go to church from childhood, when you become a Christian you will probably be warmly congratulated. It won't be long, however, before you learn that not everyone thinks the way you do. Perhaps a majority of the people you interact with most days do share a large part of your worldview, but that group does not constitute a majority of people. Christianity is the single most numerous religion in the world,

Page from a 13th-century Latin translation of Matthew

but it is still a minority of the world's population. Probably even a fair number of people who consider themselves to be Christians do not share your worldview in every respect.

How do you respond to this fact?

- Do you simply live and let live and try not to worry about it?

- Do you feel a need to convert as many people as you can to your worldview?

- Do you want to be a good influence, to be salt and light, in the hope that others will appreciate your life and will want to know why you live as you do?

- Are you tempted to change your worldview to agree with what you believe is a majority in your community, nation, or group—or do you believe that you are right and they are wrong?

You should not hold a particular worldview because you think it is popular or because you believe it will be easier to get along with others if you do. If you do this, your stated worldview will likely not be your real worldview. Your worldview should be a matter of conviction which you maintain even if your life is at stake.

Bioethicist Leon Kass has written, "We want to know just what kind of a world this is and especially what kind of beings we are and how we do and should relate to that world." This is the quest we seek to fulfill in this worldview study.

---

*[W]e look not at the things which are seen,
but at the things which are not seen; for the things which are seen
are temporal, but the things which are not seen are eternal.
2 Corinthians 4:18*

Assignments for Lesson 40

Worldview — Recite or write the memory verse for this unit.

Project — Finish your project for this unit.

Literature — Finish reading *A Long Walk to Water*. Read the literary analysis and answer the questions in the *Student Review Book*.

Student Review — Answer the questions for Lesson 40.
Take the quiz for Unit 8.

Craft Market, Malkerns, Eswatini

9 Southern Africa

Diamond mining has had a major impact on the economy and society of South Africa. The Zulu people, the largest ethnic group in South Africa, have played a crucial role in that country's history. Christians in southern Africa who don't want to tag along behind are reaching out to villages in 20 countries by means of a "Gospel Chariot." For many years South Africa followed a policy of racial segregation called apartheid, but since 1994 the country has been adjusting to new laws and a new worldview. The worldview lesson examines a basic question: What is truth?

Lesson 41 - They Say a Diamond Is Forever
Lesson 42 - This Is Our Land: The Story of the Zulus
Lesson 43 - Roll the Gospel Chariot Along
Lesson 44 - Can People Once Enemies Get Along?
Lesson 45 - Basic Issue: Truth

Memory Verse   Memorize Ephesians 2:13-14 by the end of the unit.

Books Used   The Bible
Exploring World Geography Gazetteer

Project (Choose One)

1) Write a 250-300 word essay on one of the following topics:
   - Define truth. Tell how you can discern the difference between truth, error, and deceit. (See Lesson 45.)
   - How might a "Truth and Reconciliation" Commission help the United States overcome its racial conflicts? (See Lesson 44.)
2) Write a poem or song about love and acceptance overcoming racial division. (See Lesson 44.)
3) What is a new way to reach out to others with the gospel that you think would be effective? Describe your approach and tell what would be needed for it to work. (See Lesson 43.)

Diamond Mine, Botswana

41 They Say a Diamond Is Forever

In 1870 Erasmus Jacobs pulled a baseball-sized diamond from the Orange River in South Africa that started a diamond rush. Before long, people were purchasing land claims and finding diamonds in many places across southern Africa.

Englishman Cecil Rhodes, born in 1853, was a failed cotton farmer who had dreams of greatness for Africa and for himself. He began building a fortune when he rented equipment to prospective diamond miners to pump floodwaters from their claims. Soon he started buying others' claims, which gave him the rights to the diamonds found on those claims. He bought the claims of two Afrikaans brothers named DeBeer and started the DeBeers Consolidated Mines Company. Rhodes died in 1902 but the DeBeers company, after it merged with another diamond mining enterprise, eventually controlled 90% of the diamond market around the world.

By the way, Rhodes got involved in politics in southern Africa, which at the time was part of the British Empire. He helped to establish the country that was named Rhodesia (for him), which many years later became Zimbabwe. Rhodes dreamed of a Cape to Cairo railroad that would traverse the entire continent of Africa south to north, from the Cape of Good Hope in the south to Cairo in Egypt. His will established the Rhodes scholarships at Oxford in England.

The DeBeers company was so successful at diamond mining that the market was flooded and prices fell. DeBeers began limiting the number of diamonds it put on the market each year to control the supply and thus increase the price. Into the twentieth century, most people saw diamonds as something only the wealthy and European royalty could afford. Not many prospective grooms bought a diamond engagement ring for their bride-to-be. The Great Depression of the 1930s put an even deeper hole in the diamond market, as many people began trying to sell the diamonds they already owned. Few people could afford such a luxury as a diamond.

After World War II, America had a growing middle class and enjoyed greater economic prosperity. The DeBeers company hired an advertising agency to help increase the demand for diamonds by conveying the idea that diamonds were relatively rare (they weren't) and were extremely desirable. In 1947 the agency developed the slogan, "A Diamond Is Forever." Ads encouraged the idea that a girl simply had to have a diamond engagement ring and that a boy simply had to give her one. The agency gave diamond jewelry to movie stars to wear in public appearances and encouraged the movie industry to

Unit 9: Southern Africa

Uncut Diamonds

make films in which diamonds played a prominent role. DeBeers skillfully developed the market for relatively smaller diamonds, not the huge rocks that royalty and movie stars (the American royalty to many people) owned and flashed.

It worked. The giving of a diamond engagement ring by fiancé to fiancée became standard, practically expected, not only in the U.S. but in other countries as well. The size of the diamond became an indicator of the financial status of the young man, as well as a statement of his commitment to the young woman.

A diamond ring became a necessary luxury. Other pieces of diamond jewelry found a market as well. In 1999 *Advertising Age* magazine named "A Diamond Is Forever" as the advertising slogan of the century.

This brief history of the role of diamonds in our society might sound cynical; but young men, don't get any ideas. If your intended one wants a diamond ring, get her a diamond ring. Neither of you should be materialistic, but don't worry about your effect on the world diamond market or rue the influence of marketing on our thinking, and by all means don't try to talk her out of wanting a diamond engagement ring. Honor what she means to you and think about how you can convey to her your love and your promise of lifetime faithfulness. She should be your one and only jewel, and a ring is a good way to communicate that to her and to the world—and to yourself. That is certainly what I wanted to say when I placed a diamond engagement ring on Charlene's finger after she said that she would marry me.

18th-Century Ornament from India Featuring Diamonds and Other Gems

Now known as The Big Hole, this open-pit diamond mine operated at Kimberly, South Africa, from 1871 to 1914. It is almost a mile in circumference and over 700 feet deep.

Diamond Boats Off the Coast of Namibia

Diamonds and Geography

Diamonds have a connection to geography because people find them only in certain geographic places in the world. People have to engage in the hard work of digging into the earth to retrieve these pieces of hardened carbon to begin the process that produces the beautiful jewels we see on store shelves. Diamond mining affects the geography of the mines and the people who work there.

According to the records we have, diamonds were first known in India in the 300s BC. Traders carried them to Europe by the 1200s AD. There they became the exclusive property of royalty. During that century, King Louis IX of France decreed that only the king could possess them. Over the next few centuries, other European royalty and nobility began to own diamonds, and a few people in the rising merchant class got their hands on the jewels as well. People in Brazil found diamonds there in the 1720s, and Brazil became the leading source of diamonds for the next century and a half. As mentioned above, South Africa became the major source of diamonds beginning in the late 1800s.

At first, diamonds were found in creeks and riverbeds, much like people found gold nuggets in the California gold rush. When these sources started to be exhausted, prospectors turned to underground mining as the major source. Improved technology in the twentieth century enabled people to discover many more diamond mines around the world.

Today diamond mines exist in about 25 countries, on every continent except Europe and Antarctica. Total annual recovery of all grades of diamonds from all sources runs about 100 to 150 million carats.

A recent development has been the extension of the search for diamonds into the oceans. The waters of the Atlantic Ocean off Namibia in southwestern Africa provide a good example. Experts at Debmarine Namibia (a subsidiary of DeBeers) identify the most promising places for digging within the area for which it has a license, so the process is not a general strip mining of the seabed.

Debmarine Namibia operates ships that extend large digging arms down to four hundred feet below the surface. These arms scoop up material from the ocean floor and bring it back to the ship, where other machinery sifts it. The gravel goes back to the ocean, and workers place the gems in steel briefcases, which helicopters take to shore accompanied by company officials. No human hands touch the diamonds during this entire process.

As with just about every activity that people undertake that involves the geography of our world, some controversies swirl around diamond mining. Critics charge that some land diamond mines that operate outside enforced regulations use child laborers, pay workers less than one dollar per day, and force workers to endure unsafe conditions.

Mining companies sometimes cut down trees to have easier access to mines, but this deforestation damages local habitats. Environmentalists worry

Lesson 41 - They Say a Diamond Is Forever

that mining seabeds does serious damage to the habitats of fish and other ocean life. Sea mining companies say they are careful to limit their impact on the environment and that, since most of what they dig up returns to the seabed, the damage is minimal and the habitat restores quickly. Critics question the time frame that the companies claim and say such restoration takes much longer if it ever occurs.

People decide that certain materials in the earth, such as gold, silver, diamonds, and other gems, have exceptional value. The acquisition and distribution of these materials involves human interaction with geography. Sometimes the acquisition of these materials involves the abuse of other humans. The wealth that these materials provide give people social and financial status and the ability to acquire other things they want and need.

The Bible teaches us that many things are even more valuable than precious materials from the earth.

For wisdom is better than jewels,
And all desirable things cannot compare with her.
Proverbs 8:11

Assignments for Lesson 41

Gazetteer Study the map of southern Africa and read the entry for Namibia (pages 75 and 79).

Worldview Copy this question in your notebook and write your answer: How do you determine the difference between beliefs that are either right or wrong and beliefs that are matters of opinion?

Project Choose your project for this unit and start working on it. Plan to finish it by the end of this unit.

Literature You do not have a literature title assignment this week. You can catch up on other reading or concentrate on your project.

Student Review Answer the questions for Lesson 41.

Drakensberg Mountains, KwaZulu-Natal, South Africa

42 This Is Our Land: The Story of the Zulus

What the future has in store for me, I do not know. It might be ridicule, imprisonment, concentration camp, flogging, banishment, and even death. I only pray to the Almighty to strengthen my resolve so that none of these grim possibilities may deter me from striving, for the sake of the good name of our beloved country, the Union of South Africa, to make it a true democracy and a true union in form and spirit of all the communities in the land.

"My only painful concern at times is that of the welfare of my family but I try even in this regard, in a spirit of trust and surrender to God's will as I see it, to say: 'God will provide.'

"It is inevitable that in working for freedom some individuals and some families must take the lead and suffer: The road to freedom is via the Cross."

These were the words of Albert John Mvumbi Luthuli, Zulu chief, educator, lay preacher, and anti-apartheid activist, in 1952. Luthuli devoted his life to enabling the Zulu people to enjoy freedom and equality in modern times as they lived on the land their ancestors had ruled for hundreds of years. One historian called their land one of the most attractive places for human settlement in Africa.

Natal and Zululand

On Christmas Day in 1497, Portuguese explorer Vasco da Gama sighted a coastal area of southeast Africa. He named the land Terra Natalis or Land of Christmas, after Natal ("the Birth"), the Portuguese word for Christmas. Portuguese traders later established a port further north at Delagoa Bay (now Maputo Bay in Mozambique).

Among the numerous Bantu peoples of Africa were the Nguni, who moved into what came to be called Natal in the 1500s. A leader of one group of Nguni in the 1600s, Malandela, named his son Zulu, which means sky or heaven. Zulu became leader of the Zulu people and established KwaZulu, or place of heaven, as their land. In the early 1800s, aggressive Zulu leaders used force and the threat of force to bring neighboring tribes under their rule. In time the Zulu became the largest people group in southeastern Africa. They believed the area where they lived was their land.

In 1824 the British arrived to negotiate a trade agreement with the Zulus. They established a settlement at Port Natal, the place da Gama saw and which eventually became the city of Durban. The Zulu king granted the British oversight of the port

Lesson 42 - This Is Our Land: The Story of the Zulus

and its surrounding area. This was the first British presence on what they came to see as their land.

In 1838 the Boers (also known as Afrikaners) moved out of the Cape of Good Hope area of southern Africa because of British encroachment on what they saw as their land. The Boers moved onto Zulu land, and the Zulus massacred them. The Boers reorganized and struck back, defeating the Zulus at the Battle of Blood River. The Zulus withdrew north of the Tugela River into the portion of KwaZulu that became known as Zululand. The Boers assumed control of the area south of the Tugela, another portion of KwaZulu, and called it Natal.

Other natives then began returning to Natal to resettle land they once called theirs but that they had given over to the Zulus. Then the British annexed Natal in 1843, claiming it as theirs. Many Afrikaners left to settle other parts of southern Africa, and more British immigrants came to live in Natal. Immigrants from India came to work on sugar plantations along the coast. In this racial mix, by the late 1800s Bantu people were the majority in Natal and Zululand.

Great Britain wanted to control Zululand to get its people to work in the diamond mines of the colony of South Africa and to extend its control over more of the southern Africa region. The Zulus resisted British insistence on greater control. The British invaded Zululand in 1879. After an initial defeat, British forces conquered the Zulu army; and in 1887 Britain annexed Zululand into South Africa.

The 1940s through the early 1990s were a period of racial unrest in South Africa. In 1949 Zulus attacked Indian workers in Durban. The 1980s saw numerous riots and periods of unrest in Durban that involved black Africans and Indians. Two competing black groups, the Inkatha Freedom Party and the African National Congress (ANC), often clashed for power. The South African government designated certain areas as *bantustans* (or black enclaves) into which authorities moved black residents from urban areas. In 1980 the South African government designated eleven such areas in Natal and called them collectively KwaZulu. The apartheid system ended with the new South African constitution in 1994. At that time, KwaZulu was incorporated into Natal, and the combined region became the KwaZulu-Natal province within South Africa.

Zulu Warriors (c. 1900)

The Zulu People

The Zulu are the largest ethnic group in South Africa and make up about 20% of that country's population. Some Zulus also live in Zimbabwe, Zambia, Tanzania, and Mozambique. KwaZulu-Natal is the second most populous province in South Africa. It is 86.8% black, 7.4% Indian/Asian, and 4.2% white. About three quarters of the population speak Zulu (one of the official languages of South Africa) as their native language.

Traditional Zulus grew some crops, but the major part of the Zulus' identity involved their devotion to cattle. The larger one's herd of cattle, the wealthier and more prestigious was that man to his fellow Zulus. Tribute paid to the king had to be in the form of cattle. People composed poetry that sang the praises of cattle. The Zulu used every part of cattle, including the hides for shields and garments, the tails as decorations for festive garments, and the horns for containers to hold medicine and gunpowder. Any important sacrifice involved slaughtering cattle. People saw these cattle as messengers to those in the afterlife. The person making the sacrifice would tell the animal what he wanted it to tell the man's deceased ancestor.

A traditional part of betrothal is the groom's payment of a brideprice to the bride's father. This is to compensate her family for the labor they will be losing when she becomes a wife. Traditionally the bride's father set the price, and the groom paid it in cattle. This is still done in rural areas, but today the transaction between urban Zulus takes place in cash. Sometimes a groom has to make installment payments.

Zulu Cattle Pen

Traditional Zulu Home

One practice that demonstrates the importance of cattle is that the Zulus build their huts in traditional villages around a central cattle pen. The chief's hut is at one end of the village. Many Zulus practice polygyny (a man taking more than one wife), and the primary wife's hut is closest to the chief's, with the huts for his lesser wives further away. A traditional hut has a low doorway. This makes an attacker's entry more difficult, but it also forces anyone entering to show deference and humility to the man of the hut, which is an honored Zulu tradition.

The Zulus observe many rituals and ceremonies, which include festive singing and dancing. They have long practiced the worship of deceased ancestors. Christian missionaries have had some degree of success in converting Zulus, but even after they convert, many Zulus still hold to and practice their traditional beliefs as well. Christian missionaries developed a written form of the Zulu language. The first book published in Zulu was a Bible printed in 1883. The Zulus also continue to practice traditional healing methods alongside modern medical procedures.

Lesson 42 - This Is Our Land: The Story of the Zulus

Luthuli's Life

Zulus were once fierce warriors. They gave up their autonomy to the Boers and then to the British on the land they once claimed as theirs. Not only did they lose their autonomy as a people, but they also lost their freedoms under the apartheid system of racial segregation and discrimination in South Africa. One man who helped the Zulus regain their freedoms was Albert Luthuli.

Albert John Mvumbi ("Continuous Rain") Luthuli was born in 1898 in Rhodesia (now Zimbabwe). His father, a native of Zululand, was working in Rhodesia as a missionary interpreter. However, Albert's father died when Albert was less than a year old, so his mother sent him to the family's traditional home at a mission station in Groutville, Natal. When Luthuli was an adult, he taught at a primary school in Natal. He was confirmed in the Methodist Church and became a lay preacher. He later taught at Adams College.

In recognition of his character and leadership skills, the Abase-Makolweni tribe of Zulus elected him as their chief in 1935. He led and helped his people for 17 years. During that time he became increasingly committed to and active in the movement to end apartheid (racial segregation) in South Africa. The leading group in this movement was the African National Congress. The Natal branch of the ANC elected Luthuli as its president in 1948. The group tried using peaceful protests, but the South African government brutally repressed them. When the group peacefully defied apartheid laws, over eight thousand black Africans were put in prison.

Luthuli was elected president-general of the ANC for all of South Africa in 1952. Over the next several years, the South African government persecuted Luthuli in an attempt to limit his activity. He was tried for treason, but was found innocent. He was imprisoned for a time, but released in 1957. In 1959 he was banned from attending ANC meetings and confined to his neighborhood, but he still managed

Albert Luthuli

to send written messages to ANC meetings. The South African government outlawed the ANC in 1960, but the organization continued to function underground.

Also in 1960 the Nobel Prize committee chose Luthuli to receive the Nobel Peace Prize. He was the first non-white and the first person who was not from Europe or the Americas to receive the prize. South Africa at first refused to let him go to Oslo, Norway, to receive the prize; but the government finally relented and allowed Luthuli to travel to Oslo under strict conditions in 1961.

Over the next few years, the ANC began using violence to try to accomplish their demands. Historians debate whether Luthuli opposed this change, stood by and let it happen, or gave any kind of endorsement to it. His public statements up to that time were always in support of nonviolence as the only proper means of bringing about change.

Luthuli continued to hold the office of president-general, but the ANC was not a tightly organized group during those years. People across a wide range of political views claimed to speak and act on behalf of the ANC.

Luthuli died in 1967. He did not see the end of apartheid in South Africa, but his life and work helped pave the way for that outcome.

In a way, Albert Luthuli, Nelson Mandela, and others who worked to end apartheid were like Moses in appealing to Pharaoh of Egypt to let their people go. The resistance from Pharaoh was similar to the resistance that the South African government gave to ending apartheid for many years. One difference, however, was that the Zulu wanted to be free on the land they saw as their own.

And afterward Moses and Aaron came and said to Pharaoh, "Thus says the Lord, the God of Israel, 'Let My people go that they may celebrate a feast to Me in the wilderness.'" But Pharaoh said, "Who is the Lord that I should obey His voice to let Israel go? I do not know the Lord, and besides, I will not let Israel go."
Exodus 5:1-2

Assignments for Lesson 42

Gazetteer Read the entry for Botswana (page 76).

Worldview Copy this question in your notebook and write your answer: What is something you once thought was true but now know to be untrue?

Project Continue working on your project for this unit.

Student Review Answer the questions for Lesson 42.

Sunlight Over Hills in Burundi

43 Roll the Gospel Chariot Along

You may have heard the old gospel song that has some variation of these words:

Roll the gospel chariot along,
Roll the gospel chariot along,
Roll the gospel chariot along,
And we won't tag along behind.

If a sinner's in the way we will stop and pick him up,
If a sinner's in the way we will stop and pick him up,
If a sinner's in the way we will stop and pick him up,
And we won't tag along behind.

If the devil's in the way we will run right over him,
If the devil's in the way we will run right over him,
If the devil's in the way we will run right over him,
And we won't tag along behind.

Gospel Chariot Meeting in Malawi

Some Christians in southern Africa took this song literally and devised a Gospel Chariot ministry that now uses fifteen "chariots" to evangelize and train people in twenty African countries.

Bringing People Together, Bringing Them to Jesus

In 1994 George Funk and his wife Ria, white South Africans, quit their jobs to share the gospel with their fellow South Africans. Their method involved using free materials from World Bible School (WBS), a correspondence and online ministry. The Funks contacted students who had expressed an interest in WBS materials or who had begun the course of study. They helped the students complete the course, answered their questions, brought people to the Lord, and plugged the new Christians into local churches.

Through their work the Funks bridged the racial divide that characterized life in South Africa at the time. Two years after they began their work, George studied the Bible with Machona Monyamane, a black South African who saw whites as oppressors. "I never liked church," Monyamane said. "It was a symbol of the system that oppressed us."

Gospel Chariot Driving In Malawi

Monyamane had many questions for Funk; but with each one, Monyamane said that Funk "put a smile on his face and a finger on the passage" in the Bible that provided the answer. Monyamane was converted to Christ and later became a minister in Pretoria, South Africa. Now he wants to take the gospel to the world because, as he put it, "Jesus didn't give instructions in black and white."

To increase their outreach, George and Ria built their first Gospel Chariot in 2000. Now fifteen vehicles, from small trucks to large semis, travel to small towns and large cities in twenty countries across Africa, including Uganda, Burundi, Rwanda, South Sudan, Congo, Nigeria, Namibia, Ghana, Kenya, Botswana, and Eswatini.

On each truck, one side opens to reveal a speaker's platform. An awning pulls out from the top to provide a covered speaking and sitting area. A truck can carry up to one hundred chairs. The truck also carries a PA system, an electrical generator, a baptistry, and beds for team members while they are traveling.

Often, after much advance publicity, a Gospel Chariot will come into town and an evangelistic team will conduct a campaign. Sometimes a team plants a new church with those converted, while in other situations a local congregation will nurture new Christians and follow up with seekers.

The Gospel Chariot ministry uses printed WBS materials to teach those who want to learn more about the gospel. The ministry also offers a free, six-

Lesson 43 - Roll the Gospel Chariot Along

month course of study through Nations University to equip new believers. Students enroll in this course online or at training centers in various cities. Several men who have come to Christ are now evangelists who keep the Gospel Chariots rolling along.

Funk estimates that about two thousand people per year come to the Lord through the Gospel Chariots. Amazing stories abound that tell of the Lord working. In Swaziland, a seventy-year-old man who had never been part of any church gave his life to Christ in a Gospel Chariot campaign.

In Molepolole, Botswana, the Gospel Chariot team found a good location to set up the truck, but they had to obtain permission from the tribal chief to use the site since it was on tribal land. As the team was setting up the truck, local resident Robert Reid laid down—not his fishing nets—but his meat cleaver. He left his butcher shop and walked across the street to the Chariot. "I felt pulled to the Chariot," he said later. "I was looking for truth."

Gospel Chariot Meeting In Namibia

He attended nightly services and studied with team members during the day. Three days after his initial contact, Reid obeyed the gospel. Five days after that, Reid's wife was baptized into Christ.

During the Ebola crisis in Liberia that began in 2013, Christians used the Chariots to distribute information about how to avoid contamination from the virus.

Gospel Chariot Meeting in Malawi

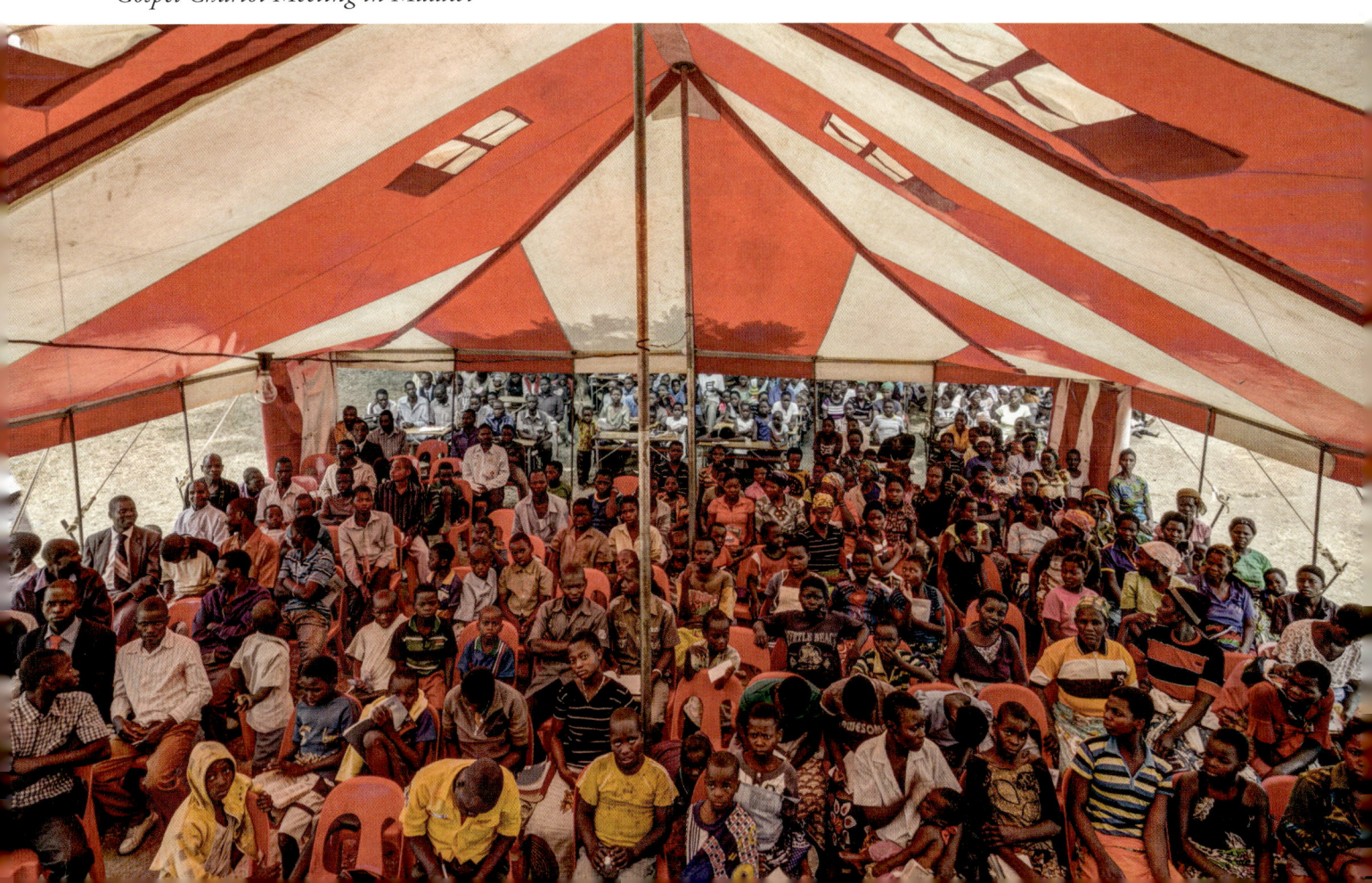

The Funks have moved to Australia. Machona Monyamane, the man who once saw the church as an oppressor, now leads the Gospel Chariot ministry along with other African Christians. These Christians, who use the Gospel Chariots to tell others about Jesus, won't tag along behind. They are like the Apostle Paul, who saw himself as an ambassador for Christ, appealing to people wherever he went to be reconciled to God.

---

*Therefore, we are ambassadors for Christ,
as though God were making an appeal through us;
we beg you on behalf of Christ, be reconciled to God.*
2 Corinthians 5:20

Assignments for Lesson 43

Gazetteer — Read the entry for Eswatini (page 77).

Geography — Complete the map activity for Unit 9 in the *Student Review Book*.

Worldview — Copy this question in your notebook and write your answer: What do you believe is a big mistake that people make by believing something that is not true?

Project — Continue working on your project for this unit.

Student Review — Answer the questions for Lesson 43.

Vineyard Near Stellenbosch, South Africa

44 — Can People Once Enemies Get Along?

Now what? The racial and ethnic groups that lived in South Africa were at odds for generations, even centuries. For over three hundred years the people in the different groups never considered themselves equals. Sometimes the groups were enemies at war with each other. During all that time there was racial discrimination, segregation, and at times, slavery. Frequently the members of one group committed unspeakable violence against members of another group.

Then in 1994 new laws officially ended racial segregation and discrimination in South Africa. Now the country had one class of citizen: South African. Segregated housing and discriminatory hiring were illegal. All adults could vote. Millions of people who had known discrimination, hatred, and violence from the time of their forefathers now had to live together in the same country with those who had been the oppressors. Would they be able to get along, work together, and prosper in the new arrangement? Could they put aside their bitter past and create a new future?

The Beautiful Land

The Republic of South Africa, which lies at the southern tip of the continent of Africa, is a beautiful land that God has blessed with abundant natural resources on and under the earth. Almost twice the size of Texas, the country of South Africa has a large interior plateau surrounded by rugged hills and a narrow coastal plain.

At the edge of the plateau is the Great Escarpment, a steep wall of rock which descends sharply from the plateau to the coastal area. The Kalahari Desert lies in the northwest region of the country, but much of the land elsewhere is rich farmland. Wildlife abounds. Mines that have produced diamonds, gold, and other riches have brought great wealth to the country.

These silhouettes in Kliptown commemorate the South Africans who waited for hours in long lines to vote for the first time in 1994.

251

Maletsunyane Falls, Lesotho

South Africa is a world away from Egypt, Morocco, and the other countries on the northern end of Africa in more ways than one. They are physically distant—air distance from Cape Town, South Africa, to Cairo, Egypt, is over 7,200 miles). The two regions are also culturally distant from each other: Northern Africa is predominantly Muslim, while southern Africa has strong European and Christian influences along with sub-Saharan tribal influences.

Nine-tenths of the country's fifty-seven million people live in the eastern half of the land and along the southern coast. Eighty-six percent of the population is Christian, about five percent are adherents of folk religions, and two percent is Muslim. South Africa has eleven official languages.

Two geographic oddities in southern Africa are the countries of Lesotho and Eswatini. Lesotho is a constitutional monarchy that gained its independence from the United Kingdom in 1966. This mountainous land, slightly smaller than Maryland, is completely surrounded by South Africa but completely independent of it politically.

Eswatini, smaller than New Jersey, is at the very northeast corner of South Africa and borders Mozambique. Its people comprise about seventy clans, and almost eighty percent of the population is rural. The country was part of the British Empire until 1988. The country is ruled by a monarch, who appoints a prime minister. One house of the legislature is popularly elected; the other is appointed. In 2018 King Mswati III changed the name of the country from Swaziland.

South Africa's Sad History

The primary significance of South Africa in the flow of human history has been its geographic location on a route that European explorers and traders took between Europe and Asia. Portuguese explorers gradually sailed farther and farther down the western coast of Africa and finally reached the Cape of Good Hope in southwestern South Africa in 1488. The Portuguese then established trading ports on the southeastern coast of Africa, from which their

Lesson 44 - Can People Once Enemies Get Along?

ships sailed further east to India, Malaysia, and other parts of Asia.

Dutch and English traders began setting up stations in South Africa in the 1600s. The Dutch East India Company granted land for Dutch farmers (boers in Dutch) to grow food to resupply Dutch trading ships. The European settlers exploited the indigenous tribes, using them as slaves and selling some as slaves in the international slave market. The Dutch also brought in people from India and Malaysia to work as slaves as well. Dutch settlers developed a variation of the Dutch language called Afrikaans. The first Afrikaans Bible was produced in 1933.

Not satisfied to share the region, the British occupied the Cape in 1795 and took over the entire Cape Colony (the region controlled by Europeans, which was most of modern South Africa) in 1806. They saw control of South Africa as key to their maintaining access to and control of India, which they considered the crown jewel of the British Empire. When the British replaced the Dutch in controlling South Africa, thousands of Dutch (called Afrikaners) moved north in the Great Trek and formed new colonies in the northeastern region of South Africa. The area north of the Orange River became the Orange Free State, and the area further north "across the Vaal" River became the Transvaal. Indigenous peoples were living in these regions, but the Dutch drove them out.

The British took over more and more land, a pattern that increased with the discovery of diamonds in 1870 and gold in 1886. Throughout the colonial period, a small white minority owned most of the land and controlled the lives of the majority black Africans. After Britain ended slavery throughout the Empire, blacks in South Africa lived on the land as laborers and sharecroppers. Despite the presence of native languages and Afrikaans, English became the language of trade and communication throughout the region.

Apartheid

The Anglo-Boer War of 1899-1902 pitted British against Dutch South Africans. The British won, but as the twentieth century progressed these two groups began working together to maintain control of the country by whites. Racial consciousness was the predominating issue, as the white minority protected its status and power at the expense of the black majority. Whites owned ninety percent of the land. The Afrikaner-led National Party won control of the government in 1948 and began the official policy of apartheid. Blacks could not live, work, or even travel in white areas without a pass; and these pass laws became an especially bitter issue for blacks. In 1960 white South Africans voted to leave the British Commonwealth and become the Republic of South Africa. Over time the white-led government took away the right of blacks to vote until elections were all-white.

By the early 1960s, South Africa was divided among four distinct groups. About nine percent of the population were "whites," mostly descendants of the British and Afrikaners. About eighty percent were "blacks," mostly members of the Bantu people. Another nine percent were "coloreds," a term South Africans used to describe people of mixed race.

During the Anglo-Boer War, the British put Boer civilians in "concentration camps," such as the Nylstroom Camp pictured below (c. 1901).

Poster by Rachael Romero (Australian, 1976)

About two percent were Asians, mostly descendants of people from India brought as indentured servants. South Africa had strict laws about what the members of the various groups could and could not do.

Many black and colored South Africans agitated against the apartheid policies. Nelson Mandela was a leader of the African National Congress (ANC). The government arrested Mandela in 1962.

In the 1970s and 1980s black South Africans began staging demonstrations and worker strikes to protest apartheid. The nation's police and security forces struck back harshly, and many people were killed in the confrontations. The explanation that the government gave was that it was trying to stop the influence of Communism in the country. Using this excuse, government agents committed horrendous acts against black citizens, often without trial. It was not uncommon for blacks to disappear and simply not be heard from again. At the same time, militant black organizations engaged in violence against whites.

The international community attempted to pressure South Africa to end apartheid. The United Nations condemned apartheid in 1962 and suspended South Africa from membership in 1974. Many nations engaged in economic boycotts against the country. The Olympic Games banned South Africa from participating beginning in 1964. South Africa became in many ways an outlaw nation.

Finally in 1990, the white government under F. W. de Klerk committed itself to ending apartheid and began repealing apartheid laws. The government removed the ban on the ANC and released Mandela from prison. The 1994 election was the first in which blacks could vote. ANC candidates received about two-thirds of the votes cast, and Nelson Mandela was elected as the first black president of South Africa. Mandela and de Klerk shared the 1993 Nobel Peace Prize.

Truth and Reconciliation Commission

How could South Africa emerge from its bitter past and avoid continued strife and a possible race war? In 1995 President Mandela established the Truth and Reconciliation Commission (TRC). South African Anglican Archbishop Desmond Tutu was the chairman. The purpose of the commission was to gather and publish the truth about human rights violations that people had committed during apartheid. The commission could determine accountability for those violations, decide on appeals for amnesty from those who requested it, and evaluate what reparations to the families of victims might be appropriate. The commission did not have the authority to prosecute alleged perpetrators.

The commission received over 22,000 statements from victims and their families. It held public meetings in several places around the country at which people could tell their stories. Most of the statements and testimony involved victims of

Lesson 44 - Can People Once Enemies Get Along?

security forces, but the stories of those who suffered at the hands of the ANC and other groups were part of the public record also. The hearings revealed the participation by Winnie Mandela, the president's wife, in some cruelties perpetrated on her political enemies. Many of these stories are heart-wrenching and stomach-turning, but they could not be swept under the rug if reconciliation and moving forward were to occur. The commission received over seven thousand requests for amnesty, conducted over 2,500 amnesty hearings, and granted about 1,500 amnesties for the thousands of incidents that had occurred during the period of apartheid.

The South African Truth and Reconciliation Commission helped the country move forward from its painful past, however imperfectly. Top military leaders and the most powerful political leaders from the period of apartheid did not cooperate with the commission's work. Some of the members of the anti-apartheid groups also resisted, claiming that they had been engaged in a "just war" and thus had nothing for which to apologize or seek amnesty. However, the public participation in the process, the public nature of the hearings, and the commission's multi-volume final report helped the country to acknowledge what had taken place and move on. Several other countries around the world have set up similar commissions to deal with their painful history, such as the tenure of a dictator or the period of Communist rule.

Nelson Mandela (shown raising his hand) and F. W. De Klerk (seated to Mandela's right) attended a ceremony in Philadelphia, Pennsylvania, in 1993. They received the Liberty Medal, created in 1988 to honor "men and women of courage and conviction who strive to secure the blessings of liberty to people around the globe."

College Students in Johannesburg (2012)

A quarter-century after the TRC took up its work, South Africa has made strides economically and in its society, but problems remain. Economists estimate that the gap between the rich and the poor is wider in South Africa than anywhere else in the world. Access to public services such as electricity, water, education, and health care has improved but is far from universal, even in urban areas. The ANC-led government has had a major problem with corruption, and other political parties have arisen to challenge the ANC's hold on power.

People around the world rejoiced when apartheid ended, but people around the world did not have to live in a post-apartheid country; South Africans did. This beautiful land, the location of great wealth and opportunity, has also been the scene of great human suffering and injustice. South Africans changed their laws; they are still in the process of changing their country's hearts and way of life.

The Apostle Paul described how Christ broke down the barrier that divided Jews and Gentiles and created one new kind of person—Christian—and thus established peace.

But now in Christ Jesus you who formerly were far off have been brought near by the blood of Christ. For He Himself is our peace, who made both groups into one and broke down the barrier of the dividing wall . . . .
Ephesians 2:13-14

Assignments for Lesson 44

Gazetteer Read the entries for Lesotho and South Africa (pages 78 and 80).
Read Nelson Mandela's Inauguration Speech (pages 270-272) and answer the questions in the *Student Review Book*.

Project Continue working on your project for this unit.

Student Review Answer the questions for Lesson 44.

From "Prang's Floral Mottoes" (1870s)

45   Truth

This lesson is the first of a series that examines some of the building blocks of a person's worldview.

Truth should be pretty easy to determine, right? I mean, something is either true or it's not, right? Consider the following statements.

1. Abraham Lincoln was the sixteenth president of the United States.
2. 2 + 2 = 4
3. ab = ba
4. Embezzling funds is wrong.
5. The extermination of Jews in the Holocaust took place and was wrong.
6. Slavery is wrong.
7. Sexual relationships outside of marriage are wrong.
8. Wearing shorts in public is sinful.
9. Jesus died for our sins on the cross.
10. God created the universe in six literal 24-hour days.
11. Man has free will.
12. Muhammad was a prophet.
13. A thunderstorm occurs when the mountain god becomes angry.
14. Democracy is the best form of government.
15. Racial segregation in public education must be maintained.
16. The sun revolves around the earth.
17. If you sail west from Europe, you will not land on a continent until you reach Asia.
18. It is impossible for humans to fly by any mechanical means.
19. The earth's climate is changing in a harmful way and humans are the cause.
20. Intelligent life exists in the universe on other planets besides the earth.

This 1864 photo of Abraham Lincoln by Anthony Berger was the basis for the image of Lincoln that appeared on U.S. $5 bills from 1914 to 2007.

Let's discuss these statements.

Statement 1 is true: *Abraham Lincoln was the sixteenth president of the United States*. It is an historical fact. It's hard to argue with this unless you deny reality altogether. But if you deny reality altogether, who are you and what is your denial?

Statements 2 (*2 + 2 = 4*) and 3 (*ab = ba*) are mathematical statements or equations. They are true, given a shared understanding of the mathematical symbols.

Statement 4 (*Embezzling funds is wrong*) is a moral and legal judgment. Given a shared understanding of the meaning of embezzlement (taking funds, especially from an institution, that don't belong to you), this statement is true.

Statement 5 gives an historical fact but also makes a judgment about it: *The extermination of Jews in the Holocaust took place and was wrong*. Most people would agree with both parts of the statement, but there are some people in the world who say that both parts of the statement are wrong. Prejudice can prevent people from recognizing and admitting the truth.

Statements 6 and 7 are moral judgments. Almost everyone today agrees with Statement 6 (*Slavery is wrong*), but this has not always been the case. On the other hand, more people used to agree with Statement 7 (*Sexual relationships outside of marriage are wrong*) than agree with it now. Moral judgments change, but they do not change what is true.

Statement 8 (*Wearing shorts in public is sinful*) is a matter of personal conscience. We do not have a statement from God that declares this to be true, but some individuals and groups believe it to be true. If someone believes it to be true, it would be wrong for that person to violate his or her conscience and do it. Opinions about this matter (and other opinions related to how people should dress) have changed over time.

Statements 9 through 13 are religious doctrines, all of which some people believe to be true. Christians believe that some of these are true (e.g., *Jesus died for our sins on the cross*) and others are not (*Muhammad was a prophet* and *A thunderstorm occurs when the mountain god becomes angry*). People who profess faith in God and belief in the Bible disagree about the truth of Statements 10 (*God created the universe in six literal 24-hour days*) and 11 (*Man has free will*).

These shoes at the Auschwitz Concentration Camp in Poland were taken from victims of the Holocaust.

Lesson 45 - Truth

Statements 14 and 15 are statements of political judgments. Many (but not all) people today agree with Statement 14 (*Democracy is the best form of government*), but our country's Founding Fathers generally did not. On the whole they feared pure democracy as little more than mob rule and preferred a republic, in which elected representatives make the laws. Many people in America in the 1950s agreed with Statement 15 (*Racial segregation in public education must be maintained*). Today practically no one does. Political judgments about what is true change.

Statements 16 through 20 give scientific conclusions, speculations, theories, and guesses. At one time people widely accepted Statements 16 (*The sun revolves around the earth*), 17 (*If you sail west from Europe, you will not land on a continent until you reach Asia*), and 18 (*It is impossible for humans to fly by any mechanical means*) as true; but now we know that they are not true. Statement 19 (*The earth's climate is changing in a harmful way and humans are the cause*) is the subject of fierce debate today. Statement 20 (*Intelligent life exists in the universe on other planets besides the earth*) is something nobody on the earth knows for sure. Some people are convinced that it is true, while others are convinced that it is not; but we do not have the evidence to know the truth one way or another. The conclusions of science are just that: conclusions (or guesses) based on the evidence available. As we have seen from history, those conclusions and guesses might be wrong. What people generally accept as scientific truth changes.

So, truth exists and falsehood exists, but there are many opinions and many changing opinions. What we "know" to be true changes. Some things we don't know whether they are true or not and might never know.

The Importance of Truth

We see from this exercise that determining what is true and what is false, what is fact and what is opinion, what is known and what is unknown, is not as easy as we might have thought. The discussion about facts, judgments, opinions, commonly held ideas, and all the rest can make your head swim; but hang in there. Cutting through the jungle of ideas to determine truth is worth the effort.

Difficult as it might be, determining what is true is a necessary endeavor. Your goal should be to have a worldview that is based on truth. We expect scientists, historians, and detectives to be relentless pursuers of truth; Christians should be nothing less. Ideas have consequences; therefore, basing your life on truth will give you a firm foundation for making decisions, dealing with what happens to you, and determining reality. You will have a much more successful life if you base it on truths such as: you should use your talents for good; you should treat others the way you want to be treated; and most people are not out to get you.

On the other hand, basing your life on what is not true will be frustrating, dangerous, and perhaps even fatal. If, for example, you base your life on the idea that the world owes you a living, or that Communism is the best way to world peace, or that gravity does not apply to you, you will face serious consequences. What is not true has consequences also. If you state something that is not true when you are a witness in a trial, you will likely face a penalty.

A Definition of Truth

A common understanding among philosophers is that truth is a proposition (or belief or thought) that conforms to reality. Alfred Jepsen says that truth is that which is constant and unchangeable; in other words, you can rely on it. We might say that truth is a system of ideas that provides a reliable way for people to interpret data and assertions in a way that agrees with reality.

In seeking to determine truth, you have to answer several questions.

Is It Truth, or Is It Something Else?

When you are considering a proposition, you have to determine whether it conforms to reality. Is it an unchanging truth? Is it a personal, group, or societal judgment, which can change? Is it a religious doctrine, which can be unchanging truth, but is not necessarily so? Is it a matter of personal conscience, which can vary with different people and can change? Is it a current finding of science, which can change? Is it merely a matter of personal preference or taste? If it is, it is not really subject to an analysis of its truthfulness.

People can have differing opinions and preferences, but they cannot have differing truths. A proposition cannot be true and not true at the same time. It is important not to elevate an opinion to the level of truth. The statement, "You have your truth and I have mine," is an attempt to redefine truth.

How Did You Learn What You Believe to Be True?

In Lesson 40 we discussed identifying the sources for your worldview. In the same way, you need to determine your authority for believing a proposition to be true or false. Did you learn it from personal observation? From logical deduction? Is your source a religious authority, such as the Bible, a minister, or a creed? Is the source for what you believe to be true your parents, your community, or a tradition that your family (or group or church) has long held? Are you considering data, or are you considering someone's interpretation of data? These factors can make a huge difference in what you perceive to be true. Only the Bible is always true.

The main thing to determine about a source is whether that source is reliable. Even if it is reliable, if it is a human source you still have to determine if the specific proposition in question is reliable. Good, honest people make mistakes, can be guilty of prejudice, and can lack knowledge. It is not enough to think, "She would never say something is true when it isn't," or "He is right about so many other things; I'm sure he is right about this." Your sources have to be solid about each statement of supposed fact.

What Difference Does This Proposition Make?

When you determine that a proposition is true, you then must determine the conclusions that flow from that proposition. What would be the consequences if you built your life on this proposition? Does this proposition lead you to be loving toward others, as the Bible defines love? Does it appear to conflict with other truths that conform to reality? Does it conform with your personal convictions? Does it challenge your convictions, and are you willing to change your convictions? How should you resolve a challenge to your convictions?

Do you need more information or more time to determine if the proposition is true? Will accepting this proposition as true help you become more like Christ, as He is revealed in the Bible? Has our general thinking and understanding about this proposition changed in the past? Is it likely to change again?

Thus we see that determining truth, error, and opinion is not so easy after all; but the quest is essential. Thus we need a firm foundation to pursue this quest effectively.

Truths About Truth

All truth is God's truth. God created the world, so whatever is true about the world comes from God. As you sort through statements, judgments, opinions, and evidence, when you determine that something is absolutely rock-solid truth, you can know that it comes from God.

Truth is true whether you know it or not. No one knows everything, and you only know a fraction of the truth. Your understanding is all you can go on, but you need to keep learning because you don't know everything and never will.

Painting of Jesus Before Pilate at the Franciscan Church and Monastery in Dubrovnik, Croatia

Truth is true whether you agree with it or not. Prejudice, tradition, fear, and other attitudes blind you to the truth. Such attitudes will not help you be who you need to be. This is why you need to be as completely honest with yourself as possible. A huge difference lies between something being true and your wanting it to be true.

Truth is truth and error is error, but some wrongs have less severe consequences than others. In other words, some truth is more central and essential than other truth. You can be mistaken about wearing shorts in public and still go to heaven; but being mistaken about the identity of Christ has eternal consequences.

It is not true that the only truth we can know is what someone can prove by the scientific method. We can know some truth by the preponderance of evidence. For instance, we can know that Jesus Christ died on the cross for our sins, even though we cannot prove it by scientific evidence.

The way we hold the truth can be an issue. To believe the truth but to express it in a judgmental, harsh, and condescending way is wrong. There is a wrong way to be right.

Jesus Is the Basis of Truth

In the Gospel of John, Jesus said, "I am the way, and the truth, and the life; no one comes to the Father but through Me" (John 14:6). In other words, Jesus claimed to be truth—the reliable, accurate basis on which all people can base their lives and understand the world in which we live. Jesus' life and teaching conform to reality. His words, His life, and His heart are true. The beginning point of truth is Jesus.

When Jesus stood before Pilate, He said, "For this I have been born, and for this I have come into the world, to testify to the truth. Everyone who is of the truth hears My voice" (John 18:37). Pilate responded by asking, "What is truth?"

Bible scholars have long debated what Pilate meant by this question. At the very least, Pilate showed that he had an imperfect grasp of truth, the very thing that Jesus claimed to be. How sad that an important official in the mightiest empire of the day did not have a good grasp of truth.

Deciding to trust the doctrines of Scripture is not a blind leap of faith—a wish that Scripture is true. Trusting the Bible as true means accepting those teachings as conforming with reality. This is a firm place to stand as you seek to grow in your grasp of truth.

The Pursuit of Truth Is Worth It

God is a God of truth who cannot lie (Psalm 31:5, Titus 1:2). Truth exists. Your job is to identify what is eternally, steadfastly true—what conforms to reality—and to live your life on that basis. This involves taking a hard look at what you believe, what the world says is true, and what the authority or authorities that you hear say is true. This requires being honest, humble, and courageous enough to live on the basis of truth, regardless of what others say or do and regardless of what you might need to change.

Don't settle for your current understanding, and don't float in such a way that you never come to any conclusions. Just because the search for truth is difficult does not mean that the search isn't worth it. Because truth is so important, we should expect the search to be challenging. The process of learning the truth and living by the truth is worth the effort; in fact, no lesser effort is worthy of you.

Jesus had much to say about truth because determining truth can be a challenge and because falsehood can be deceptive and attractive.

So Jesus was saying to those Jews who had believed Him, "If you continue in My word, then you are truly disciples of Mine; and you will know the truth, and the truth will make you free."
John 8:31-32

Assignments for Lesson 45

Worldview Recite or write the memory verse for this unit.

Project Finish your project for this unit.

Student Review Answer the questions for Lesson 45.
Take the quiz for Unit 9.

Pena Palace, Lisbon, Portugal

10 Southern Europe

The Basque people are an ethnic group associated with a specific geographic location: northwestern Spain and southwestern France. Intense debate has taken place over the naming of the Republic of North Macedonia because of the neighboring province of Macedonia in Greece. We tell the heartbreaking story of the Greece-Turkey population exchange of 1923. The microstates of Europe are geographic anomalies, throwbacks to the small city-states of medieval times. The worldview lesson examines Paul's sermon in the Areopagus (often called the sermon on Mars Hill) in Acts 17 as a lesson in God's perspective on human geography.

Lesson 46 - The Basques: One People in Two Countries
Lesson 47 - What's In a Name? The Saga of North Macedonia
Lesson 48 - You Go Here and You Go There
Lesson 49 - Microstates: Vestiges of Earlier Times
Lesson 50 - Paul's Sermon on Human Geography

Memory Verse   Memorize Acts 17:26-27 by the end of the unit.

Books Used

The Bible
Exploring World Geography Gazetteer
The Day the World Stopped Turning

Project (Choose One)

1) Write a 250-300 word essay on one of the following topics:
   - Many people have been forced to move because of war, economic dislocation, or for some other reason when they have not done anything wrong. If this happened to your family, what would you want to take with you, what would you leave, and what would be the meaning of those items?
   - How would you build relational bridges to another ethnic or cultural group? Why would you want to do so?
2) Imagine that you have been elected president of the tiny (imaginary) country of Mylandia. Write and deliver your inaugural address, stating your priorities.
3) Deliver Paul's sermon from Acts 17 to your family. Prepare different family members to respond in the ways people did when Paul delivered the sermon.

Literature

The Day the World Stopped Turning takes the reader to the beautiful Camargue region in the south of France. There we meet Kezia and Lorenzo, who share a unique and beautiful friendship and a rich parallel history. This historical novel flashes back to the World War II experiences of their two families.

Kezia's family is Roma, ever-outsiders as they travel and work from village to village. Lorenzo's family farms a beautiful piece of land. Because of Lorzeno's special needs, few take the time to know and understand him. When the storm of war reaches their lives, they find refuge in their friendship, facing loss and suffering together. This beautiful and thoughtful story explores relationships, the experience of being an outsider, gain and loss, and how events shape people and people shape events.

Michael Morpurgo was born in England in 1943. He started writing stories for children while he was a teacher who enjoyed reading aloud to his classes. He has written over 150 books for children. He and his wife founded a charity called Farms for City Children which has operated for decades and given tens of thousands of children the opportunity to experience a working farm. In 2018 he was knighted by Queen Elizabeth II for his services to children and charity.

Plan to finish *The Day the World Stopped Turning* by the end of Unit 11.

Bay of Biscay at Donostia-San Sebastian, Spain

46 The Basques: One People in Two Countries

Question: Who lives in Spain?
Answer: The Spanish, of course!
Well, it's not that easy.

The country of Spain is actually a collection of ethnic regions, each with its own distinct culture. They have their own dialects, and, in some cases, their own language. In fact, less than half of the population of Spain, about 45%, considers themselves Spaniards. About 28% are Catalonian, who live mainly in the region of Catalonia. Other groups include the Galicians, Castilians, Andalusians, Aragonese, and Lusitanians. About 2% of the people who live in Spain are Roma people—whom many call gypsies. They have endured prejudice and discrimination from others, but some have blended into their surrounding culture and even been elected to parliament.

History and Geography

This cultural diversity is the result of two main reasons: history and geography. Historically, many people groups have invaded the Iberian Peninsula, on which Spain and Portugal lie. Phoenicians and Carthaginians established settlements and trading posts on the peninsula. Invaders have included Iberians (a mixed group from Europe that didn't share a single language or culture), Celts, Romans, Vandals, Huns, and Visigoths. A large number of Jews settled in Iberia, and Muslim invaders gained control of almost all the peninsula in 711.

Geographically, a large central plateau averaging about a half-mile above sea level dominates the peninsula. One mountain range cuts across it, and other mountain ranges border it. The south of Spain has a desert area, complete with palm trees and tropical vegetation. Geographers consider this region a continuation of the Sahara in northern Africa. To the north, the Pyrenees mountain range creates a natural border between Spain and France. These mountain ranges and distinct geographic areas create places where people groups have settled and maintained their distinct identities. City-states within the regions tended to dominate them.

In 1469 Ferdinand of Aragon married Isabella of Castile. Their marriage united these two powerful regions, and their rule eventually brought about the unification of all of Spain under one throne—at least on the surface. In 1492, Ferdinand and Isabella ordered all Jews to leave the country unless they were willing to convert to Christianity. That same year, the Spanish defeated a Muslim force, and this for the most part removed the Islamic influence in Iberia. These developments led to a period of peace in Iberia and provided Ferdinand and Isabella the freedom to sponsor the westward journey of the Italian navigator Christopher Columbus, who hoped to find an easier route to Asia to take advantage of trading opportunities there.

Basque Country

However, unification has not settled easily on all of Spain. This is especially true for one group, the Basques. Basque Country consists of four historic provinces in northwest Spain (darker orange on the map at right) and three in southwest France (lighter orange). The region lies on the Bay of Biscay and includes the western end of the Pyrenees.

We are not sure exactly where the Basques came from or when, but they appear to have come from elsewhere in Europe perhaps over 3,000 years ago. Once they settled in what became their mountainous homeland, they resisted invasion by others. Their isolated homeland helped them maintain a distinct culture, including the Basque language. This language, which they call Euskara, is the only non-Indo-European language in Europe and might be the oldest language still in use on the continent.

Geography provided the Basques isolation; it also influenced how the Basques supported themselves. The land did not enable widespread farming efforts, so many became miners and herders of sheep and cattle. Positioned on a bay in the Atlantic Ocean, Basques also became shipbuilders, sailors, and fishermen.

Basque Shepherd

Lesson 46 - The Basques: One People in Two Countries

Leaving Home

The Basque role in seafaring became legendary. Basque whalers and codfishers roamed as far as the east coast of Canada, perhaps before Columbus headed west but for sure shortly thereafter. Archaeologists have found artifacts of Basque settlements in eastern Canada; however, the Basques were not interested in living there permanently. They came in the spring each year and used their land bases to dry the cod, process the whales, and perhaps do some trading, before heading home in the fall.

Some members of Columbus' first crew were Basque sailors. A few years later, when the last of the five ships in Ferdinand Magellan's original fleet finally arrived back in Spain in 1522, completing the first known trip around the world, the commander of the ship was Juan Sebastian Elkano—a Basque. When the Frenchman Jacques Cartier began his exploration of Canada in 1534, he encountered Basque ships in the waters around Newfoundland as well as in the St. Lawrence River. Basque whaling activity in Canadian waters continued until 1626.

Other Basques went even further afield in later years. Having experience with mining, some migrated to California in the 1850s to try to capitalize on the gold rush there. Only a few people struck it rich in gold mining and they weren't Basques, so Basques became sheep and cattle herders in California. Then some Basques returned to fishing and whaling, while other Basques spread into other areas of the American West. An especially large group settled around Boise, Idaho. The Basque district in that city is called the Basque Block, and an annual festival celebrates the Basque heritage of Boise.

Twentieth Century Experiences

In the 1930s, Spanish dictator Francisco Franco wanted to force conformity to his central authority. He attempted to eliminate distinctive cultural expressions, including the use of local languages such as the Basque Euskara.

Geography Swapping

France and Spain were enemies in the Thirty Years War that ravaged Europe in the 1600s. Even after the war was over, France and Spain kept fighting each other. Finally in 1659, the two countries decided to negotiate a peace treaty on the Ile de Faisans (French), or the Isla de los Faisanes (Spanish), or Konpantzia (Basque)—in English the place name is Pheasant Island.

The 660-foot-by-150-foot island (shown above) sits in the middle of the Bidasoa River that serves as part of the border between France and Spain. In that location the negotiators concluded the Treaty of the Pyrenees. One provision of the treaty called for the two countries to share control of the island. France would rule it for six months and then Spain would rule it for six months.

The two countries have swapped control of the island in this fashion ever since. An historical marker is on the island, but no one lives there. No visitors are even allowed on the island except during special heritage days. The only responsibilities involved with overseeing the island are keeping the place looking nice and chasing off illegal campers.

The Basques found this irritating enough, but then things got worse. During the Spanish Civil War of 1936-1939, Franco wanted to provide an opportunity for his ally Adolf Hitler to practice using his new air attack capabilities. Since Franco found the Basque separatist tendencies troublesome, he ordered an air attack on the Basque city of Guernica on April 26, 1937. The bombing was merciless on civilians gathered in the marketplace for market day. When the attack ended, thousands of people lay dead and injured and the town of Guernica was virtually destroyed.

When Franco died and Spain became a constitutional monarchy in the mid-1970s, the central government began permitting greater autonomy in the separate regions. This has included the use and acceptance of Euskara. However, since the central Spanish government banned the language for so many years, not all Basques know how to speak it. Only about 27% of Basques use Euskara.

In the late twentieth century, a Basque separatist movement arose that employed violence and terrorism to try to force the Spanish government to allow them to withdraw from Spain. Some eight hundred people lost their lives in terrorist attacks before the group announced its abandonment of violence. Three groups with distinct political outlooks exist within Spanish Basque country. One group is content to live with a degree of autonomy under Spanish rule. A second would prefer to have more independence but are not willing to use pressure and violence to achieve it. A third group, relatively small, would be willing to use violence again to achieve independence.

Sign in Euskara, Spanish, and English at a Beach in Basque Country

Basque Dance at a Folk Festival in Bilbao, Spain (2010)

Lesson 46 - The Basques: One People in Two Countries

Language plays a major role in a people's identity. Castilian Spanish is the official language of Spain, but some groups are not happy with that dialect having this honor.

French Basques have generally become more integrated into French culture than is the case on the Spanish side of the Pyrenees. By far most of the Euskara-speakers live in Spain.

Thus the Basques, about 5.5% of the population of Spain, have a complex history. They have been largely isolated, but they have played a role in developments around the world. Their current life is less dramatic than it has been at times in the past, but they are still proud of their distinctive heritage and identity even as they have become more integrated into the global culture. Geography has helped to shape their identity. It has helped them develop characteristics that they have kept alive both in Basque Country in Spain and as they have emigrated to other places around the world.

The Basque region has been calm in recent years, but not so the Catalan region of Spain. In 2017 a movement arose in Catalonia demanding separation from Spain. Catalonia is a wealthy region with significant manufacturing and trade. Many Catalans believe that the central Spanish government draws away financial resources that in their view should flow to the Catalonians.

In Romans 15-16 Paul mentions several geographic places where he had traveled: Macedonia, Achaia, Illyricum, Jerusalem, and Asia (the western part of modern Asia Minor). He hopes to go to Rome, where the recipients of this letter lived, and he mentions one other place he wants to visit:

Therefore, when I have finished this, and have put my seal on this fruit of theirs [a contribution for Christians in Jerusalem that he had been collecting], I will go on by way of you to Spain.
Romans 15:28

Assignments for Lesson 46

Geography — Study the map of Southern Europe and read the entries for Gibraltar, Portugal, and Spain (pages 82, 87, 94, and 98).

Worldview — Copy this statement in your notebook and write your answer: List some Greek philosophers you have heard of.

Project — Choose your project for this unit and start working on it. Plan to finish it by the end of this unit.

Literature — Begin reading *The Day the World Stopped Turning*. Plan to finish by the end of Unit 11.

Student Review — Answer the questions for Lesson 46.

Ohrid, North Macedonia

47 What's In a Name? The Saga of North Macedonia

The story begins once upon a time, but for Matthew Nimetz it was no fairy tale. At times it was more like a nightmare. Nimetz spent a fourth of his life trying to resolve an international diplomatic conflict that aroused intense emotions on both sides of the issue. We'll get back to Nimetz shortly to explain his role in the dispute, which has roots that go back for centuries.

Once upon a time, in the land we call the Balkans, lay the kingdom of Macedon. Its first great leader was Philip, who extended his rule into Greece. His son, Alexander (called the Great), became an even more powerful leader and created a kingdom that stretched from Egypt, through the Middle East, and into the western part of what we know as India.

In his youth, Alexander studied under the Greek philosopher Aristotle. Alexander loved Greek culture and the Greek language. He and his successors planted their language and culture throughout the lands where they ruled.

Rome took control of Macedon in 146 BC and called it Macedonia. Thereafter the region was subject to many rulers and to being divided among many other countries. Slavic people later settled the northern part of Macedonia, developed a distinct culture, and became associated with the many nations of the Balkans to the north, although they still claimed the heritage of Alexander and ancient Macedonia.

In 1912 an international agreement following the Balkan Wars divided Macedonia among Bulgaria, Serbia, and Greece. After World War I, the Paris Peace Conference created the Kingdom of Serbs, Croats, and Slovenes, which included the portion of Macedonia held by the Serbs. The rest of ancient Macedonia continued as the province of Macedonia in Greece. In 1929 King Alexander I

Lesson 47 - What's In a Name? The Saga of North Macedonia

(notice the name!) of the Kingdom of Serbs, Croats, and Slovenes declared that his country was now the kingdom of Yugoslavia, which means Land of the Southern Slavs.

The Axis Powers invaded Yugoslavia in April of 1941 during World War II. Communists led by Josip Broz Tito and supported by Communist Russians threw off Nazi rule in 1944 and established a Communist Yugoslavia. The new country included six "people's republics": Croatia, Montenegro, Serbia, Slovenia, Bosnia-Herzegovina, and what Tito called the People's Republic of Macedonia. This Macedonia included the northern part of ancient Macedonia as well as adjacent lands. At the time, the U.S. State Department called Tito's naming it Macedonia "unjustified demagoguery representing no ethnic or political reality" and suggested that the name might become an excuse for aggression against the province of Macedonia in Greece, which would unite all of ancient Macedonia under Communism. Tito's brutal rule held the volatile ethnic and religious mix of Yugoslavia together, but mistrust and hatred among the groups in the country simmered under the surface.

As Communism disintegrated in Europe, Yugoslavia began to come apart in 1991. Slovenia and Croatia seceded from Yugoslavia in June of 1991. The Republic of Macedonia followed in December. Long-suppressed differences flared, and the Balkan region was the scene of significant ethnic fighting during the 1990s. Other areas became independent also, and by 2003 Yugoslavia ceased to exist.

The newly-independent Macedonia applied for membership in the United Nations, but Greece objected and blocked their joining because of its name. To most Greeks, the name Macedonia was Greek and belonged to Greece. Greece agreed to the diplomatic compromise of referring to the new nation by the temporary name of Former Yugoslav Republic of Macedonia (FYROM), and that is how it became part of the UN in 1993.

President and Mrs. Tito of Yugoslavia with President and Mrs. Nixon at the White House (1971)

The Greek Province of Macedonia and the Former Yugoslav Republic of Macedonia

To Greeks, the name Republic of Macedonia was illegitimate, an attempt to take something that wasn't really theirs. Greeks saw it as an attempt by Slavs to steal a significant part of Greek culture and identity. On the other hand, the people of the FYROM believed that they had a legitimate heritage from ancient Macedonia that enabled them to use the name. FYROM wanted to join NATO to improve their security in an unstable part of the world and to join the European Union (EU) to help their weak economy. Greece, however, already a member of both organizations, vetoed FYROM's membership in both. A leader of FYROM during this volatile period didn't help matters when he named the airport in Skopje, the capital city, the Alexander the Great Airport and had statues of Alexander erected throughout the city. Some Greeks continued to believe that FYROM wanted to take over their province of Macedonia.

Despite this lingering name issue, which caused occasional diplomatic sparks to fly, Greece and FYROM developed business and tourism ties. Interestingly, people crossing the border in either direction would see a sign welcoming them to Macedonia!

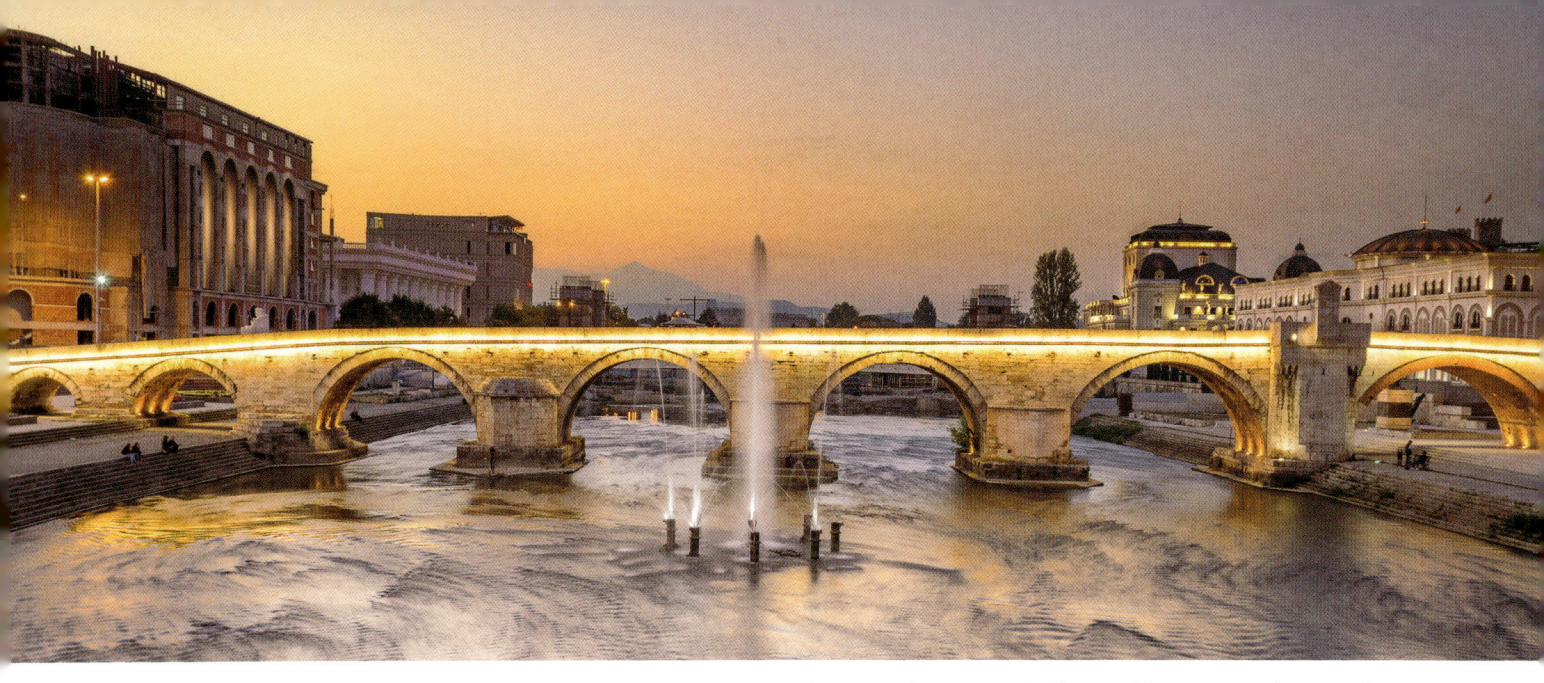

Bridge over the River Vardar in Skopje, North Macedonia

A Long-Term Diplomatic Mission

At this point, we reintroduce Matthew Nimetz. Born in 1939, Nimetz is an attorney, government official, and diplomat. He has held positions in several presidential administrations and has served on several nonprofit boards. In March 1994 President Bill Clinton appointed Nimetz as his special envoy to resolve the name dispute. Nimetz served in this role until September of the following year. As a result of his work, Greece and FYROM signed an interim agreement that resolved several issues, but not the name issue.

Starting in 1999, Nimetz was the special representative of the UN Secretary-General for resolving the name dispute. He continued to hold other positions and did not work full-time on this single project, but he devoted considerable effort to it. For his work he received a salary of $1 per year.

Resolution

In 2018, after twenty-four years of Nimetz's work and changes in the governments of both countries, Greek Prime Minister Alexis Tsipras and FYROM Prime Minister Zoran Zaev signed an agreement to resolve the issue. FYROM agreed to change its name to the Republic of North Macedonia (North Macedonia or NM for short) and declared that it would no longer make any ethnic or historic claims to Alexander the Great or ancient Macedonia. However, the country will call its language Macedonian and its citizens Macedonians. Greece agreed to accept that change and to drop its opposition to NM becoming a member of NATO and the EU. The airport had already been renamed Skopje International.

Opposition to the proposed agreement was strong in both countries and led to numerous protest rallies. Many Macedonians believed that they were abandoning their heritage. Many Greeks insisted, as one protest banner proclaimed, "There is one Macedonia and it is Greek!" However, a referendum in Macedonia approved the agreement although about sixty percent of voters did not participate. The Greek parliament approved the agreement in a close vote. North Macedonia began the process of changing signs, currency, passports, and other official documents to reflect its new name.

Greece then became the first NATO member to give its approval for NM to become the 30th member of the mutual defense group (every current member has to give its agreement for a new member to join). North Macedonia became a member of NATO in March of 2020 and will also likely become a member of the EU.

Standing in the background and looking on with disapproval is Russia. Russia believes that the move extends the influence of the United States and Western Europe further into the Balkans, an area where Russia believes it has strategic interests. The nearby countries of Georgia and Ukraine have also expressed interest in becoming members of NATO and the EU, which would push Western interests and military presence even closer to Russia.

But we can hope that, after two decades of work by Matthew Nimetz and with some political risk-taking by the leaders of the two countries, the Republic of North Macedonia and Greece, unlike much of the rest of the Balkans, will live happily ever after.

The Apostle Paul spent considerable time in Macedonia. His ministry in Philippi played an important role in the spread of Christianity.

North Macedonia formally became part of NATO on March 27, 2020, in a ceremony at the U.S. State Department. Because of the COVID-19 pandemic, the ceremony was part virtual (see iPad at right) and part socially distanced.

So putting out to sea from Troas, we ran a straight course to Samothrace, and on the day following to Neapolis; and from there to Philippi, which is a leading city of the district of Macedonia, a Roman colony; and we were staying in this city for some days.
Acts 16:11-12

Assignments for Lesson 47

Gazetteer Read the entries for Greece, Italy, and North Macedonia (pages 88, 89, and 93).

Worldview Copy this question in your notebook and write your answer: What do you think Paul meant when he said that in God, "we live and move and exist (or "have our being," Acts 17:28)?

Project Continue working on your project.

Literature Continue reading *The Day the World Stopped Turning*.

Student Review Answer the questions for Lesson 47.

Greek and Armenian Refugees Near Athens (1923)

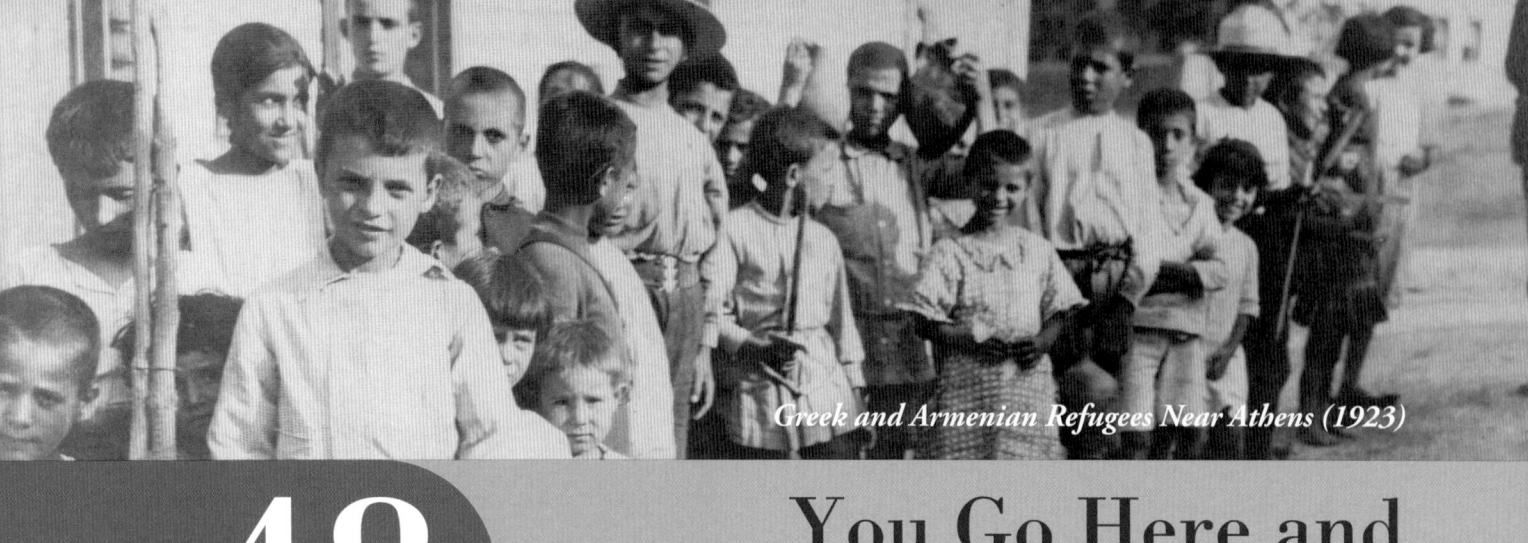

48 You Go Here and You Go There

Hundreds of thousands of people packed up what they could carry, said goodbye to their neighbors, left their homes, made their way to the port, and sailed to their new home villages, where they did not know the language, the people, or the customs.

Meanwhile, hundreds of thousands of others packed up what they could carry, said goodbye to their neighbors, left their homes, made their way to another port, and sailed to their new home villages—sometimes moving into the very houses left by the other group—where they also did not know the language, the people, or the customs.

If this seems strange, in one sense it is but in another sense it is not. What is even more strange is that these people had no choice in the matter but moved as a result of their respective governments' decrees. Furthermore, the nations that these governments led had been enemies for centuries. These hundreds of thousands of forced immigrants were leaving where they did not want to leave and going where they did not want to go. What might be even more surprising is that this population exchange, which involved about two million people, took place in the twentieth century.

The reason for it was primarily religion.

Moving People as Government Policy

Throughout history, some government leaders have decided that one way to implement their goals is to treat people as pawns or cattle by forcing them to leave where they are and move to another place.

When Assyria conquered the Northern Kingdom of Israel in 722 BC, the Assyrians forced the people of Israel into exile away from their homeland and scattered them throughout the Assyrian Empire. In return, Assyria repopulated the land of Israel with their own pagan people as a way to control that area and prevent further rebellion.

The Babylonian conquerors of Judah took the same first step in 606-586 BC when they forced thousands of the people of Judah into captivity in Babylon. The Persians, who in turn defeated the Babylonians, allowed the Jews to return to their homeland, and many did.

History tells of many other forced expatriations such as these. For instance, in 1492 Catholic Spain expelled Muslims and Jews from its borders. In the late 1800s, Russia forced many Jews to leave their homes and go somewhere else—anywhere, just out of Russia.

Lesson 48 - You Go Here and You Go There

A sad chapter in American history was the policy of forcing native nations to move onto reservations. The Trail of Tears, in which the government removed the Cherokee from their traditional homelands in North Carolina, Georgia, and Tennessee and required them to walk to reservations in Oklahoma, is only the best known example of many. Americans who wanted Cherokee lands (especially in the area where gold had been discovered in Georgia) in some cases moved into the houses that Cherokee had left. The reservations where natives had to move exist today in many places throughout the United States.

In World War II, the Soviet Union took control of Poland. After the war, to exert even greater control over Polish territory, the Soviet Union forced thousands of Poles to leave eastern Poland and settle in what had been eastern Germany, and then repopulated eastern Poland with Russians.

Thus, in the context of history, the harsh people movement described at the beginning of this lesson was merely another incident in the long line of government policies that have required one or more groups to leave one place, often their ancestral home, and move to another place.

The Flight of the Prisoners by James Tissot (French, c. 1900) depicts Jews leaving Jerusalem after its destruction by the Babylonians.

The Conflict Between Greece and Turkey

The hatred between Greece and Turkey, two geographic neighbors separated by the Aegean Sea, goes back literally for thousands of years. Invaders from Greece apparently attacked the city-state of Troy on the western coast of what is now Turkey at some time in antiquity. Homer's *Iliad* is a fictionalized account of this battle. Later, when Persian armies attacked Greek city-states, they did so through Anatolia (modern Turkey). In some of his first conquests, Alexander the Great, seeking to further the influence of Greek culture, moved through Anatolia and took control of it.

Rome dominated both the Greek peninsula and Asia Minor (another geographic term for Anatolia). In the Christian era, Constantine established a new Rome, called Constantinople, on the European side of Turkey, to the west of the Bosphorus strait.

Muslims on the Greek Island of Crete (c. 1900)

The Orthodox tradition of Christianity came to dominate Christian practice in Greece, in the area north of Greece on the Balkan peninsula, and in the area around Constantinople. This city (earlier known as Byzantium) was the capital of the Byzantine or Eastern Roman Empire.

After the rise of Islam, Muslim armies conquered Constantinople in 1453 and established the Ottoman Empire. At the height of their power, the Ottoman Empire extended into Greece and further north in the Balkans. For centuries the Ottomans occupied Balkan lands and oppressed Balkan peoples, including Greeks. Some people in these areas converted to Islam, but many Europeans resented the Ottoman presence in Europe. The conflict between Muslim and Orthodox peoples continued and grew.

Over time, the Ottoman Empire gradually lost power under less able leaders, as groups within it wanted freedom and as other forces attacked it. For over a century, people referred to the Ottoman Empire as "the sick man of Europe." Deciding to fight the sick man, Greece in the early 1820s fought a successful War of Independence from the Ottoman Empire. The conflict was especially brutal and bloody for both sides. One observer called it "a mutual war of extermination." The war began with the Greek massacre of twelve thousand Turkish civilians at Tripolista. In retaliation, Ottoman armies killed twenty-five thousand Greeks on the island of Chios (the numbers of deaths in these massacres are matters of debate). Things went even further downhill from there, and the memories from this war did not fade. Another Greece-Ottoman war erupted in 1897.

Thus the people exchange described at the beginning of this lesson did not come out of nowhere, but instead was one more painful interaction between these two long-standing enemies. Why did Orthodox Greeks live in predominantly Muslim Anatolia? The Orthodox Byzantine background was one reason. Another was that Greece was mired in poverty while economic opportunities were

Lesson 48 - You Go Here and You Go There

available in Anatolia. Why were Turkish Muslims in Orthodox-majority Greece? One reason was that the Ottoman Empire had practiced an expansionist policy there and Muslims had moved into Greece to live there under the Ottoman government.

In 1912 a conflict known as the Balkan War began when Greece joined with Serbia and Bulgaria to drive the Ottomans out of the Balkans. Hundreds of thousands of Muslims fled to Istanbul and displaced many Greek Orthodox from their homes there.

World War I and After

In World War I the Ottoman Empire aligned with the Central Powers of Germany and Austria-Hungary. The Ottoman government deported from Anatolia thousands of Greek Orthodox, whom they accused of being collaborators with the Allied Powers of Great Britain and France. The Central Powers lost the war and the Ottoman Empire collapsed. Soon after the end of that conflict, Greece invaded and seized western Anatolia. Meanwhile, Kemal Ataturk was working to build a new Muslim state to be called Turkey. Some Turkish Muslims took revenge on local Orthodox citizens.

Simmering conflicts between the Allied Powers and the Central powers continued for several years. For one thing, the United States rejected the Treaty of Versailles, so the U.S. separately had to reach peace agreements with the countries against whom it fought; but other unresolved issues and conflicts remained. The newly-formed country of Turkey, which controlled part of the former Ottoman Empire, continued its unsettled relations with Greece. Greece invaded western Anatolia, and Turkey struck back in yet another war, this one extending from 1921 until 1922.

Finally, on July 24, 1923, Great Britain, France, Italy, Japan, Greece, Romania, and the Kingdom of Serbs, Croats, and Slovenes reached an agreement with Turkey in the Treaty of Lausanne (Switzerland). The Allies recognized the existence and boundaries

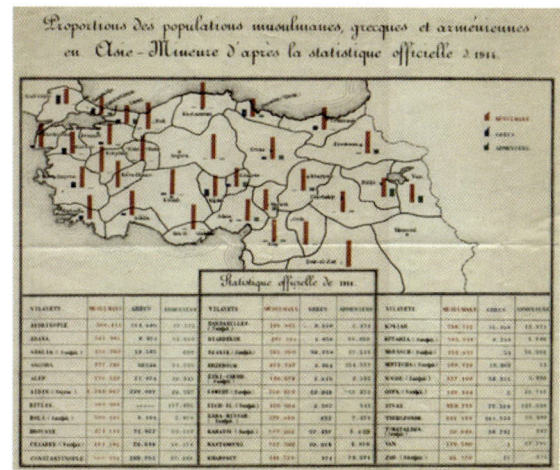

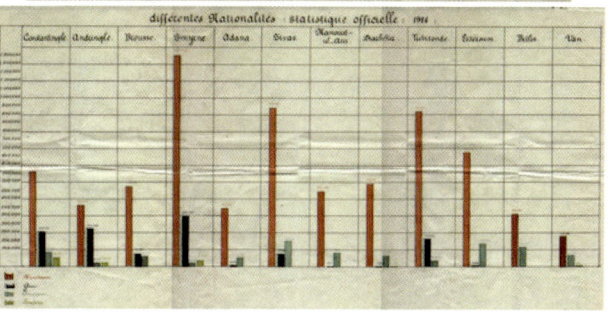

This 1914 report from the Ottoman Empire shows the relative populations of Turks (red bars), Greeks (blue bars), and Armenians (green bars) in different regions.

of Turkey. Turkey gave up any claims to Arab lands. The Allies dropped their insistence on independence for Kurds within Turkey and on Turkey giving up lands in Armenia.

The treaty also established the commitment of Greece and Turkey to initiate an exchange of peoples. Turkey wanted to be an all-Muslim state and expel Greek Orthodox persons, and Greece wanted to be rid of the Turkish Muslims within its borders. It might be hard to imagine world powers agreeing to this kind of mutual forced expatriation, but they did and it would not be the last time.

Greece had a recent precedent for this action. In 1919, Greece and Bulgaria signed an agreement calling for the voluntary exchange of minorities in the two countries. About 37,000 Greeks lived in Bulgaria, and about 150,000 Slavs of Bulgarian ethnicity lived in Greece. This exchange took place, but how voluntary it was is another matter. The precedent was set.

The American College for Girls in Constantinople attracted Greek, Armenian, and Turkish students. This 1920 photo shows students playing basketball.

The Exchange

The exchange between Greece and Turkey was the first time that two countries agreed to a compulsory transfer of populations to deal with minorities. The standard was specifically based on religion. Turkish citizens who were Greek Orthodox had to leave Turkish territory, and Greek citizens who were Muslims had to leave Greece. The exchange involved about a half-million Greek Muslims who went to Turkey, and about 1.3 million Orthodox Turks who went to Greece. Turkish Muslims who lived in the part of Thrace controlled by Greece (at the northern end of the Aegean Sea) were exempted. A small number of Greek Orthodox in Turkey also were not included in the census that determined who the government compelled to leave.

Cold numbers, even large numbers like these, do not tell the whole story. In each country, in different parts of the country, hundreds of thousands of people faced their forced departure with deep sadness. By far most of the people did not rejoice to be leaving a land that had caused them distress; instead, they were leaving lifelong friends, businesses and farms they had developed, and culture and geography they knew and loved. Older people probably had the hardest time because of the length of time they had lived in their homes and because of the physical hardships of packing, journeying to ports, making the voyage, traveling to their new homes (mostly by foot or wagon), finding their way in a new culture and environment, getting to know new neighbors, and figuring out how they were going to live.

Those who were children when the exchange took place carried deep, often hard memories into their old age, when interviewers sought their stories around the end of the twentieth century or the beginning of the twenty-first. In many if not most cases as the years passed, these exiles were able to blend into their new lands, develop friendships, and learn to live well. But many tried to go back for visits as often as they could to the places they were born and spent their early years. They eventually blended in to their new homes to a great extent, but to many people who had always lived in those towns and did not have to move, those who had come were always the *muhacer*, Arabic for migrant, even in later generations. Some natives refused to let their children marry into muhacer families. Some abandoned properties and places of worship still remain unused.

This 1922 photo shows the "American Bakery" in Constantinople. The languages shown on the windows include Armenian, Ladino (a form of Spanish written in Hebrew), English, Turkish, Greek, and Russian.

Kayakoy, located in southwestern Turkey, was home to residents of Greek ancestry until 1923. In the decades since they left as part of the population exchange, the town has fallen into ruin.

Later Conflict

The exchange did not eliminate ill will between the two nationalities. Many Greeks and Turks still feel prejudice toward each other. The sharpest clash between the two populations has occurred on the island of Cyprus in the Mediterranean. Eighty percent of Cypriots are of Greek descent, and almost all of the rest are Turkish. Throughout history, occasional clashes have erupted on Cyprus; and both groups have wanted the island to be recognized as part of their home country.

Britain governed the island as a protectorate from 1878 until 1960, after which time conflicts began anew. In 1983 a Turkish Cypriot leader proclaimed the Turkish Republic of Northern Cyprus, constituting about the northern third of the island. Turkey is the only country in the world to recognize this region as a sovereign country. Violence has calmed and people live out their everyday lives, but the unsettled political situation remains.

On occasion Greece and Turkey have worked together. Greece and the Ottoman Empire worked together to rid the Aegean Sea of pirates in 1839. Greece supported Turkey's membership in NATO in 1951. In August of 1999, when an earthquake struck Izmit, Turkey, Greek rescue teams were among the first to arrive. A month later, an earthquake struck Athens, Greece; and Turkish teams responded.

In 2019 and 2020, Greece and Turkey were in conflict again, this time over the rights to energy reserves off the disputed island of Cyprus. Thus geography and human activity continue to intertwine.

Unit 10: Southern Europe

Twice a Stranger

People have come to associate the phrase "twice a stranger" with those whose government forced them to take part in this population exchange. They became strangers to their homelands, but as they did they also became strangers in the lands to which they moved.

Greece and Turkey are geographic neighbors, but their leaders and their people have often treated each other as enemies instead of neighbors. The geographic factor of where they were born, coupled with their religious beliefs, completely upset their lives and the lives of later generations.

Jesus addressed the question of how His disciples should relate to their enemies.

Nicosia, the capital of Cyprus, is divided into northern (Turkish) and southern (Greek) sections by a mixture of walls and barricades.

But I say to you who hear, love your enemies, do good to those who hate you, bless those who curse you, pray for those who mistreat you.
Luke 6:27-28

Assignments for Lesson 48

Gazetteer Read the entries for Albania, Bosnia and Herzegovina, Croatia, Kosovo, Montenegro, Serbia, and Slovenia (pages 83, 85, 86, 90, 92, 96, and 97).

Geography Complete the map skills assignment for Unit 10 in the *Student Review Book*.

Worldview Copy this question in your notebook and write your answer: How would you try to convince someone who believes in a totally different religious system that the gospel is true?

Project Continue working on your project.

Literature Continue reading *The Day the World Stopped Turning*.

Student Review Answer the questions for Lesson 48.

San Marino

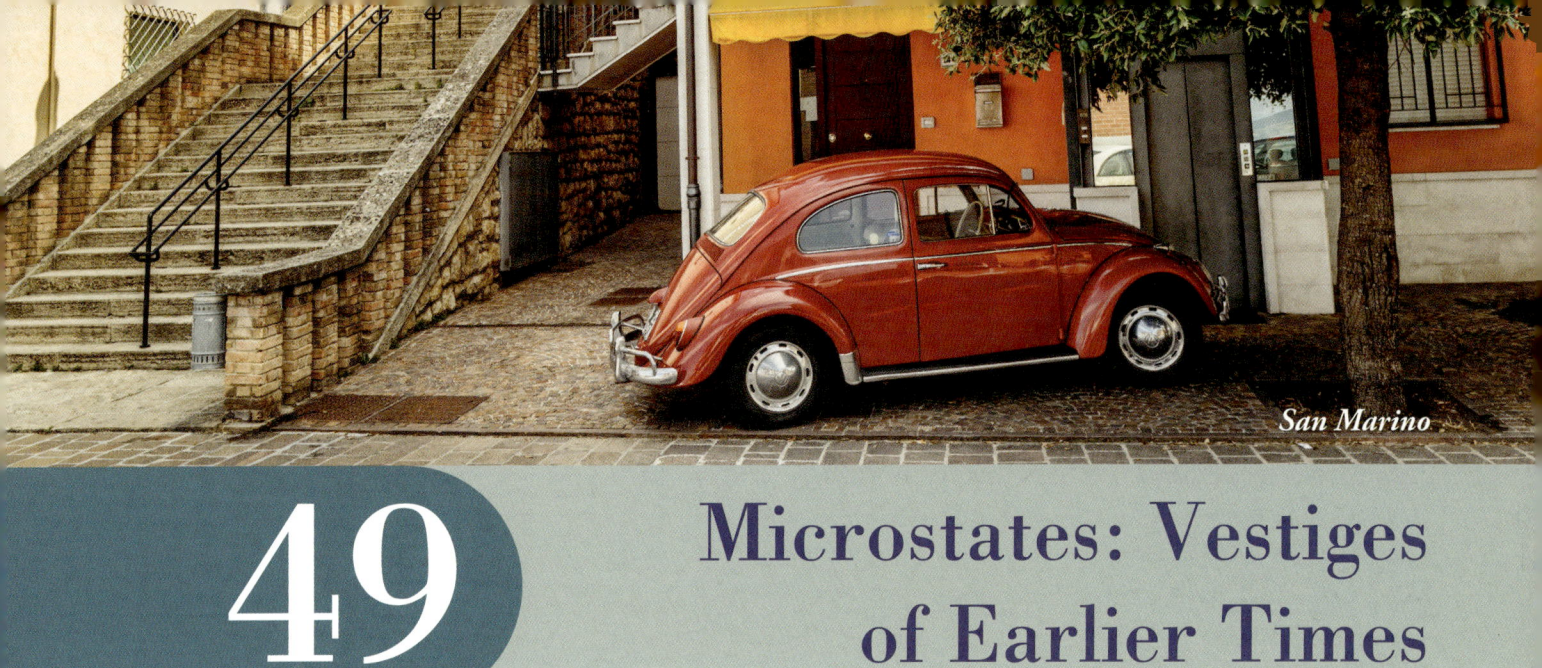

49 Microstates: Vestiges of Earlier Times

Let's take a fun trip to some fascinating places you may have never heard of. Some of the countries in this lesson lie outside of the region of southern Europe. We included them here to bring together in one lesson the countries of Europe designated as microstates.

Microstates? How Did We Get Microstates?

For much of the history of the world, people organized units of government that covered small areas. Historians usually call these city-states. Tribal units in Africa are another kind of government that covers small areas. Large areas such as China, even areas that shared a single language, were for centuries ruled by many local kings or chieftains. Kingdoms and empires such as Assyria, Babylon, and Rome were exceptions and usually achieved great size by conquering other, smaller kingdoms that never abandoned their individual identities.

Europe was no different in this. The large nation-states we know today such as Russia, France, Germany, Spain, and Italy did not exist for centuries. We noted in Lesson 46 that Spain achieved unity by bringing together several different ethic groups and their land areas on the Iberian Peninsula. At one time what is now Italy consisted of over two hundred small kingdoms, duchies, principalities, and other such areas. After many years of work, a parliament of representatives from many local states on the Italian peninsula declared the unification of Italy in 1861.

At the same time, over three hundred German-speaking kingdoms, duchies, principalities, and so forth existed for centuries, some large and some tiny. Otto von Bismarck, the chancellor of the most powerful one, Prussia, worked for years to bring many of these small governments together. He declared the unification of Germany in 1871.

However, a few tiny states continue to function in Europe to this day. Most are in mountainous areas of southern Europe, another example of how geography can keep people separate. Let's visit what geographers call microstates and see how these tiny countries maintain their identity of smallness and yet also connect with the regions around them and the world at large.

The Most Serene Republic of San Marino

We start in the Apennine Mountains of northeastern Italy. As the story goes, in 301 AD a

281

Christian stonemason named Marinus led a group of believers to this area to escape persecution by the Romans. Over the centuries, despite repeated takeover attempts by rulers of other Italian states, the people of San Marino maintained their independence. They established a republican government in the 1300s, which makes it the world's oldest republic still functioning today. Leaders there wrote the basis for their current governing document in 1600. In the early 1800s, Napoleon and the Congress of Vienna both recognized San Marino's sovereignty.

During the period when many people were working to unify the Italian peninsula, San Marino provided sanctuary for unification leader Giuseppe Garibaldi and some of his co-workers from rulers of small kingdoms who did not want to give up their local power to a unified nation. In return, after unification the Italian government allowed San Marino to remain independent. The pope recognized the independence of the country in 1862.

So let's go there—but first we have to figure out how to get there. There is no airport. A highway leads into San Marino from Rimini, the nearest Italian city; and Rimini offers helicopter service to San Marino during the summer.

As we approach San Marino, the outstanding feature we see is Mt. Titano. Rising 2,425 feet above sea level, the mountain has three peaks. On each peak stands a medieval tower. They even have names: La Guaita, Cesta, and Montale. The capital city, also called San Marino, is located on the side of the mountain. Mt. Titano and the San Marino Historic Center comprise a UNESCO World Heritage Site.

San Marino is the third-smallest country in Europe. It is all of 24 square miles, about one-third the size of the District of Columbia. San Marino is about as Italian as it can be without being part of Italy. Italy completely surrounds it, the official language is Italian, and it sends about ninety percent of its exports to Italy.

Almost all of the approximately 34,000 citizens, the Sammarinese as they are called, are Roman Catholic; but the country has no official state religion.

Life is good in San Marino. Average life expectancy is 83.3 years, the fifth highest in the world, and per capita income is also one of the world's highest. You should be pretty safe there; the public security force numbers about fifty. The country is 97.7% urban; in other words, there is very little farmland. The soil is fairly rocky, but the mild climate and ample rainfall make life easier for the few farmers who are there.

The voters elect a 60-member parliament, called the Grand and General Council, every five years. The council then selects two members to be captains-regent. These are the heads of state, who serve six-month terms and cannot be reelected for three years.

As difficult as it might be to get there, over two million tourists manage to find their way there each year. Tourism and the sale of stamps and coins provide significant revenue for the country.

About done in San Marino? Then let's head west (somehow).

The Principality of Monaco

Five miles west of Italy on the Mediterranean coast lies Monaco. France surrounds it on three sides. The coastal areas of France that lie on either side of it are part of the French Riviera. We'll have a bit easier time getting to Monaco. A highway and a railroad line run through it as they go between Italy and France. The city of Monaco has a port, and we can always take our trusty helicopter. Monaco has a heliport.

Monaco is the second smallest country in the world. It encompasses about three-fourths of a square mile, or about three times the size of the National Mall in Washington, D.C. (We're working our way to the smallest country; it's next). Monaco sits on hills and rocky headlands at the foot of Mt. Agel in France. In places the country only extends about two hundred yards from the coast. The two cities of Monaco (the capital) and Monte Carlo occupy cliffs overlooking the sea. Monaco has no forests and almost no farming.

Monaco

The country has a population of about 39,000 and is the most densely populated country in the world. The citizens are known as Monegasques. The official language is French, but we will hear English and Italian also. A number of locals speak a dialect also called Monegasque that is based on French and Italian. Many wealthy people from other countries come to live there. One big reason is that Monaco has an attractive tax system. Monaco has the highest life expectancy in the world: 85.6 years for men and 93.5 years for women. About 90% of the citizens are Roman Catholic. The Roman Catholic Church is the official state religion, but the constitution guarantees freedom of religion.

The country's biggest industry is tourism. Can you imagine why? The average January temperature is about 50º, in summer it rarely gets above 90º, and it only has about sixty days with rain each year. One feature that draws people are the world-famous road races held in Monaco each year, but the biggest attraction are the gambling casinos. These are just for visitors. Monegasques may not even enter a casino unless they work there. Colorful postage stamps generate a fair amount of revenue for the government; and the country does have a small industrial sector.

Ancient Phoenicians utilized the natural harbor there in the 700s BC. Greeks established a colony there in the 500s BC. The name Monaco derives from the Greek *monos* (one) *oikos* (house), meaning people living alone or as a single habitation. Romans and Carthaginians used it as a trading post. Genoese traders came around 1191 and built a fortress there in 1215. Many people wanted to rule the place, but the Grimaldi family took control in 1297. Over the succeeding years others ruled from time to time, but the Grimaldis regained control in 1419. France seized it during the French Revolution, but the Congress of Vienna returned it to the Grimaldis under the protection of Sardinia in 1814. Sardinia gave it back to France in 1860. The next year France kept half the territory but gave independence to the rest, which is what we know as Monaco now. The resolute royal Grimaldi family celebrated 700 years of rule in 1997.

Monaco is a principality, which means that the chief of state is a prince, the head of the Grimaldi family. He lives in a castle, part of which dates from the 1700s. The prince chooses a minister of state to oversee the government. By law the minister of state must be a French citizen. Voters elect a 24-member National Council to five-year terms.

OK, now it's time to board your yacht (without having visited a casino, of course) and head for Rome.

Swiss Guards, Vatican City

The State of the Vatican City

You might be surprised that a few blocks within a city can be a country, but they are. First, the history.

In ancient Rome, Vatican Hill was the site of the public gardens and circus of Emperor Nero. According to tradition, it was the site where Peter and many other Christians died as martyrs. About 325 AD, after Emperor Constantine came to faith in Christ, he ordered a basilica (a special church building) to be built on Vatican Hill over the site believed to be where Peter had been buried. This structure stood for over one thousand years until 1506, when Pope Julius II undertook the building of a new St. Peter's Basilica on the same site.

In church history the Roman Catholic pope claimed governing authority over many areas of Europe. He ruled many places in Italy that were collectively known as the Papal States. When the unification of Italy took place, the pope at the time refused to accept this loss of temporal authority. After a decades-long standoff, in 1929 the pope and the Italian government signed a treaty. In the treaty, the pope recognized the country of Italy (then ruled by Benito Mussolini's Fascist government) and gave up claims to the Papal States. In return Italy recognized the State of the Vatican City, which includes Vatican Hill.

Vatican City (often called the Vatican) comprises 109 acres within the city of Rome on the west bank of the Tiber River. It is the smallest country in the world. The Vatican is the home of the Holy See (a word that means seat), which is the headquarters of the Roman Catholic Church. The head of the church and of Vatican City is the pope. The government of Vatican City is called an ecclesiastical elective monarchy. This means that the Church through its College of Cardinals chooses the pope, who becomes monarch of the City. The Vatican also has oversight of a few other buildings in Rome and the pope's summer residence in the village of Castel Gandolfo in the Alban Hills outside of Rome.

The Vatican has a population of about one thousand, half of whom are priests and nuns who work at the Vatican. Almost five thousand people, most of whom live outside of the city, work there. Vatican City carries out many typical functions of government, such as printing stamps, minting coins, and producing license plates; providing street lighting, street cleaning, telephone service and water service; and maintaining a bank, an observatory, and a post office. Vatican Radio began in 1931 with the help of Guglielmo Marconi, the inventor of radio, and now broadcasts news and religious programming in about forty languages. The Vatican also produces television broadcasts of special events involving the pope.

The Swiss Guards provide protection for the pope, as they have since the late 1400s. Many people at that time recognized Swiss soldiers as the best in the world. Swiss soldiers served the rulers of many European countries. In 1506, following a suggestion by a Swiss bishop, a contingent of guardsmen became the first permanent security force for the pope. Today members of the 100-man Guard must be single, male Roman Catholics between 19 and 30 years of age and citizens of Switzerland.

People around the world know Vatican City as the home of beautiful buildings, priceless works of art, and important museums and libraries. UNESCO designated Vatican City as a World Heritage Site in 1984.

Now on to Malta.

The Republic of Malta

What is it like to live on a small island in the middle of everything? You can find out on Malta.

The Republic of Malta is an archipelago of five strategically located islands, three of which are inhabited, south of Sicily in the middle of the Mediterranean Sea. The total land area is 122 square miles (about twice the size of the District of Columbia), of which the main island, also called Malta, takes up 95. Most of the 457,000 inhabitants live on the eastern half of Malta Island, and about half of the island's population lives in the capital city of Valletta.

Much of the land is flat and rocky with some dissected plains (plains intersected with long depressions) and many coastal cliffs. The highest point is the Dingli Cliffs on the western coast, which rise to a plateau 850 feet above sea level. People have cultivated about a third of the land for farming. Farmers in some areas have terraced the land such that the terraces look like giant steps.

No surprise, a major industry is tourism. Over one million tourists come each year. The country imports about three-fourths of its food needs; and with limited freshwater sources, it relies on desalinization facilities for much of its water.

About 90% of the people are Roman Catholic, and the Roman Catholic Church is the official state religion. Schools teach their lessons in the two official languages of Malti and English; and all schools, public and private, teach Catholic Church doctrine.

Malta has been involved in world affairs for thousands of years. Artifacts indicate the presence of Phoenicians, Carthaginians, Greeks, and Romans. Bible students know that Paul and his companions suffered a shipwreck on Malta during his voyage to Rome, as recorded in Acts chapters 27 and 28. In 1848 James Smith, a Bible scholar and expert yachtsman who was familiar with the travel lanes of the Mediterranean, published *The Voyage and Shipwreck of St. Paul*. After careful research, Smith determined that Luke's account in Acts was an extremely accurate description of the waters around Malta as well as its coastal area. Most scholars believe that a bay on the northeastern coast of Malta is where Paul's ship ran aground; thus today it is called Saint Paul's Bay.

By the way, after Paul's shipwreck, the residents of Malta underwent a change of worldview regarding Paul and a snake. When a viper fastened itself on his hand, the people took this to be a sign that he was a

Valletta, Malta

murderer. However, when Paul shook the viper off and was not harmed, the people started to believe that Paul was a god (Acts 28:3-6).

Great Britain claimed Malta as part of its empire in 1800, and the Congress of Vienna endorsed this claim in 1815. Britain established its military headquarters for the Mediterranean on Malta. It was a strategic naval base for the Allies during World War I. In World War II, the British presence helped the Allies patrol the sea lanes between Italy and Africa. The islands' rocks and inlets provided places for ships to anchor as well as submarine bases. The military used underground passages on the islands as bomb shelters. Fighter planes that were based there defended shipping convoys. Because of heavy Nazi bombing in 1942, Britain's King George VI awarded the island the George Cross for the courage and endurance of the Maltese people. After the war, NATO established its Mediterranean military headquarters there. The British undertook to disassemble its empire following the war, and Malta achieved independence in 1964. In 1979 both the UK and NATO withdrew their military forces from the country.

The Malti language has roots in the Phoenician tongue but also has Arabic influence because of a strong Muslim presence there in the Middle Ages. Malti is the only official Semitic language in the European Union and the only Semitic language written in Latin script (Semitic languages include Hebrew, Aramaic, and Arabic). The language has many loan words from English, Italian, and French.

Whew! A lot has happened on these little islands. Now on to the Iberian Peninsula for the next stop on our microstate tour.

The Coprincipality of Andorra

Nestled in the Pyrenees Mountains that separate France and Spain and that mark the northern border of the Iberian Peninsula, Andorra is slightly larger than Malta but only has a population of about 77,000. The mountains, which rise as high as 9,665

Farm Buildings in Andorra

feet above sea level, served to keep Andorra isolated for many years. Today the mountains, the glaciers on them, and the valleys and rivers below them offer beautiful scenery for the eight million visitors who come each year. Good highways bring tourists into the country, but it still has no railroad and the nearest airport is in Spain. Andorra is not only difficult to get to; it is difficult to become a citizen there. To do so, you must be a resident for twenty-five years and you must renounce your previous citizenship.

Andorra has two heads of state. The country is a parliamentary coprincipality. According to the accepted story, Charlemagne liberated the region from the Muslims in 803. The Spanish count of Urgel (the adjacent region of Spain) ruled it beginning in the 800s. Later the count gave it to the bishop of Urgel. Disputes among French and Spanish counts and bishops continued for centuries.

From 1278 until 1993, Andorra was governed by the French chief of state and the Spanish bishop of Urgel. Under this arrangement, every other year Andorra paid the French head of state about two dollars and gave the bishop of Urgel about eight dollars as well as six hams, six cheeses, and twelve hens. Because of its dual French and Spanish heritage, Andorra has two postal systems, one French and the other Spanish.

In 1993 Andorra adopted its first constitution. Under that document the people elect a 28-member legislature, the General Council of the Valleys, which appoints a head of government or prime minister. The president of France and the bishop of Urgel are coprinces whose roles are largely ceremonial.

Lesson 49 - Microstates: Vestiges of Earlier Times

This arrangement technically makes both of them monarchs, but no one seems to be concerned about it.

The official language is Catalan, which many consider a Romance language as are French, Spanish, Italian, and Romanian. Some Andorrans also speak French, Spanish, or Portuguese. The population is predominantly Roman Catholic, although Andorra has no official state religion. Only Roman Catholics may get married in Andorra, and the Church maintains all public records.

Because the country is so mountainous, most of the limited farmland is pasture for sheep, although farmers grow some crops.

Family life is very important for the people of Andorra. A typical three-level family farmhouse has stone walls and a slate roof. A barn or toolshed is on the lower level, the living room and kitchen are on the main floor, and bedrooms are above. Historically the eldest son in a family inherited all or most of the property, so younger children frequently moved out of the country. Now more people of Andorran descent live in France or Spain than in Andorra itself.

Now we head northeast to visit our final European microstate, which is only about four hundred miles from where we started in San Marino.

The Principality of Liechtenstein

This landlocked country, slightly smaller than the District of Columbia, has had people living in it since well before the time of Christ. It became part of the Roman Empire and later of the Holy Roman Empire (the often not-very-powerful patchwork of countries that existed from 800 to 1806). In 1699 Johann-Adam of the house of Liechtenstein in Vienna purchased the county of Schnellenberg from the Empire; he bought the adjacent county of Vaduz in 1712. In 1719 the Holy Roman Emperor Charles VI joined the two counties and declared them to be an imperial principality. Johann-Adam thus became prince of Liechtenstein, and a member of this dynasty has ruled the country ever since.

However, no prince of Liechtenstein set foot in it until 1842. It is the only remaining principality of the Holy Roman Empire.

Sandwiched between Switzerland and Austria, most of its 39,000 citizens live on the western side near the Rhine River, which is its western border. The eastern two-thirds feature a range of the Alps. The highest peak is 8,527 feet above sea level, and much of the country lies above 6,000 feet. Liechtenstein is considered 85% rural, and it has a rich diversity of woodland plants and wildlife. While snow covers the bare mountain peaks, forests and flowers grow on the mountain slopes; and significant farming takes place near the Rhine. An industrial sector began developing in the 1950s and continues to grow, although the country has almost no natural resources of commercial value and so industries must import virtually all raw materials. The climate is mild. You'd never guess, but tourism is a major industry as is the sale of postage stamps (do we see a pattern here?). The official language is German, but many of the people speak a dialect of German known as Alemannic. About three-fourths of the population are Roman Catholic. Liechtenstein has not had an army since 1868; it depends on Switzerland for its security.

The prince of Liechtenstein owns one of the finest privately held art collections in the world. It includes the world's largest Faberge egg (a jeweled egg crafted in czarist Russia). The country is home to additional art museums also.

Balzers, Liechtenstein

An Overview

Each of these six microstates has a unique story that goes back centuries. At the same time, their stories do have common threads. For the most part they have attracted wealth because of friendly tax laws. They have benefited from tourism (and stamp collecting!). The Roman Catholic Church has a strong presence in each one.

Two factors have contributed to their longevity. One is geography: four are mountainous and one is an island. They do not have abundant natural resources such as gold that other, larger countries typically want. Another factor is the political skill required to adapt to changing circumstances. One adaptation by five of them is the institution of representative government. The strength of the Roman Catholic Church has helped the Vatican to survive.

The microstates are independent, but at the same time they are also dependent on other countries. For instance, they depend on nearby countries for military protection. In addition, they have found it advantageous to enter into economic alliances with other countries. Without military protection and economic activity, these countries would find it difficult to survive. In their own way, the microstates of Europe are examples of how people have interacted with geography and with each other.

Shout for joy, O heavens! And rejoice, O earth!
Break forth into joyful shouting, O mountains!
For the Lord has comforted His people
And will have compassion on His afflicted.
Isaiah 49:13

Assignments for Lesson 49

Gazetteer Read the entries for Andorra, Malta, San Marino, Vatican City, Liechtenstein, and Monaco (pages 84, 91, 95, 99, 109, and 111).
Read the excerpt from *A Little Pilgrimage to Italy* (pages 273-274) and answer the questions in the *Student Review Book*.

Project Continue working on your project.

Literature Continue reading *The Day the World Stopped Turning*.

Student Review Answer the questions for Lesson 49.

Areopagus and Modern Athens

50 Paul's Sermon on Human Geography

The large crowd fell silent as the stranger in town stood to speak. This was the culmination of weeks of controversy and animated discussion as people tried to fit the stranger's message into their worldview but found it difficult to do so. Now the stranger had a hearing before the group of men who would decide whether his message deserved serious consideration or whether it was just another wild idea that passed through the city from time to time.

When the apostle Paul spoke before the Areopagus in Athens, Greece, as recorded in Acts 17:22-31, he was explaining the new message that he had been presenting in the city. Since that day, people have often seen his sermon, delivered in the midst of that highly respected and long-standing body, as a prime example of how to present the gospel to a pagan audience. For our purposes in this curriculum, we want to see what his message has to say about the study of human geography.

This lesson is the first of several that show how the Bible utilizes geography to teach lessons and the relevance of Scripture to the study of geography.

The Setting

Athens had been the intellectual capital of the Mediterranean world for centuries. By Paul's day, the golden age of Athens under the leadership of Pericles and the lives of the great philosophers Socrates, Plato, and Aristotle were long past. Rome was the undisputed world superpower now, but most people in the Mediterranean world still saw Athens as the hub of intellectual activity.

Paul had come to Athens because Jewish opponents in another Greek city, Berea, had stirred people up against his message. A group of friends had escorted him to Athens and left him there until Silas and Timothy joined him. Paul had faced opposition in Thessalonica and Berea, but that did not stop him from continuing to teach others the truth that had changed his life.

The scenes that Paul saw as he walked through Athens stirred his spirit within him as he considered the many idols in the city. Sadly, this intellectual

View of the Acropolis from the Areopagus in Athens

capital of the world had failed to grasp the central truth of the universe: the reality of the one true God. The Athenians were for the most part pagan polytheists groping for the spiritual truth that lay behind the world they knew. Paul wanted them to know the truth that God had shown to him, the truth that sets men free.

The Areopagus was a geographic feature of the city. The word means Hill of Ares. Ares was the Greek god of war, the equivalent of the Roman god Mars. The Romans knew the site as Mars Hill, which is the way that the King James Version of the Bible renders it. The Council of the Areopagus was a body of men who passed judgment on the ideas that people discussed in Athens. The council decided whether any new teaching was intellectually healthy and should be permitted. The council had originally met on Mars Hill; but in Paul's day, except in a few rare instances, it met in the Royal Porch in the agora or marketplace in Athens. Even though it met in a different location, the council kept its original name.

The Greek World of Ideas

Athenians had been surprised at Paul's teaching in the Jewish synagogue and in the marketplace in Athens because he had been expressing some new ideas, namely that Jesus was the Son of God and that Jesus had come back to life after being dead. These ideas were so unusual to Athenians that some thought Paul was preaching about two new deities: Jesus and Resurrection.

Both concepts were strange to the Greeks. The idea that the one true God could come to earth in the form of a man was completely foreign to the Greek worldview. In their minds many gods existed; and although their gods sometimes took human form, the idea of incarnation, God truly and fully becoming man, was alien to their thinking. Resurrection was also impossible. The Greek philosopher Aeschylus had written, "Once a man is dead and the ground drinks up his blood, there is no resurrection."

Some Epicurean philosophers and some Stoic philosophers had heard what Paul had been teaching, and they couldn't figure him out either. Epicurean and Stoic philosophies were two of the leading schools of philosophy in Athens at that time. Zeno of Cyprus (335-263 BC) taught in the Stoa Poikile (Painted Hall) in Athens, and thus his philosophy became known as Stoicism. To Zeno, Reason was the god who guided the universe. To live in accordance with reason (or Reason, personified) meant living a reasoned, rational life. The goal of mankind, Zeno

Lesson 50 - Paul's Sermon on Human Geography

said, should be to live a resigned, passive life, putting aside all impulses and emotions.

Another philosopher, Epicurus (341-270 BC), had taught that the highest good mankind could pursue was to avoid pain and suffering. Some of their opponents ridiculed this idea and claimed that Epicureans only wanted to live for pleasure; but actually, the movement taught the value of living a simple life free from the cares of this world.

Both of these philosophies, and the many philosophies and religions that men have devised before and since, have been human attempts to determine what we are to make of this world in which we live and how we are to live in it. We have an amazing world, full of fascinating people, places, things, truths, and ways that things work; but the central question with which many people wrestle is, what is the ultimate purpose of it all? The Epicureans, Stoics, and many others have attempted to give an answer to that question. Paul appeared to be offering one more idea about ultimate purpose, and his idea was one that many Greeks simply could not fathom. What he taught was so different from the typical philosophies and beliefs of that day that people had a hard time understanding what he was saying. As is so often the case when people don't understand (or don't want to understand) a new message, those who heard him attacked the messenger.

To the Epicureans and the Stoics, Paul was a babbler, someone who picked up stray ideas like rags and tried to blend them into a philosophy that did not make sense to them. Paul's message was not about using a little more reason or pursuing a little more pleasure while maintaining belief in the same gods. Instead, his message offered a completely new view of the world, a new way for people to live in it, and a new idea about where it was all headed. As a result, skeptics in the city took him before the Council of the Areopagus to try to determine the exact nature of his strange new teaching.

The Athenians were much like many people today who want to hear and discuss new ideas. The 24-hour news cycle and social media bombard us with speculation, the results of modern scientific research, and the latest political activity. People love to hear this kind of thing and to offer their own perspectives on all of this. At the end of the day, however, just talking about the news doesn't accomplish much. We need to analyze the news from a Biblical worldview.

Detail from The School of Athens *by Raphael (Italian, 1511)*

The Sermon

Paul began by complimenting the Athenians for being a religious people. They had a deep interest in religious matters, as evidenced by the many idols and temples in the city. However, their religiosity was unfocused and in the end mistaken because they believed in many deities—except, ironically, the one true God. Paul noted the altar dedicated "To an unknown god," and then proceeded to introduce the Athenians to the God they did not know.

Paul spoke from a God-centered worldview. He said that one God, the only true God, created the world and everything in it. God is the source of the world in which we live. Whereas Athenians believed that they sprang from the soil of Attica, Paul told them that God is really the giver of all life. The Athenians placed great importance on making sacrifices to the gods, but since the one true God created everything, He did not need anything from man.

The apostle then referred to the Biblical story of Creation when he explained that from one man (and one woman) has come every nation, every ethnic group there is. God made mankind to live "on all the face of the earth"; in other words, to interact with the world. This, as you remember, is the theme of human geography. So Paul told us that God actually created the subject of human geography.

But there was more to Paul's message than just the fact that God created the world and the people in it. God has appointed for the nations their times—the periods in which they have lived, their rises and falls—and their locations, the "boundaries of their habitations." In other words, all that we study in this curriculum, all that you study in any curriculum, and everything that happens in personal, family, national, and international affairs is part of God's plan and under His sovereignty. This is what human geography means from the perspective of faith in God.

God made humans to interact with the world: to work, invent, create, achieve, and relate to one another. As wonderful as those activities are, however, they are not our highest purpose. They are not worthy of us as people made in God's image who have an eternal destiny. God made mankind to seek Him and to know Him. This is not an impossible quest because He is not far from every person; that is to say, finding Him and knowing Him are possible. After all, as the poet Epimenides of Crete said, in Him we live and move and have our being. And, as the Greek poet Aratus said, we are all the offspring of God.

The psalmist said, "It is He who has made us, and not we ourselves" (Psalm 100:3). When the Greeks and others devised belief systems that attempted to explain the origin of the earth and mankind, they were in a sense trying to create themselves, to devise their own creation myth. What they needed to learn, and what people today also need to learn, is that we originated with God, not in a human-imagined Big Bang or as the by-product of a war among the gods. The two pagan Greek poets whom Paul quoted understood the truth to this extent.

Seeking God gives meaning and context to our quest for knowledge in this world and about this world. Any other perspective or worldview misses

Carved Depiction in Berlin of Paul's Sermon

Lesson 50 - Paul's Sermon on Human Geography

this meaning. As evidenced by their quest for knowledge and their understanding of architecture and other elements of the world in which they lived, the Athenians understood something about how to interact with the Creation. Paul wanted to teach them how to interact with the Creator.

As God's offspring, we have a spiritual nature. We can think, reason, and create. We are responsible for how we live. Overall, however, we people (including the Greeks) have failed to live as we should. Those religious Athenians said they were seeking truth. That was the purpose of the Socratic method of seeking truth by asking questions. But the Athenians had not found it yet. They were still groping. Paul wanted to direct their quest toward the proper goal. Trying to find our way by probing our own understanding is not the way to find the truth. We need to trust in the Lord, the One who is outside of ourselves, and not lean on our own understanding (Proverbs 3:5-6).

The Athenians' central failing was their perception of God. We ought not to think that humans can represent the one true God by an idol made of gold, silver, or stone. Idols represent false gods. Since we are God's offspring and since we are spiritual beings, God must be more than an object that people make from physical materials, even precious materials. Those materials from the earth have their proper purposes and uses, but those purposes do not include bearing the divine likeness. Human beings are made in the image of God; nothing else is.

Paul told the Athenians that God was willing to overlook the times of ignorance (and the Greeks confessed their ignorance by having an altar to a God they didn't know), but now they could know the truth because Paul was telling them the truth. The time had come for all people everywhere, including the wise and accomplished Greeks, to repent.

The time of ignorance was past, and the time of accountability had come. God has a day ahead of us in which we shall face His judgment. This means that our world has a destiny; it is headed somewhere.

Paul's Sermon in Greek on a Plaque at the Areopagus

Time is not just going through ever-repeating cycles. The interactions of man with his world, of man with man, and of man with God all have a point, a day of reckoning. God has appointed the judge that all people will face on that day, the man He raised from the dead. So Paul circled back around to the idea of resurrection and explained its meaning with regard to Jesus and its significance for Him as judge of the world.

No god in the whole system of Greek deities had the responsibility to judge the world. The One whom God raised from the dead has this responsibility, and He will carry it out.

The Response

When Paul finished, people had various reactions.

Some mocked him for talking about resurrection.

They knew that Paul was Jewish, and the typical Athenians would have seen Paul's message as foolishness and something that didn't fit at all into any Greek philosophy (1 Corinthians 1:18-25).

Some seemed open to further discussion. A few joined him and believed, including one of those on the council of the Areopagus.

People have had these different responses to the gospel—mocking rejection, a desire to know more, and acceptance and faith—ever since.

The Meaning of Paul's Sermon for Human Geography

Paul's sermon touched on many aspects of human geography. He delivered it in the marketplace of a great city, surrounded by the works of man, in the midst of an accomplished culture and in response to religions and philosophies that people had created.

The geographic setting of Athens on the Greek peninsula in proximity to the Mediterranean Sea had a major influence on the city's history. Paul referred to works of literature that came from the minds of people. He spoke of Jesus, who was crucified on a hill and buried in a cave. Athens contained a multitude of ethnic groups: Greeks, Jews, God-fearing Gentiles, and many others. Those who believed Paul's message became members of a new people group: Christians.

Paul spoke to people who had a particular worldview, and he introduced another worldview to them. That new worldview said (and continues to say) this: Mankind's greatest achievements in this passing world will ultimately be meaningless without a submission to and relationship with the one, true, eternal God. Greek philosophy was impotent to change people and to confront the realities of sin and death, but Jesus conquered these realities. The amazing achievements of ancient Athens and the amazing achievements of modern man will one day rejoin the dust of the earth on which they were achieved. As we live in this world, we anticipate living forever in the heavenly city with the one true Maker of all.

As Paul expressed it:

. . . He made from one man every nation of mankind to live on all the face of the earth, having determined their appointed times and the boundaries of their habitation, that they would seek God, if perhaps they might grope for Him and find Him, though He is not far from each one of us.
Acts 17:26-27

Assignments for Lesson 50

Worldview Recite or write the memory verse for this unit.

Project Finish your project for this unit.

Literature Continue reading *The Day the World Stopped Turning*.

Student Review Answer the questions for Lesson 50.
Take the quiz for Unit 10.
Take the second geography, English, and worldview exams.

Cliffs of Moher, Ireland

11 Western Europe

The beautiful geography of the Lake District in England makes it a special place worth preserving. This was a goal for Beatrix Potter. Many people associate tulips with the Netherlands, but they actually originated in a different place thousands of miles away. The tulip and flower industry developed in the Netherlands beginning in the 1500s. Weather conditions were a significant factor in the Battle of the Bulge that took place in Europe during World War II. Following the war, the Marshall Plan helped rebuild countries that the war had devastated. The worldview lesson considers the meaning of faith.

Lesson 51 - Not a Fairy Tale
Lesson 52 - The Flower That Made the Netherlands Famous
Lesson 53 - When Weather Helped Make History
Lesson 54 - Helping a Continent Be Strong and Free
Lesson 55 - Faith

Memory Verse   Memorize Romans 12:1-2 by the end of this unit.

Books Used   The Bible
Exploring World Geography Gazetteer
The Day the World Stopped Turning

Project (Choose One)

1) Write a 250-300 word essay on one of the following topics:
   - Describe a beautiful place you enjoy and tell why preserving it is important to you. See Lesson 51.
   - What things encourage your faith in God, and what things discourage it? See Lesson 55.
2) Lay aside the idea that you are too old for this, find some Beatrix Potter books, and look at the illustrations that this successful author, artist, businesswoman, and conservationist made that reflect scenes in the Lake District.
3) Create a display that shows growing things we usually associate with particular geographic locations, such as cactus in the American Southwest, cotton in the American South, kangaroos and other indigenous animals in Australia, cinnamon in Ceylon, and rooibos tea in South Africa. You might assemble pictures and objects with explanations of the items and where they grow, or make a map.

Stockley Bridge, Lake District National Park

51 Not a Fairy Tale

Beatrix Potter was born in London, England, in 1866. The daughter of a wealthy couple, Beatrix was educated at home by her governesses. The Potter family took their holidays in the country, often in Scotland, but as she got older in the beautiful Lake District of northern England.

Beatrix grew up loving folk and fairy tales. She also displayed a talent for drawing and painting. As a child she wrote and illustrated stories about the pets that she and her younger brother kept. In 1901 Beatrix wrote and illustrated a story, *The Tale of Peter Rabbit*, which she had privately printed. The next year, the major London publisher Frederick Warne republished the book in a professional edition, which was an immediate success. In 1903 Beatrix wrote and illustrated two more children's books, and the tide was unleashed. Her books were enormously popular. The illustrations that she painted for her books portrayed the lakes, mountains, forests, and other settings that she had come to love in the Lake District.

Working closely with Beatrix in the publication of her books was Warne's son, Norman. They fell in love and became engaged in 1905, but Norman died a month later of leukemia. Beatrix continued on with her plan to purchase Hill Top Farm, a small working farm in the Lake District just outside the village of Near Sawrey. There she retreated to write, paint, and learn farm management.

Four years later Beatrix Potter purchased Castle Farm, which was across the road from Hill Top Farm. Potter wanted to own land in the Lake District and thus be able to preserve it from commercial development. Helping her in these purchases was the local attorney William Heelis, who shared Beatrix's love of the land. Beatrix and William married in 1913 when she was forty-seven; William was a few years younger. They moved into the house known as Castle Cottage on Castle Farm.

The Potters became deeply involved in their community's life. For instance, she founded a trust to improve health care, and she worked to oppose commercial development of the Lake District, such as the idea to introduce hydroplanes on Lake Windermere. Beatrix developed a keen interest in breeding and raising Herdwick sheep. She became an admired and respected breeder and consistently won prizes at local shows. On her farms she raised cattle, pigs, chickens, turkeys, and ducks in addition to the Herdwicks. She also raised rabbits for meat during World War I.

Beatrix and William continued to purchase farms and other land in the Lake District from their professional incomes. In addition to her book sales,

Hill Top, Beatrix Potter's Home

early on Beatrix had the brilliant business insight to license tie-in products to her stories. Because of her failing eyesight, her increased interest in and responsibility for farm and land management, and her continued care for her parents in London, Beatrix's last "little book" for children appeared in 1918. At her death in 1943, she left fifteen farms and over four thousand acres in the Lake District to the National Trust. William died about nineteen months later.

Beatrix Potter's agricultural and conservationist legacy lives on today in the Lake District. Sheep raising is the primary farming activity there. The district welcomes over fifteen million tourists every year, but the National Trust and local governments closely protect the natural beauty of the area from industrial exploitation and tawdry tourist development. The Beatrix Potter Gallery, a museum which displays many of the original paintings she produced for her books as well as other artifacts, is located in the seventeenth-century building that was once her husband's law office in the town of Hawkshead.

The Beauty of Geography

What makes a geographic setting beautiful in our eyes? Is it the greenery of hills, fields, and forests? Is it the variation of level or rolling lands contrasting with hills and mountains? Do the waters of lakes, rivers, or oceans generate soothing thoughts? Do blue skies and white clouds join with green lands

Unit 11: Western Europe

to create an unforgettable image? Are we intrigued by the thought that people have lived in or traveled over the same land for perhaps thousands of years?

Whatever it is that we find beautiful in a given setting, what we are admiring is geography. God gave us the earth as our home, to sustain us, for our enjoyment, and as evidence of His power and care. We must constantly balance our need for food and economic development on one hand with the need to preserve the land as a resource for future generations on the other. We have needs and desires today that the land can meet, but we also must keep the long term in mind so that we conserve resources for generations yet unborn.

The Lake District of England

The Lake District comprises an area of about nine hundred square miles in the northwest of England, an area which includes the Lake District National Park. The district is in the county of Cumbria, which contains part of the area once known as Cumberland. In the district lie the tallest mountains on the island of Great Britain, although the tallest, Scafell Pike, is only about 3,210 feet high. It also contains several picturesque lakes, including Britain's largest and longest lake, Windermere. One reason for the lush greenery there is the plentiful rainfall of over one hundred inches per year. Winds

These sheep are grazing in the Lake District. Their owners mark their wool with dyes to identify them from a distance.

Lake Windermere

---

from the west pick up moisture from the ocean and rise with the mountains where the moisture cools, condenses, and then falls.

Riches lie beneath the ground also. The district has been the site of lead and copper mining, but slate used for building material is its best known mineral resource. The quarries of the district have produced the materials for many buildings and for the thousands of miles of stone walls that line the roads and divide the pastures in the region. Craftsmen made many of these walls by stacking stone without mortar.

Humans have traversed and lived on this land since ancient times. For instance, Castlerigg Stone Circle may be older than Stonehenge and is one of many stone circles in England. The circle might have been a place for gatherings or for the offering of pagan sacrifices. Castlerigg is different from Stonehenge in that you can walk up to the stones at Castlerigg and touch them. Also in the Lake District are ruins from the time of the Roman occupation of Britain.

The Lake District played a prominent role in the literary movement called Romanticism, which flourished in the late 1700s and the first half of the 1800s. This literature emphasized an individual's unique thoughts and feelings as opposed to objective descriptions of people and events. Romantic poetry celebrated nature and often created a dreamy escape from the hustle of everyday urban life. Samuel Taylor Coleridge, William Wordsworth, and Robert Southey (the last two of whom served as poet laureate of Great Britain) were three poets associated with the Romantic Movement and were known collectively as the Lake Poets.

Today the Lake District is known for its tourist activities, such as boating, hiking, long auto drives along winding country roads beside stone walls and beautiful farms, visiting the farms, and visiting historic sites. The many charming villages are filled with shops, restaurants, and hotels. The Lake District is a place where people have worked with geography to produce a place of beauty that millions of people enjoy. Perhaps one day the Lake District will capture your heart as it did Beatrix Potter's, William Wordsworth's, and mine.

Village of Hawkshead

Praise the Lord from the earth...
Mountains and all hills;
Fruit trees and all cedars;
Beasts and all cattle;
Creeping things and winged fowl....
Psalm 148:7a, 9-10

Assignments for Lesson 51

Geography — Study the map of Western Europe and read the entries for the Channel Islands, the Isle of Man, and the United Kingdom (pages 100, 103, 107, and 114).
Read the excerpts from *A Guide Through the District of the Lakes* (pages 275-276).

Worldview — Copy this question in your notebook and write your answer: What does the idea of faith mean to you?

Project — Choose your project for this unit and start working on it. Plan to finish it by the end of this unit.

Literature — Continue reading *The Day the World Stopped Turning*. Plan to finish it by the end of this unit.

Student Review — Answer the questions for Lesson 51.

Zaanse Schans, Netherlands

52 The Flower That Made the Netherlands Famous

Images of the Netherlands: Windmills. Wooden shoes. Paintings by the Dutch Masters such as Rembrandt. And of course, tulips. Vast fields of swaying, colorful tulips.

People have long associated tulips with the Netherlands. There's just one interesting thing about this image:

Tulips didn't originate in the Netherlands.

Withdraw your mental image from the windmills of the Netherlands to a place thousands of miles away from northern Europe, to a barren, inhospitable region of mountains and valleys north of the Himalayas, where China, Tibet, Russia, and Afghanistan come together. These are the Tien Shan Mountains of China and the Pamir Mountains of Russia. These mountain ranges have usually served as barriers to human interaction between East and West. Because of snow, the mountain passes are largely impassable eight or nine months of the year.

Few people live here and little vegetation grows here; but in the poor, sandy soil of the valleys there grows a hardy flower we know as the tulip. Their stem is shorter than what we see in the Netherlands, and they are usually red although they do have variations in color and shape. Nearly half of the 120 known species of tulips grow wild here.

From East to West

This region is the homeland of the people we call the Turks. At some point the Turkish people began moving west, across the steppe to the regions of the Caspian Sea, the Caucasus, the Black Sea, and Anatolia. Tulips might have spread west naturally, but almost certainly the Turks carried tulip bulbs with them as they moved.

People in these areas of the world have long venerated the tulip as a symbol of beauty and perfection. By 1050 the people of Persia (largely modern Iran) and Baghdad found the tulip worthy of praise in story and poetry. Before the end of that century the Seljuk Turks came west and conquered Anatolia, wresting it from the Byzantine Empire. The Seljuks saw themselves as carrying on the greatness of Rome. One of their cherished accomplishments was the cultivation of the tulip.

The Seljuks met defeat in the early 1300s, and smaller kingdoms of the Turks arose. The leader of one of these was Osman of Sogut, whom Europeans called Ottoman. He gained power and ruled a large area. His rule even extended to the gates of Vienna. Ottoman was a harsh and cruel ruler, but he loved beauty and gave tulips the highest position of honor.

Ottoman Dish with Tulips and Peonies (c. 1550)

Among medieval Muslims, gardening was a favorite pastime because to them gardens symbolized paradise. To them, the tulip was the flower of Allah. When Babur established the Mogul dynasty in northern India, he planted tulips in his garden. In 1389, when Ottoman Turkish warriors fought Christian armies at Kosovo, the shirts that the Turks wore under their armor were embroidered with tulips. After Ottoman Turks led by Mehmed captured Constantinople in 1453, the victorious Muslims planted numerous gardens in the city that featured tulips.

By the early 1500s, the tulip was an honored flower in the Ottoman Empire. Some 1,500 varieties grew in Istanbul, with such names as The Light of Paradise, The Matchless Pearl, and Diamond's Envy. A century later, the city boasted some eighty flower shops and three hundred men whose profession was to grow and cultivate flowers, whom the Turks called florists.

Discovery in Europe

By contrast, as late as the first quarter of the 1500s, the tulip was unknown in Europe. Some think that the governor of the Portuguese colony in India might have brought one or more with him on his return to Portugal in 1529. The ambassador of the Holy Roman Empire to Istanbul commented on the tulip in a written document in the 1550s. One or more tulips are known to have been growing in a garden in Augsburg, Germany, in 1559; but in the 1560s European botanists who knew about the tulip saw it as a novelty.

However, once the flower became known in Europe its popularity spread quickly because people could carry the bulbs easily from place to place. Tulips were growing in Vienna by 1572, in England by 1582, and in Frankfurt by 1593. Wealth was increasing in Europe as a result of overseas trade and the discovery of silver in Spanish South America. Many wealthy families enjoyed spending some of their money to create beautiful, elaborate gardens. When different species of tulips are planted close together, insects mix the pollen and new hybrids multiply. These hybrids resulted in glorious new color combinations in the petals. Some color variations are the result of a virus that can infect the flowers.

Not everyone was content simply to look at tulips. Some people crushed the bulbs and used them as medicine. And since the tulip bulb resembles an onion, some people thought that the thing to do with a tulip bulb was to eat it! (During World War II, many Dutch people ate tulip bulbs because of food shortages. Newspapers printed recipes for tulips.)

In 1562 a trading ship docked at Antwerp in the Low Countries (Belgium, the Netherlands, and Luxembourg) and introduced tulips into Northern Europe. Some were eaten, some were planted. Tulips first blossomed in the Low Countries in 1563 and spread from there. The poor, sandy soil of the Low Countries, reminiscent of the soil in which tulips first grew, proved to be a fertile home for the flowers. But it was the work of one scientist and tulip-lover that had the most impact in spreading the appreciation for and production of tulips.

Lesson 52 - The Flower That Made the Netherlands Famous

Carolus Clusius

Charles de L'Ecluse was born in France in 1526. Born into a Catholic family, as an adult he became a Protestant and from then on used the Latin form of his name, Carolus Clusius. He studied botany (at the time considered a branch of medicine) at the university in Marburg, Germany. Clusius traveled widely throughout Europe and developed a long list of acquaintances and correspondents. He could speak nine languages.

Clusius learned of the Antwerp tulips and might have seen them as early as 1564. For a time he served in the court of the Holy Roman Emperor in Vienna, Austria. He distributed tulip bulbs and seeds (they can grow from seeds) to his contacts all over Europe. In 1592 Clusius accepted a teaching appointment on the medical faculty at the University of Leiden in the Netherlands. He was not the first person to grow tulips in the Dutch Republic (that had been done by 1573), but he did develop a system of classification for tulips and did more than anyone to spread tulips and knowledge about them throughout Europe.

The Flowering of Interest in the Tulip

Interest in tulips became a fascination and then an obsession. The increased wealth in the Netherlands due to trade enabled many people to plant elaborate gardens. It was not at all uncommon for thieves hired by one wealthy landowner to steal tulips from the gardens of other wealthy landowners. Requests for bulbs overwhelmed Clusius, and his own garden was a target for thieves. Nevertheless, he continued to correspond with people all over Europe and to send them bulbs.

Protestant refugees escaping persecution in Catholic countries often came to the Netherlands. A significant number were wealthy and wanted to invest their money in promising opportunities. Tulip production was one of those activities. Poetry and artwork began to feature the flower.

Flower devotees came to see the tulip as the ultimate flower. Gardeners all over Europe tried to outdo one another with the colors of their hybrids. By the 1630s some one thousand varieties existed in Europe, about half of these in the Netherlands. During this same time, in Paris, France, the tulip surpassed the rose in terms of popularity and demand; and the fashions of the French court influenced the fashions of royalty and nobility all over Europe. Some Netherlands merchants sold bulbs to the Ottoman Empire! Some experts published books with paintings of different styles of tulips.

The increased demand led to increased prices for tulip bulbs. In the 1630s a typical carpenter in the Netherlands could earn 250 guilders a year. Clusius's annual salary when he began teaching at the University of Leiden in the 1590s was 750

Bust of Clusius in Leiden, the Netherlands

guilders. A middle-rank merchant could earn 1,500 guilders per year. During the 1630s, a single bulb might sell for one to two thousand guilders. A house sold in 1633 for three rare tulip bulbs. A farmhouse and land sold for a package of bulbs. Tulips thus had become a medium of exchange. This was not the case for all tulips; less-desired hybrids sold for much less.

The beauty of tulips was not the only thing that people desired; some saw them simply as a way to make easy money. Professional flower growers multiplied. Traveling salesmen—some of questionable dependability—offered bulbs for sale. Some poorer Netherlanders believed that a small investment in growing tulips could produce a big profit.

The Tulip Bubble and Its Bursting

In late 1636 and early 1637, the price of a single bulb of a rare variety of tulip could double in just over a week. Documented examples include a bulb that a trader bought for fifteen guilders which he sold for 175 guilders. One that cost 45 guilders sold for 550; another, even rarer type that a trader bought for forty-five guilders sold for nine hundred. Bulbs might change hands ten times in one day. The price of two thousand guilders per bulb was not uncommon; some went for four thousand to five thousand each. Some traders sold bulbs that were still in the ground. A futures market developed in which people invested on the basis of what they thought bulbs would bring at a certain time in the future.

In early February 1637, the estate of a tulip trader who had died held a private sale and public auction of the tulips he had owned. Single bulbs went for thousands of guilders each. The man's orphaned children (his wife had died earlier) saw the sale bring them a total of 90,000 guilders. They had gone from being poor orphans to wealthy children.

Then, starting in mid-February 1637, the bottom fell out of the tulip market all over the Netherlands within a matter of days. People panicked and sold

Still Life with Flowers
by Hans Bollongier (Dutch, 1639)

their bulbs at any price. A bulb worth thousands one day could not find a buyer the next. A bed of tulips that the owner had bought for six hundred guilders only brought six when he tried to sell it. Some lost almost all of their net worth, which had actually only consisted of tulip bulbs in the ground.

What happened? Several factors. The crazy demand outstripped supply. During the period of the craze no new varieties came on the market (why experiment with possible losers when winners were everywhere?). The speculation centered on the most expensive varieties, and there were simply not enough cheaper bulbs for people to buy. In other words, the money supply and the bulb supply became exhausted.

Moreover, many of the bulbs traded during the height of the craze were actually worthless. The emotion of the market had driven prices up, but then the owners learned that what they had was not worth anywhere near what they had paid;

Lesson 52 - The Flower That Made the Netherlands Famous

thus they could not recoup their investment with any future tulips from that bulb or lot. Many had made their purchases with borrowed money (what the 1929 New York Stock Exchange called buying on margin), so when they could not repay what they owed they defaulted. The market had been emotion driven and not logic driven, so what went up had to come down.

Was the tulip craze crazy? Yes, but it was not the last time such frenzied buying took place. Dahlias in France became the rage in 1838. Dutch gladiolas brought a similar response in 1917. Land prices in Florida shot up in the 1920s, only to come down in the years following. Your mother or grandmother might remember department store customers fighting each other for Cabbage Patch Kids dolls in the early 1980s.

The Dutch Flower Industry Today

Flower sales in the Netherlands bottomed out and recovered after the 1636-1637 mania. Today the Netherlands still serves as the world headquarters for the flower industry, which produces about five percent of the country's gross domestic product.

However, the Dutch flower auctions that were long a staple of the flower market face threats as the market—and the world—change. Some growers have moved to Africa, where they find a warmer and less expensive climate. The Internet has enabled long-distance direct sales to large chain stores and long-distance bidding for flowers in the Netherlands market. Some smaller growers have gone out of business in the new, competitive climate.

The Tulip Folly by Jean-Léon Gérôme (French, 1882) depicts a scene during the tulipomania of the 1630s. A Dutch nobleman watches soldiers trample flower beds in an attempt to limit supply and keep prices higher.

Thus tulips, as with so much else in the world, have a geographic factor. They originated in one geographic area that has certain characteristics, spread across a large geographic area of the globe, and wound up in yet another geographic area where they could flourish in the same kind of soil in which they began. Because this last area had wealth and was a center of trade and learning, tulips spread widely from there. But now that traditional flower center is facing competition from yet another geographic area where people are growing tulips and other flowers. The geography of the earth and the activities of humans combine for fascinating results.

Jesus said the beauty of the flowers that God created surpassed even Solomon in all his glory.

And why are you worried about clothing? Observe how the lilies of the field grow; they do not toil nor do they spin, yet I say to you that not even Solomon in all his glory clothed himself like one of these.
Matthew 6:28-29

Assignments for Lesson 52

Gazetteer Read the entries for Ireland, the Netherlands, and Switzerland (pages 106, 112, and 113).

Worldview Copy this question in your notebook and write your answer: What do you admire about Abraham's faith?

Project Continue working on your project for this unit.

Literature Continue reading *The Day the World Stopped Turning*.

Student Review Answer the questions for Lesson 52.

Members of the U.S. 101st Engineers in Luxembourg (January 1945)

53 When Weather Helped Make History

Seventy years later, Paul Rogers still remembered the weather.

"It was the worst place you could be. It snowed, and it got cold. And I mean cold. People don't know what cold is. Nobody could get to us, or they didn't even try because it was so cold," Rogers said in a 2014 interview.

Rogers was part of Easy Company, 506th Parachute Infantry Regiment, 101st Airborne Division, United States Army, during World War II. Historian Stephen Ambrose wrote a book about the outfit, *Band of Brothers*, in 1992.

German Soldiers in Luxembourg (December 1944)

A native of Missouri, Rogers enlisted in the Army in 1942 and requested an assignment with the parachute infantry. His first jump was on D-Day, June 6, 1944. He landed in a tree. Three months later, he parachuted into the Netherlands. A few days later, mortar shrapnel injured his right arm, which necessitated eight weeks in a hospital in England. He returned to the front in eastern France in late fall.

A Desperate Attack Late in the War

As 1944 was ending, the German army was on the run, retreating before the advancing Allies. After landing in northern France on D-Day, the Allies had moved south to liberate Paris from the Germans. Then they turned east and headed toward Germany, pushing the German army back.

The Allied plan was to invade Germany, capture the German capital of Berlin, and end the war in Europe. Allied forces had already defeated the Germans in northern Africa, invaded Italy, and liberated Rome. Allied victories in northern Europe had led to widespread speculation that the defeat of Germany was only a matter of time.

307

But German leader Adolph Hitler was not ready to give up. He designed a counter-offensive that called for German forces to break through the Allied lines and continue on to capture the port of Antwerp, Belgium. This would divide British and American forces and take a key supply port out of Allied hands. Hitler hoped that this push would stop the Allied advance and turn the tide of the war.

Hitler was so committed to this plan that he ordered troops fighting against Russia on the eastern front to redeploy west to take part in the assault. As a consequence of this move, Russian troops were able to advance with less resistance toward Berlin. Hitler also ordered all able-bodied German men between sixteen and sixty to report for duty to strengthen the German war effort.

For the place of this attack, Hitler chose the Ardennes Forest, in the center of the American line. The assault extended from southern Belgium into Luxembourg. The Ardennes is a heavily forested, hilly region, features that Hitler thought would hide the German advance. The high plateau there has numerous valleys which made tank movements difficult.

The weather in the Ardennes in the fall of 1944 had been miserable: rain, fog, snow, sleet, and cold. Army vehicles had turned the unpaved roads into quagmires. Men who were there reported that the mud was literally knee-deep. The supply of gasoline for vehicles was running short. The Allied advance toward Germany was slow but still moving.

The Action Begins

Then on December 16, 1944, the German counter-assault began. German forces pushed the center of the Allied line back several miles. (My father, Wesley Notgrass, was a U.S. Army soldier who was in the battle. He said that high-ranking officers told them they were not to say they were retreating; they were to describe it as pulling back.) The resulting west-pointing bulge in the Allied line gave the battle its name. Some Germans, dressed in

The author's father, Wesley Notgrass, eating snow in Belgium around the time of the Battle of the Bulge.

American uniforms, driving stolen American jeeps, and speaking good English, moved through the area and spread misinformation among American troops, which caused significant confusion.

As the opposing armies struggled, maneuvered, attacked and repositioned, the weather got worse. It was the hardest winter Europe had experienced for decades. Eight inches of snow covered the ground. The temperature was below freezing for long periods of time, sometimes below zero. My father reported that at one point the thermometer dipped to twelve degrees below zero. Imagine being in a tent or foxhole or moving through a forest in such conditions! Paul Rogers recalled, "It was not only 29 days on the ground, but 29 days in the ground" (in other words, in foxholes). Many of those who were not in the front lines were able to sleep in houses or other buildings, often in the cellar.

Of course, both armies had to suffer through these conditions. Men were poorly clothed and equipped just to be outside in that kind of weather, let alone fighting and trying to keep their equipment functioning. Some had no gloves and wore gunny sacks on their feet. Troops had to start the engines on tanks and other vehicles every few minutes to keep the oil from freezing. Many men had problems with

Lesson 53 - When Weather Helped Make History

their feet for the rest of their lives. Some Germans wore white uniforms to hide their activity.

"Nuts"

In their advance, German forces surrounded the town of Bastogne (pronounced Bass-TONE) in Belgium, near where Rogers and other members of the 101st were positioned. The Germans bombarded the town heavily and caused many American deaths and injuries.

On December 22, the German commanding officer sent a written message by his staff to the American commander in Bastogne, General Anthony McAuliffe. The message demanded that the Americans surrender to avoid total annihilation. McAuliffe sent back a one-word written reply: "Nuts."

The skies cleared on December 23, and American planes began to resupply the troops in Bastogne. General George Patton's Third Army relieved the city on December 26.

This U.S. Army map shows the situation on December 25, 1945. The Germans are on the east (right) side of the map. You can see the "bulge" in the Allied lines in the middle with Bastonge surrounded inside the bulge.

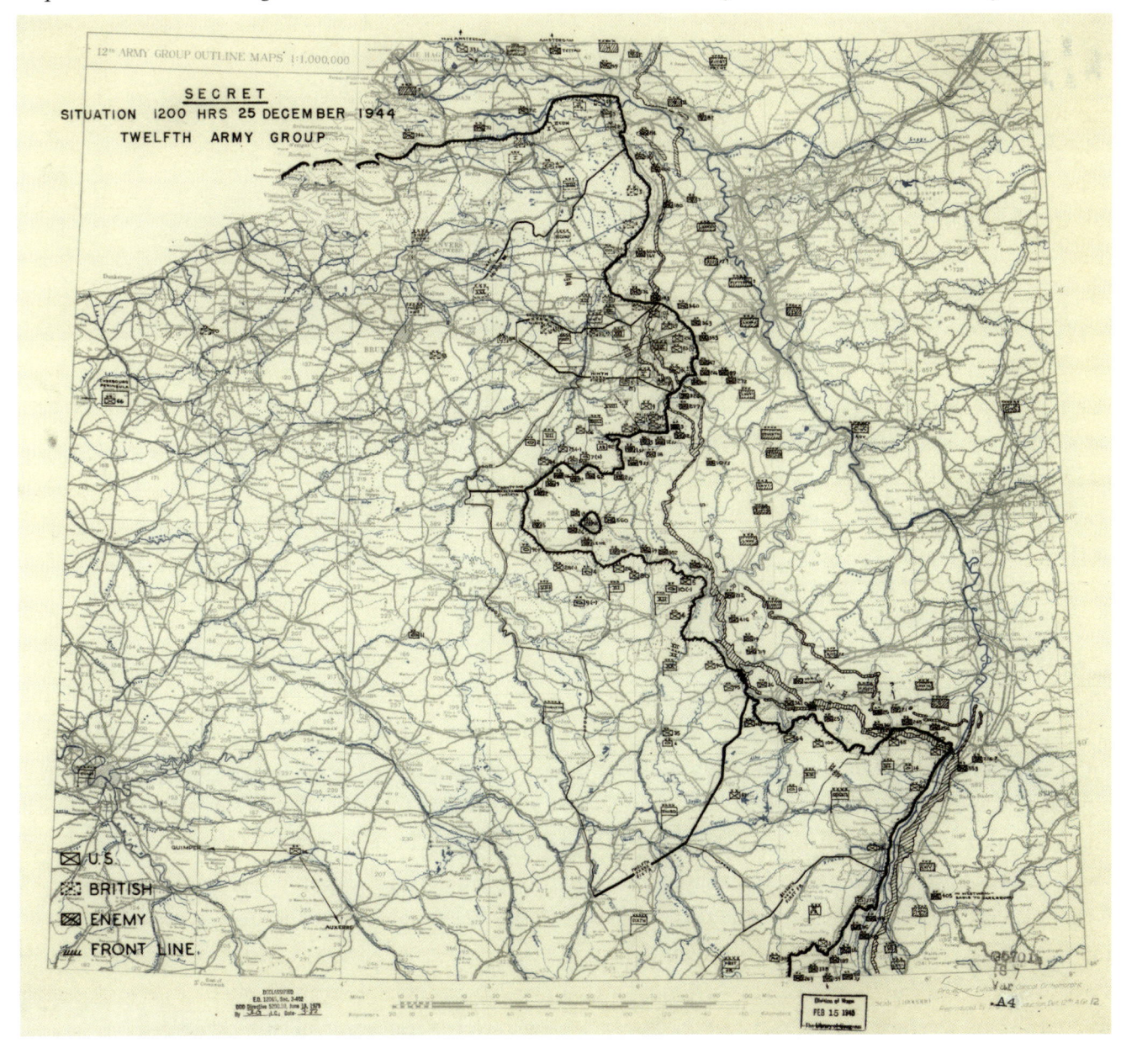

Christmas

The Christmas holiday was on everyone's mind, especially on Christmas Eve and Christmas Day. However, neither side tried to arrange a cease-fire. Some of the heaviest fighting occurred on Christmas Day.

The setting was visually stunning. One lieutenant recalled, "If it hadn't been for the fighting, that would have been . . . the most beautiful Christmas . . . . The rolling hills, the snow-covered fields and mountains, and the tall, majestic pines and firs really made it a Christmas I'll never forget in spite of the fighting." Wherever churches were able to hold services on Christmas Eve or Christmas Day, soldiers attended those services in large numbers.

Outside of Bastogne on Christmas Eve, a squad of Americans were hiding in a barn behind German lines. German soldiers discovered them and captured them. A German captain took the prisoners into the house, had them shave and clean up, and invited them to the party the Germans were preparing. The older Belgian couple who lived there set a full, beautiful table. After the meal, the soldiers joined in singing "Silent Night," the German captors in German, the American prisoners in English. The next day, German soldiers ordered the Americans to head toward a POW camp in eastern Germany.

Allied Victory

The German attack faltered. Clearing skies in late December enabled Allied planes to take to the skies and start attacking German positions. Meanwhile, the German air force (Luftwaffe) was largely out of fuel, thanks in large part to Allied bombing runs into Germany. On January 3, 1945, the Allied counter-attack began shrinking the bulge and pushing the Germans back. By the end of January, the Allies were again advancing full-force to the east. It was the beginning of the end for the Germans. They surrendered to the Americans on May 7, 1945, and to the Russians the next day.

Remains of a Foxhole Near Bastogne (2015)

The Battle of the Bulge was perhaps the greatest single battle ever that involved U.S. troops. Opinions vary as to how many troops were involved (what did it mean to be involved?), but one count estimates 600,000 Germans, 500,000 Americans, and 55,000 British. One casualty count put the total at 80,000 dead or wounded for the Allies, 120,000 for the Germans.

Rogers' Easy Company continued east and was in Germany when the Nazis surrendered. The unit originally had 140 men. By the end of the war, forty-nine men in the unit had lost their lives. Over 360 men rotated into the unit as replacements or transfers. Easy Company had a casualty rate of 150 percent, which means that about 210 of the men who served in it were killed or wounded.

Lesson 53 - When Weather Helped Make History

Many of the men of Easy Company gathered for reunions for decades after the war. Their children continued to organize reunions in more recent years. Paul Rogers died in 2015 at the age of 96. To his last days, he remembered his involvement in the battle when weather helped make history.

In the book of Job, as God humbled Job for his lack of wisdom and understanding, He mentioned snow in the context of war.

Have you entered the storehouses of the snow,
Or have you seen the storehouses of the hail,
Which I have reserved for the time of distress,
For the day of war and battle?
Job 38:22-23

Assignments for Lesson 53

Gazetteer — Read the entries for Austria, Belgium, France, and Germany (pages 101, 102, 104, and 105).

Worldview — Copy this question in your notebook and write your answer: What is the difference between a worldview of faith and a worldview of doubt and fear?

Project — Continue working on your project for this unit.

Literature — Continue reading *The Day the World Stopped Turning*.

Student Review — Answer the questions for Lesson 53.

Scout Camp in the Netherlands (1950)

54 Helping a Continent Be Strong and Free

Millions of people did not have enough to eat.

Farmland, factories, roads, and bridges were destroyed.

Millions were unemployed.

Many feared an imminent threat of foreign invasion.

And these were the crises in the nations that had WON the recent war.

Europe had not fully recovered from the Great War of 1914-1918 when a second world war engulfed the continent from 1939 until 1945. The Nazi onslaught, followed by the conquering Allied invasion, resulted in economic chaos across the length and breadth of Europe that continued into 1947.

Devastation from War

The two world wars destroyed a large part of the Europeans' ability to farm, work, and trade. The people of the victorious Allied nations wanted desperately to resume their normal lives, but they lacked the resources to do so. For example, the standard of living (the availability of goods and services) in France after World War II was about half of what it had been before the war. Industrial production in Italy after the war was about one-fifth of what it had been before the war; agricultural output there was one-half of prewar levels (Italy had been an ally of Germany at the start of the war, but it surrendered in 1943 and joined the Allies). The agricultural harvest of 1947 in Europe was particularly poor and left people even more desperate for food. Officials in France expressed concern about their limited supply of bread.

In addition to all of the economic hardships, political questions simmered. For instance, American, British, French, and Russian military forces occupied and controlled defeated Germany; but the Allied nations had no long-term plan for what would happen there.

What To Do About Germany and the Soviets

Many European political leaders and much of the general public in Europe thought that Germany should somehow pay for starting another war that resulted in so much suffering, death, and destruction. However, history taught that "making Germany pay" would cause more problems than it would solve. The victorious nations in World War I had required Germany to pay dearly after that

312

Lesson 54 - Helping a Continent Be Strong and Free

conflict. The Treaty of Versailles that had ended that war had placed the blame for the war fully on Germany, stripped Germany of its military forces, and forced Germany to pay reparations (payments of money) to the victorious nations. The burden of these payments had helped prevent Germany from recovering economically after that war. Germany eventually was unable to pay the reparations, and the victorious nations cancelled them.

A large number of Germans deeply resented this punishment. Adolph Hitler played on this resentment to justify his seizure of power and his aggression against other countries. Allied nations knew that harsh treatment of Germany after this second world conflict might result in yet another conflagration. But how could the nations of Europe recover and not live in fear of another war?

In addition to these realities in Europe, a new factor had emerged in the geopolitics of the continent. Geopolitics refers to the influence of geography on political activities. To the east of Europe lies Russia. In 1917 Communists had seized power in Russia and neighboring countries and had set up the world's first Communist government, the Union of Soviet Socialist Republics (U.S.S.R.), also called the Soviet Union. The country was sometimes called Russia because that country was by far the largest part of the U.S.S.R.

The Soviets had joined the Allies against Germany in the second war, not because of any Russian commitment to freedom and democracy but because the Soviet Union feared being attacked by Germany and saw an opportunity to expand its power by conquering nations that Germany had previously seized. During and after the war, the Soviets had in fact taken over European nations to its west, including Poland, Hungary, Finland, Romania, Bulgaria, and the eastern part of Germany. This was the line of countries that in 1946 Winston Churchill called the Iron Curtain. Communists had openly expressed the goal of world domination. Many people feared that the weakened nations in western Europe might be vulnerable to Soviet conquest as well. Once an ally, the Soviet Union had become the new enemy.

A Period of Instability

Communist political parties had already gained strength in France, Britain, and other European nations. In some countries, an alarming number of Communists had won election to the national parliament. Germany had taken over Greece in 1941; when the Nazis withdrew from Greece, Greek Communists threatened to take over. Only assistance from the United States had helped Greek military forces defeat the Communists. The possibility of Communist takeovers in other European countries was no mere theoretical possibility. It had already happened in eastern Europe. Might it also happen in France, Britain, and other countries in western Europe?

World geopolitics were in a period of change and instability. Great Britain had suffered tremendously during the war, and it was no longer able to support its extensive world empire. The crisis in Greece had arisen because Britain could not provide continued assistance to the free government there. After years of turmoil, India, the "jewel of the British Empire," became independent in August of 1947. The loss

In March of 1947, people in Krefeld, Germany, gathered to draw attention to their need for "kohle" (coal) and "brot" (bread). Notice the destroyed building in the background.

of empire meant the loss of many of Britain's trade arrangements within the empire. Only two great powers existed in the postwar world that had the ability to influence significantly what happened in other parts of the world: the United States and the Soviet Union. These two countries had opposite motivations and goals, and Europe lay between them.

The Truman Doctrine

In March of 1947, President Harry Truman described to Congress the situation in Greece. He spoke of "the gravity of the situation which confronts the world today," referring to Greece and also another part of the world, Turkey, which needed help to modernize its economy and raise its people's standard of living. Truman's proposal to give aid to countries that were resisting domination by domestic rebels or foreign invaders came to be known as the Truman Doctrine. Congress soon approved emergency aid to Greece and Turkey. A growing number of people in the United States believed that the U.S. needed to do more to help Europe resist Communist domination.

The goal of United States foreign policy was not the defeat of Soviet Communism and a rollback of their control over countries they had come to dominate. Instead, American policy was one of containment: containing Soviet influence where it existed at the time. The U.S. accepted a balance of power between the two countries and the reality of spheres of influence that each controlled.

The Marshall Plan Proposed

Secretary of State George Marshall was a retired U.S. Army general. He had been the Army Chief of Staff, the highest ranking officer in the Army, during World War II. Marshall understood the situation in Europe and all of the dangers the continent faced. On June 5, 1947, Marshall delivered the commencement address at Harvard University. He began by stating "the world situation is very serious." In the speech he proposed a program in which the United States would provide assistance to the countries of Europe, to help the Europeans get back on their feet and to be able to resist the Communist threat. This assistance would include factory parts, food, money, and other items. But Marshall wanted Europeans to take the initiative, develop a proposal, and present it to the United States. His goal was for Europe to return to prewar conditions in four years. Marshall said the program was intended to fight "hunger, poverty, desperation, and chaos."

In response to Marshall's proposal, representatives from sixteen European countries met in Paris to discuss what they needed. It was the first time that representatives of the nations of Europe had examined the economic condition of the entire continent. However, their discussions did not go smoothly at first. For instance, much of what they did was simply to come up with a list of what they each wanted, which resulted in a proposal of about

U.S. Secretary of State George Marshall (1947-1949)

Lesson 54 - Helping a Continent Be Strong and Free

$28 billion in aid, an amount the United States was unable and unwilling to provide.

In addition, centuries-old geopolitical rivalries among the European nations resurfaced. Would British factories be rebuilt first, or French? When would factories in Allied-controlled Germany receive help? Would Germany be able to resume steel production, or would the other European nations keep Germany at a weakened level while they rebuilt and rearmed? The European nations and the United States invited the Soviet Union and the Soviet-controlled countries in Eastern Europe to take part, but the Communists refused to participate, as everyone in the West thought they would. The Soviets charged that the American proposal was just a way for the U.S. to extend its "imperialistic" control over Europe.

Deliberations of Congress

Congress refused to approve the Europeans' initial proposal. One factor in the deliberations of Congress was that in 1946 the Republican Party had gained majorities in both the Senate and the House of Representatives for the first time since 1932. Many Republicans opposed sending additional assistance to Europe. They thought that the United States had been involved enough in international affairs, and they wanted to turn their attention to long-neglected needs within the U.S. itself.

A geopolitical reality for the United States was that the Atlantic and Pacific Oceans had protected this country from the extensive physical destruction and loss of life that Europe and Asia had suffered. The U.S. lost over 418,000 military and civilian personnel because of the war, but the loss of life worldwide numbered in the tens of millions. American involvement in the war had broken the country's long-standing pattern of isolation from foreign entanglements, and many Americans wanted to see the country return to its previous isolation.

Despite the initial rejection of the European aid request, Congress continued to discuss some

This woman is collecting a signature from a man in North Carolina to "urge speedy action by Congress on the Marshall Plan" in 1948.

form of aid through 1947. President Truman made another speech to Congress in December of that year, urging approval of the package. During this period, many people in the U.S. were becoming concerned that Communist agents were infiltrating the U.S. government and other parts of American society. In late February 1948, an event took place in Europe that made the request for assistance even more urgent. Communist forces directed from the Soviet Union took control of Czechoslovakia. This added yet another country to the Soviet bloc and extended the Iron Curtain even further west. Congress approved the European Recovery Program (usually called the Marshall Plan), and Truman signed it into law on April 3, 1948.

The Impact of the Marshall Plan

Assistance began flowing to Europe immediately and made a huge difference in specific ways. The U.S. eventually shipped wheat, tractors, cotton, tires, airplane parts, and much more, including materials to rebuild factories, roads, and bridges, as well as some direct financial aid to seventeen countries in Europe. The part of Germany still controlled by the Allies, called West Germany, participated also.

Europe began a remarkable recovery. People began having sufficient food. French harbors, seventy percent of which had been destroyed in

Many different artists created posters highlighting the goals and effects of the Marshall Plan. The poster at left by F. J. F. Nettes says "Together Work, Together Strong" in Dutch. The poster at right by Gottfried Honegger-Lavater says "Inter-European Cooperation for a Higher Standard of Living" in French.

the war, were rebuilt. Tractors, seed, and fertilizer revitalized European agriculture. Mining and industrial equipment helped put people back to work. Germany produced twice as many cars in 1953 as it had in 1936. One polio victim in France received an iron lung from Denver. Above all else, the Marshall Plan restored the confidence of Europeans that they could endure materially and be free politically. The influence of Communist parties in western European countries fell to insignificance. Communism had nothing that most Europeans found attractive once the people had what they needed to live in freedom and without want.

The United States economy grew also. Europeans were once again able to trade with the U.S. This meant that Europeans sold goods to the U.S., and thus Europeans had the money to buy goods from the U.S. The American economy had begun to recover from the Great Depression during the war through its production of wartime materials, but the increased postwar trade with Europe helped the U.S. economy to grow even more.

The policy of containment limited the Soviet Union's sphere of influence where it was in Europe. However, this did not mean that the Soviets quit trying. Another geopolitical crisis developed in June of 1948. The Soviets controlled East Germany, while the Allies controlled West Germany. In addition, the two sides maintained divided control over Berlin, the German capital that lay completely within East Germany. On June 24, 1948, the Soviets blockaded all land access to West Berlin. In response, President

Lesson 54 - Helping a Continent Be Strong and Free

Truman ordered an airlift of all necessities to keep West Berlin functioning and free. The Berlin Airlift continued until May of 1949, when the Soviets lifted the blockade. Thus, the United States defused another confrontation with the Communists.

The Marshall Plan aid program changed dramatically in 1950, when North Korea invaded South Korea. The U.S. government directed more of its resources to this new war effort, which presented an even more alarming illustration of the Communist threat. European countries also began focusing on military preparedness in case the Soviet Communists moved on Europe. Marshall Plan assistance continued until 1952.

Aid from the United States to Europe under the Marshall Plan totaled $13 million. It wasn't everything that European leaders requested, but it was enough to make a difference. Perhaps the most important effect of the Marshall Plan was that the people of the United States had the opportunity to provide humanitarian aid to those in need, as they have done many times before and since.

Long-Term Effects

Germany remained divided between East Germany, ruled by Communists, and West Germany, which became a free democracy in 1949. The failures of the Communist system eventually led to the fall of Communism in Eastern Europe beginning in 1989. Germany officially reunited in 1990. The Soviet Union dissolved in 1991.

Ten European nations, along with the United States and Canada, formed the North Atlantic Treaty Organization (NATO), which coordinated military activities among member nations. NATO stood as a defense against the possibility of Communist invasion of any participating country. The NATO charter says that member countries will see an attack on one member country as an attack on all member countries. NATO has since grown to 30 countries, including some former allies of the Soviet Union.

European countries began to lay aside their rivalries and work together to coordinate trade and production. Trade tariffs were lowered or eliminated. This trend of cooperation eventually led to the creation of the European Common Market in 1957, which became the European Union (EU) in 1993. Most of the EU uses a common currency, the euro. The EU is not a perfect arrangement. In 2016 Great Britain began the process of withdrawing from it. However, cooperation is better than renewed armed conflict, of which Europe has seen plenty in the previous generations.

The Marshall Plan did not directly fight Communism. Instead, it strengthened the nations of Europe, and this helped Europe achieve economic and political stability, which helped Europe to resist Communism. The doctrine of containment lay behind American involvement in the later geopolitical conflicts in Vietnam and Korea, where Communists wanted to extend their control over free areas.

The Marshall Plan redefined geography. The Atlantic Ocean was no longer a barrier to cooperation. The Atlantic Community, which includes countries on both sides of the ocean and is exemplified by NATO, came into existence. The United States and Europe, which once had kept each other at arm's length, began working together for mutually beneficial aims.

Geopolitics, international negotiations, and the deliberations of Congress came together in the Marshall Plan, which addressed a situation that literally involved life and death for multitudes of people and for the nations in which they lived.

Isaiah gave a word from the Lord to Israel regarding the nature of true religious devotion. It is not just going through the motions of fasting, but instead:

Is this not the fast which I choose,
To loosen the bonds of wickedness,
To undo the bands of the yoke,
And to let the oppressed go free
And break every yoke?
Is it not to divide your bread with the hungry
And bring the homeless poor into the house;
When you see the naked, to cover him;
And not to hide yourself from your own flesh?
Isaiah 58:6-7

Assignments for Lesson 54

Geography — Complete the map skills assignment for Unit 11 in the *Student Review Book*.

Project — Continue working on your project for this unit.

Literature — Continue reading *The Day the World Stopped Turning*.

Student Review — Answer the questions for Lesson 54.

55 Faith

Imagine a group of children on their first day at summer camp. At the end of the day, the camp director leads everyone to a shallow creek. Across the creek stretches a rope bridge. Planks held in place by ropes make the walkway, and a network of ropes serves as the sides and handrails.

First-time campers are terrified at the thought of even stepping onto the bridge, not to mention walking across it. The bridge seems weak and unstable to them, and walking on it seems far too risky to attempt.

Camp counselors, who have worked at the camp for several summers, stride fearlessly onto the bridge. They get to the halfway point, turn around, and call to the reluctant campers to join them. The first-timers look at each other in amazement and challenge each other to be the first to step out.

Finally one rookie takes a tentative step onto the bridge. It sways a bit but holds fast. Then she takes another step, and another. Her best friend follows her, frightened at first but gradually more confident. After these two campers return to solid earth, other campers venture out. Soon all the campers have walked onto the bridge. By the end of their week at camp, the campers think little about walking across the bridge. For one thing, they are used to it; for another, they know that the mess hall awaits them on the other side.

What the counselors knew that the campers did not know, is that no one had ever fallen from the bridge. In addition, the counselors knew that state inspectors review the bridge before camp opens every year, and workers replace any sections that are questionable in the least. The bridge was strong enough to hold them.

(This story is an adaptation of one told by Bob Hendren.)

The Definition of Faith

Hebrews 11:1 defines faith as the assurance of things hoped for; in other words, the conviction of things not seen. The rest of that chapter is a list of heroes of the Bible who lived just that way: they acted on the basis of their assured hope and conviction, even though they could not see the outcome of their efforts immediately. This is how the writer of Hebrews wanted his readers to live. In the face of persecution, they needed not to turn back but to keep their eyes on the unseen Jesus, who Himself had lived by faith in God. It is in this way that the people of God gain His approval.

The prime example of faith in the Bible is Abraham (see Romans 4). When God told the childless Abraham that his descendants would be as numerous as the stars, Abraham believed Him (Genesis 15:1-6). On the surface of things, it appeared that God's word to Abraham was impossible; but Abraham trusted that God would do what He said; and He did. Abraham also trusted when God told him to sacrifice Isaac. Abraham obeyed God even though it appeared that this would destroy the promise. Hebrews says that Abraham believed that God would bring Isaac back from the dead (Hebrews 11:19). Abraham did not have everything laid out before him as an assured result, but he believed God and God kept His promise.

It is true that Abraham slipped in his faith when he had a child by Hagar. Even the faith of Abraham, the model of faith, was not flawless. God kept His promise to give Abraham a son by Sarah (Genesis 16:1-6). This tells us that we can depend on God's promises not because of the unwavering nature of our faith but because of the nature of the One who made the promises.

The story about the rope bridge at camp illustrates several truths about faith.

(1) As the Bible uses the word, faith means trust. Only rarely does the Bible describe faith as an intellectual assent to a set of doctrines. Jude 3 speaks of "the faith which was once for all handed down to the saints." This verse refers to "the faith" as the body of doctrinal truths of Christianity, but even in Jude they are truths to which disciples are to commit themselves entirely and to trust as true, not just a list of propositions to which they give intellectual agreement.

The young campers could see that people walked out on the bridge. As they stood on the bank watching, they might have thought, "I believe that bridge would hold me and not fall," but it was only when they stepped out onto the bridge that they would truly be trusting it. As James says, "Faith, if it has no works, is dead" (James 2:17).

16th-century engraving of Abraham's sacrifice by Phillips Galle and Frans Floris

The same is true regarding believing in God and Jesus. You might stand on the sidelines and think, "I agree that God exists and that Jesus is His Son," but it is only when you step out, when you live on that basis and trust that truth for your life, that you have faith, in the Biblical sense of the word.

(2) The important aspect of faith is the object of faith, not the strength of one's faith. A strong faith in a weak bridge might get a person in trouble, but even a weak faith in a strong bridge will get a person across the creek.

Being absolutely sure of a philosophy or religion that is in fact wrong will not result in a successful life here or eternity hereafter. On the other hand, even a tentative first step onto a strong bridge will take you in the direction you need to go. As you walk in faith (trust), your faith will grow. You cannot expect to have a mature faith with your first step. Just as wisdom and understanding grow, faith grows with the experience of living by it.

(3) A person's faith is imperfect. Faith and doubt can coexist in the same person. A man whose son was plagued by an evil spirit once came to Jesus wanting help. "If you can do anything, take pity on us and help us," the man pleaded. Jesus replied, "'If you can'? All things are possible to him who believes." The man responded, "I do believe; help my unbelief" (Mark 9:22-24).

This desperate father's confession, "I do believe; help my unbelief," speaks for the hearts of many people. You can believe and not believe at the same time; in other words, a person's faith is imperfect. Even when you step out onto the bridge, you might still have doubts. The solution is not to retreat in doubt or to be paralyzed until you can be 100% absolutely sure. The solution is to act in faith. The goal of following Jesus is not to have a flawless faith at the start, but to begin to believe in the perfect Object of faith.

(4) Faith is not "giving God a try." Believing in Jesus is not a shot in the dark, a last gasp, a shrugging the shoulders, "Why not? I've tried everything else" approach. Faith is not thinking that if it doesn't work out to your liking you can always go back to your previous philosophy or worldview. Faith means giving your all, really trusting God with everything, stepping out onto the rope bridge with no turning back. Faith means placing your life in the hands of Jesus. Faith means putting your trust in Jesus as Lord and Savior, counting on His death on the cross as the source of your forgiveness. If you do this, you will find salvation and wholeness. There is no escape clause with Jesus. The bridge will hold you up, but you really have to step out onto it to experience it holding you up. The good news is that Jesus is strong enough to save you and to take you where you want and need to go.

Influence of Faith on Worldview

Faith determines whether you see Creation as the handiwork of God or the result of chaos. Since you were not present at the Creation, you have to decide which explanation you trust to be true. God says that He created the world.

Faith determines how you live. You can live selfishly, taking advantage of people, fudging on the truth, manipulating people with tears and threats, and using people for your pleasure and your own sinful purposes. Or, you can live by trusting that God's ways of kindness, love, and patience are right, that He will provide, and that telling the truth is the best way to live, even in situations when living this way does not appear to be best.

You can live with the unbelieving worldview that people are out to get you and so you have to be defensive, or you can trust that God will take care of you and so you can be kind and generous.

You can choose to work and strive (and perhaps cut corners) in order to have and to acquire, which is trusting in yourself; or you can use your life for God's purposes and, while still working, trust that God will provide. Jesus described the worldview that He wanted His followers to have: "Do not store up for yourselves treasure on earth. . . [and] do not be worried about your life, as to what you will eat or what you will drink"; instead, "Seek first His kingdom and righteousness, and all these things will be added to you" (Matthew 6:19, 25, 33).

Faith influences how you see prayer. Faith is not trusting in prayer, but it is trusting in the One to whom you pray—again, faith in the right Object.

Faith involves trusting that God will work things out for good, according to His will (Romans 8:28), as opposed to believing that you must take matters into your own hands. Faith also involves knowing when to act in faith as opposed to intellectually believing but not believing enough to do anything (e.g., saying "I believe that bridge would hold me up," but never stepping onto it). This requires wisdom. God in His wisdom allows us to be in situations where we must make the decision to act or not, and this helps us grow in faith.

The opposite of faith is fear. Instead of trusting that God's way is right, you fear in real-life situations that God's way won't work. Instead of stepping onto the bridge, you stay on the bank. Instead of living with all of the possibilities that faith in God opens up to you, you remain a prisoner of your own fears and self-imposed limitations. A prisoner bound by fear has a view of the world that is different from that of a person who believes that he is free to live by trusting in the Creator of the universe.

In his letter to the Christians at Ephesus, Paul prayed for his readers to have a distinctively Christian worldview, with Christ dwelling in their hearts through faith.

For this reason I bow my knees before the Father, from whom every family in heaven and on earth derives its name, that He would grant you, according to the riches of His glory, to be strengthened with power through His Spirit in the inner man, so that Christ may dwell in your hearts through faith; and that you, being rooted and grounded in love, may be able to comprehend with all the saints what is the breadth and length and height and depth, and to know the love of Christ which surpasses knowledge, that you may be filled up to all the fullness of God. Now to Him who is able to do far more abundantly beyond all that we ask or think, according to the power that works within us, to Him be the glory in the church and in Christ Jesus to all generations forever and ever. Amen.
Ephesians 3:16-21

Assignments for Lesson 55

Worldview — Recite or write the memory verse for this unit.

Project — Finish your project for this unit.

Literature — Finish reading *The Day the World Stopped Turning*. Read the literary analysis and answer the questions about the book in the *Student Review Book*.

Student Review — Answer the questions for Lesson 55. Take the quiz for Unit 11.

Seljalandsfoss Waterfall, Iceland

12 Northern Europe

The Sami people of northern Scandinavia—especially in Finland—live on the edge in many ways. The country of Estonia is a prime example of life in a digital world. Residents of the Faroe Islands have a distinctive history, culture, and lifestyle. We study the importance of land surveying and make special note of the Struve Geodetic Arc in Europe. The worldview lesson examines the basic question of the existence of God and the significance of that fact.

Lesson 56 - Living on the Edge: The Sami of Finland
Lesson 57 - Is This What Tomorrow Looks Like?
Lesson 58 - They Do Things Their Way: The Faroe Islands
Lesson 59 - Surveying the Matter: The Struve Geodetic Arc
Lesson 60 - The Existence of God

Memory Verse Memorize Hebrews 11:6 by the end of this unit.

Books Used The Bible
Exploring World Geography Gazetteer
Kidnapped

Project (Choose One)

1) Write a 250-300 word essay on one of the following topics:
   - "Deciding to Believe" See John 7:17 and Lesson 60.
   - Discuss the pros and cons of technology, specifically the trends in Estonia that Lesson 57 discusses, such as digital signatures, debit cards, personal identification numbers, online voting, and so forth.
2) Research the profession of surveying, especially focusing on its traditional and modern tools. Take a simple survey of the property on which you live.
3) Research George Washington's work as a surveyor: how he did it, how it helped him, and other topics.
4) Imagine you were taking a trip to the Faroe Islands. Find out how much it would cost, what special preparations you would need to make (such as taking enough warm clothes!), and what you would like to do there.

Literature

Kidnapped narrates the thrilling adventures of teenage orphan David Balfour from the lowlands to the highlands of Scotland in the turbulent mid-1700s. David has plenty of troubles of his own when his uncle refuses to grant his rightful inheritance. After his uncle has him kidnapped and supposedly removed from the scene, David becomes involved in a much bigger struggle. He falls in with the Jacobites, Scottish rebels who sought for decades to restore the British throne to the House of Stewart (sometimes spelled Stuart).

Robert Louis Stevenson's large body of literary works has lasting influence and popularity. Two of his best-known works, *Treasure Island* and *A Child's Garden of Verses*, attest to the diversity of his interests and talent. He was born in Scotland in 1850. He traveled widely, often combining a love for adventure with trying new climates for his lifelong struggles with his health. He bought an estate on the Pacific island of Samoa in 1890, where he died in 1894.

Plan to finish *Kidnapped* by the end of Unit 13.

Note: Because a grasp of the historical background will be helpful in understanding the novel, please read the literary analysis of *Kidnapped* in the *Student Review Book* before reading the book.

56 Living on the Edge: The Sami of Finland

They are not Lapps, and they do not live in Lapland. They call themselves the Sami, and they call their homeland Sapmi. The Sami are the traditional reindeer-herding people of the northern Scandinavian peninsula. They are perhaps the last indigenous people in Europe.

The Sami have lived in this region for many centuries, so you might think they have a settled way of life. In fact, in some ways they have lived and continue to live on the edge.

On the Edge of Europe

Sapmi lies on the extreme northern edge of Europe, largely within the Arctic Circle. For many years the Sami lived as nomads in a place known for long, cold, dark winters. They lived a simple life by following reindeer herds. Today a large number live in villages and maintain herds in those places.

Reindeer play an important role in Sami life. They provide people with milk and meat; hides that are used for tents, shoes, and clothes; antlers and bones that the Sami make into tools, weapons, and crafts; and sinews that they use in sewing. The Sami language has an estimated one thousand words related to the appearance and behavior of reindeer.

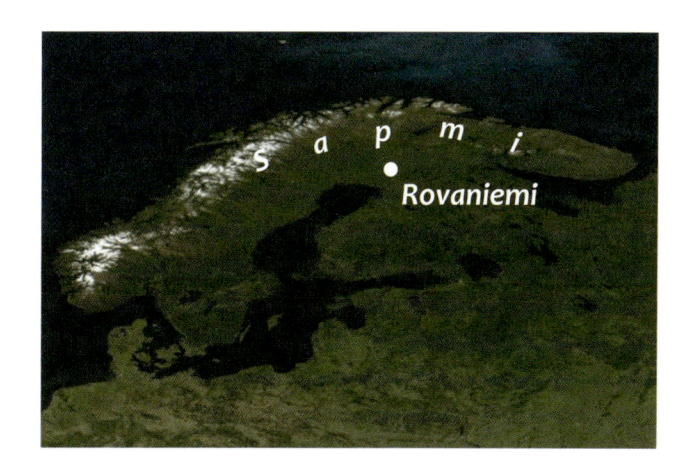

Since the countries of Scandinavia did not draw their boundaries in a way that respected the Sami homeland, Sami people are spread across Norway, Sweden, Finland, and the Kola Peninsula of Russia. Estimates vary as to the number of Sami in each country. There are perhaps 30,000-40,000 in Norway, 20,000 in Sweden, 9,000-10,000 in Finland, and 2,000 in Russia. In this lesson we focus on the Sami of northern Finland.

On the Edge of Their Way of Life

As has been the experience of many indigenous people in many places around the world, the Sami have been the victims of discrimination and a lack of respect for their culture on the part of the majority ethnic group of the country. For many

325

Traditional and Modern Examples of Sami Shelter

years schools in Sapmi could not teach in Sami languages. As a result, fewer and fewer Sami could speak and read their language. However, in 1995 the Finnish constitution granted the Sami the right of self-government with regard to their language and culture. Now the Sami must by law receive services in their own language.

Land rights are another major concern for the Sami. The government owns about 90 percent of the land in Finnish Sapmi, so for the most part they do not have the rights that come with property ownership. For instance, if the government wants to harvest timber in a Sami area in a way that would threaten a reindeer habitat, there is little that the Sami can do to oppose it.

Extensive pit mining is already taking place in Sapmi to obtain gold, nickel, copper, and other minerals. The government has proposed building a railroad through Sapmi to connect with a port in Norway on the Scandinavian coast (Finland itself does not have a northern coastline). This would enable goods to be shipped more easily along the Northeast Passage to the north of Europe, taking advantage of the northern water routes being frozen for a shorter period during the year. Advocates of the railroad say it will create jobs and bring economic growth to the region. Opponents say that it could destroy the Sami way of life by dividing reindeer migration routes and grazing land.

Finland on the Edge

Finland itself has spent many years on the edge between competing world powers. Sweden ruled Finland from the 1100s until 1809, when czarist Russia took it over. When the Communist Revolution of 1917 ousted the tsar, Finland declared its independence from Russia.

However, the Soviet Russians still wanted Finland. As a sideshow of World War II in Europe, the Soviet Union invaded Finland in 1939. The Winter War lasted into 1940 and Russia gained about ten percent of Finnish territory. But the Soviet invasion bogged down in the terrible Finnish winter, and the Soviets did not conquer the entire country.

The Finnish government allied itself with Germany to get help to prevent a Soviet takeover.

Finnish Soldiers Making Breakfast (1939)

Lesson 56 - Living on the Edge: The Sami of Finland

However, the Finns retained some of their independence. For example, they refused to hand over Finnish Jews to the Nazis.

In 1944 the tide of the war turned against Germany. The Soviet Union ordered Finland to expel the Germans, which it did. Before the Germans left the northern city of Rovaniemi, and after the residents had evacuated, the Germans set the city on fire and destroyed 90% of it. Finnish architect Alvar Aalto oversaw the rebuilding of the city. The new street layout when seen from the air looks like the head of a reindeer, antlers and all.

The rebuilt and bustling Rovaniemi, a little south of the Arctic Circle, is now the capital of the Sapmi region of Finland. Five years after the war, Eleanor Roosevelt wanted to visit the Arctic Circle. The Finns built a log cabin for her just outside of Rovaniemi. In later years, several world leaders visited the cabin, including Soviet premier Leonid Brezhnev and Israeli prime minister Golda Meir. Developers built a rural-style wooden village around the cabin, which became Santa Claus Village, a tourist destination for over a half-million people every year. Every Christmas the Rovaniemi postal system receives 700,000 letters to Santa from all over the world.

During the Cold War, Finland had to walk a tightrope between going its own way and avoiding Soviet domination. Finland did not accept aid from the Marshall Plan because the Soviets pressured them not to. Finland did not join NATO because of Soviet pressure, but neither did it join the Warsaw Pact, the Communist response to NATO that consisted of the Soviet Union and Soviet-dominated

Rovaniemi, Finland

Reindeer Sleigh Rides in Rovaniemi

East European countries. Finland's policy moves led to the creation of the term Finlandization, which means not questioning the actions of a powerful neighbor in exchange for a degree of independence.

With the crumbling of the Soviet Union and its satellite states, Finland has moved politically closer to the West. It became a member of the European Union in 1995, and it conducts joint military exercises with NATO.

The Role of Helsinki

The capital of Finland is Helsinki, a city on the southern coast of the country. That coast lies on the Gulf of Finland, an arm of the Baltic Sea. On the edge between Western democracies and the Communist Soviet Union, Helsinki has hosted several important international meetings and summits.

The 1975 Helsinki Accords were the result of meetings between representatives of the United States and the Soviet Union and their allies. The Accords guaranteed the Soviet Union that Western nations would respect the existence of the Communist governments in Eastern Europe. On the other hand, the Accords stated guarantees about human rights that Soviet dissidents later used to pressure Moscow into easing political repression in the U.S.S.R. Sixteen years later, the Soviet Union was gone.

In 1990 U.S. President George H. W. Bush and Soviet leader Mikhail Gorbachev met in Helsinki after Iraq had invaded Kuwait. Soviet-supported Communist governments in Europe had already begun falling. At the summit in Helsinki, Gorbachev supported the U.S.-led Gulf War that pushed Iraqi troops out of Kuwait. Gaining Soviet support for an American military activity was a major diplomatic victory for the U.S.

In 1997 President William Clinton and Russian president Boris Yeltsin met in Helsinki. Other meetings between lower-level officials took place

Lesson 56 - Living on the Edge: The Sami of Finland

there in subsequent years, and in 2018 President Donald Trump and Russian president Vladimir Putin met in Helsinki.

The Sami people have faced and continue to face many difficulties. However, they have worked hard to maintain their ethnic identify and to care for the land they inherited from their ancestors. The prophet Micah described the messianic ideal when everyone can live in peace on their own land.

Donald Trump and Vladimir Putin in Helsinki (2018)

Each of them will sit under his vine
And under his fig tree,
With no one to make them afraid,
For the mouth of the Lord of hosts has spoken.
Micah 4:4

Assignments for Lesson 56

Gazetteer — Study the map of Northern Europe and read the entries for Finland and Iceland (pages 115, 120, and 121).

Worldview — Copy this question in your notebook and write your answer: How do you know God exists?

Project — Choose your project for this unit and start working on it. Plan to finish it by the end of this unit.

Literature — Begin reading *Kidnapped*. Before you begin, read the literary analysis in the *Student Review Book* to learn the historical background to the novel. Plan to finish the book by the end of Unit 13.

Student Review — Answer the questions for Lesson 56.

Tallinn, Estonia, on the Baltic Sea

57 Is This What Tomorrow Looks Like?

Quick. What might be the Internet access capital of the world?

Silicon Valley, California? Singapore? London?

Try Tallinn, Estonia. Really.

The Setting

Estonia is a small country, about twice the size of New Jersey. Its population is about 1.3 million people, roughly the same as Dallas, Texas. Some 400,000 of them live in the capital of Tallinn. The country is located on the Baltic coast, with Latvia to its south, the Baltic Sea to the west, the Gulf of Finland to the north, and huge Russia looming to its east.

Much of Estonia has a low, wet, coastal landscape. Its highest point is 1,043 feet above sea level. Many rivers flow through Estonia, the longest being 90 miles in length. Because of ocean currents and resulting winds, the country enjoys more moderate temperatures than Russia.

The Background

Estonian national identity goes back to medieval times. Clans led by elders lived there, and Viking warriors began coming through in the 800s. Tallinn dates to the 1200s when a nobleman built a castle in that location. It was a key member of the Hanseatic League of trading cities in northern Europe. The wealth that came as a result of trade financed elaborate church buildings and well-appointed merchants' homes.

However, Estonia has been dominated by other countries, especially Russia, for much of its history. It was part of the Russian Empire when the Communist Revolution occurred in 1917. Estonia declared its independence the next year. Then during World War II the Soviet Union took back control in 1940. Some 60,000 to 70,000 Estonians were killed or deported, and the Communists moved Russians into the country. In the back-and-forth military

Trail Through Mukri Bog

Lesson 57 - Is This What Tomorrow Looks Like?

action that followed, Germany gained control for a time. After the war, Russia again ran the country and made it part of the Union of Soviet Socialist Republics.

The U.S.S.R. eventually crumbled, and in 1991 Estonia again declared its independence. The country had been a victim of Communist inefficiencies. It was dreadfully behind in technology. It was not uncommon for someone who applied for a landline telephone to have to wait ten years to get one. At independence, each citizen received from the government a payment—equal to a little over ten U.S. dollars—to start life anew.

The New Estonia

The first generation of free Estonian leaders were mostly young. For example, the first prime minister who took office after the Soviet era was 32 years old. The national leaders decided to look forward and not backward in rebuilding their country. They knew that digital technology was where the world was heading, and they wanted to be a leader in it.

In 2000 Estonia passed a law that gave digital signatures equal validity with written signatures. This opened the door to a society and an economy that is mostly paperless. Since no Estonian had used a checkbook in the Communist days, banks quickly went to issuing debit cards. Filing one's annual income tax return takes a few minutes and can be done online. An Estonian can register and start a business in about twenty minutes online. About one-third of the country votes online from home. Citizens can access 99% of the country's public services online at any hour.

In the early 2000s, Tallinn created a free, citywide Wi-Fi network. Every child that is born receives an eleven-digit identification number that he or she uses in many aspects of living, conducting business, and relating to the government—sort of a Social Security number on steroids. The country's elementary schools (all schools are wired for the Internet, of course) teach children computer programming—not just computer literacy or computer skills, but programming.

Estonia is connecting with other countries through its online accessibility. Finland and Estonia have an agreement through which each country accepts the other's digital ID cards. This enables access to a person's medical records wherever they are needed since the records are stored—you guessed it—online. In addition, Estonia has established a policy that encourages foreigners to set up businesses in Estonia without actually moving there. It's called virtual residency or e-residency. Registration involves paying a fee to Estonia, but each company pays taxes in whatever country it owes them. The companies then pay monthly expenses for accounting and administration in Estonia. One big advantage of this is that companies registered in Estonia can do business throughout the European Union, which Estonia joined in 2004. The country became a member of NATO the same year. Estonia has also created the e-visa, which makes travel to Estonia easier for "digital nomads," people who work remotely in different places around the world.

Contactless Payment System on a Public Bus in Estonia

Unit 12: Northern Europe

Trouble and Response

Are you thinking, "What if someone commits a cyberattack on this countrywide system?" Someone did.

In 2007 the government of Estonia decided to move a World War II-era statue of a Soviet soldier from the middle of Tallinn to a nearby cemetery. Pro-Russian demonstrations, some of them destructive, followed for days (about one-fourth of the country are ethnic Russians). Then the Estonian Parliament, many public services, and several banks suddenly went offline. Estonia believes Russia did it; Russia denies it.

Not long afterward, Estonia created a NATO cyberdefense center that conducts drills to prepare for future cyberattacks. It also established a data embassy in Luxembourg: a storage building that houses backups of all Estonian data and that enjoys the same right of sovereignty as any nation's embassy does. Estonia also began utilizing blockchain technology, a distributive database that supposedly cannot be hacked.

Skype Office in Tallinn (2013)

Another development in Estonia that had worldwide impact was the software developed in 2003 by a group of friends for the Internet calling platform Skype. We can measure the success of this development in many ways, not least of which is this: Microsoft bought Skype in 2011 for $8.5 billion. Many of those who profited from the sale invested heavily in Internet start-up companies in Estonia, which attracted additional foreign investment. This "Skype effect" has been significant for the country's continued growth.

Mohni Island, Part of Lahemaa National Park

Lesson 57 - Is This What Tomorrow Looks Like?

The Old Continues, Too

E-businesses and digital interaction with government agencies are not all that happen in Estonia. Another major industry is the definitely undigital activity of harvesting peat from the extensive bogs in the coastal areas of the country. Peat is composed of decayed vegetation; people use it for fuel and fertilizer. Peat ranks second after petroleum as an energy source in Estonia and has the advantage of being renewable. People also obtain cranberries and marsh tea from Estonian bogs.

Old things are popular in Estonia also. Tallinn still boasts many medieval structures along winding streets and alleys with names hundreds of years old. Historic Old Town, which sits on a hill, is a popular tourist attraction. Tourists can see majestic buildings from previous eras in other cities of Estonia also. Lahemaa National Park on the Gulf of Finland is another popular destination. About six million foreign tourists visit Estonia every year—that's over four times the population of the country!

Estonians created in the image of God have used their talents to rebuild their country when Communist rule had left them pretty much starting from zero. When Nehemiah visited Jerusalem he found the city's wall in ruins, but his encouragement led them to take on the job of rebuilding the wall.

I told them how the hand of my God had been favorable to me and also about the king's words which he had spoken to me. Then they said, "Let us arise and build." So they put their hands to the good work.
Nehemiah 2:18

Assignments for Lesson 57

Gazetteer Read the entries for Estonia, Latvia, and Lithuania (pages 118, 122, and 123).

Worldview Copy this question in your notebook and write your answer: Fyodor Dostoevski wrote that if God does not exist, all things are lawful. How is this true or not true?

Project Continue working on your project for this unit.

Literature Continue reading *Kidnapped*.

Student Review Answer the questions for Lesson 57.

Sheep and Village on Streymoy Island

58 They Do Things Their Way: The Faroe Islands

They do things their way on the Faroe Islands. Given their geography, they almost have to.

The Faroes are a group of 18 islands—17 of them inhabited—to the north of Scotland and situated between Norway and Iceland in the North Atlantic Ocean. In other words, they are out there, not really close to any land mass.

The land area of the Faroes is about half of the land area of Rhode Island. In other words, they are small. And yet, they are not so small and so out there that nobody lives there or goes there. The islands are home to about 50,000 people. Double that number of tourists go there each year, half of whom are cruise ship passengers who stay for only a few hours.

The Faroese have developed a way of living that works for them. Here are a few examples.

Sheep View 360

When Google Street View had not mapped the roads of the islands, a team of Faroese developed an unusual Plan B to let the world see what beauty lies on the island.

The team devised a harness with a 360-degree camera, and then mounted all of this on the back of five of the 80,000 sheep that live on the islands. Sheep are a major part of their economy and culture (the word Faroe comes from the Old Norse word for sheep). They turned the sheep loose to walk down the pathways of five of the islands and recorded what a traveler could see from the back of a sheep while noting GPS coordinates of the places pictured. Then the team uploaded the images and data to Google Street View.

The goal of the effort was to help get the word out about the islands in the hope that more people will want to visit.

Voluntourism

Getting tourists to volunteer to help maintain the land is an idea that worked for the Faroese. The Faroe Islands were closed for a weekend in 2019 before the heavy tourist season began. Through social media and other avenues, the islands' tourism office recruited 105 people from 25 countries to come for that weekend to perform some maintenance and upkeep on ten projects on the islands.

A major concern was improving the pathways through the mountains that many tourists hike during their visits. The voluntourists performed such tasks as removing loose stones from the paths and driving stakes to mark the pathways clearly. The footpaths are generally public access (though some

Lesson 58 - They Do Things Their Way: The Faroe Islands

landowners charge fees to those who come through), but the pastures nearby are private property where trekkers are not to tread.

Safe, identifiable pathways are important in this land located at 62 degrees north latitude, where grass is thin and recovers slowly. In terms of tourism, good hiking routes can attract tourists, and the owners of pastures can earn some extra money by offering bed and breakfast accommodations, sack lunches, and guided tours. In terms of the environment, the Faroese want to insure the maintenance of the nesting sites for snipe, skuas, oystercatchers, guillemots, fulmars, puffins, and other birds that find at least a temporary home there.

Government

The Faroe Islands are autonomous except in the ways they aren't, and they are Danish but with a Norwegian background. It works for them.

As best we can tell, Norwegian Viking explorers settled the islands in the 800s. Norway ruled the islands until 1380, when the thrones of Norway and Denmark were combined. After that point, the Faroes became more closely associated with Denmark.

The Danish government gave the islands to a Danish nobleman to rule in 1655, which he and his heirs did (not altogether successfully) until 1709, when the Danish government once again assumed direct control. During the 1800s Denmark exercised even closer control over the islands. In 1816 the Danish government made them a province of Denmark.

The islands' geographic location had an impact during World War II. During the war, Germany occupied Denmark but Great Britain occupied the Faroes. This prevented Danish control of the islands, but it also enabled Faroese fishermen to profit from selling fish to the British at relatively high wartime prices.

Path and Village on Mykines Island

After the war, both Danish and Faroese political leaders saw the wisdom of giving the islands greater independence. Denmark accomplished this with the Home Rule Act of 1948. The Faroes have their own parliament and their own flag. They rule themselves except in the areas of justice, defense, and foreign affairs. However, Denmark provides about one-third of the Faroese government's funding, and the Faroes hold two seats in the Danish parliament. Denmark is a member of the European Union, but the Faroe Islands are not.

Geography makes for some interesting arrangements. The small country of Denmark holds authority over the tiny Faroe Islands as well as Greenland, the largest island in the world.

Roykstovan (center) and St. Olav's Church (left) on Streymoy Island

Fishing

Fishing is the largest single component of the Faroes' economy (tourism is second). About 95% of Faroese exports by volume is fish.

The little Faroe Islands have become the leading exporter of fish to Russia, although the pathway to this status is complicated. The European Union imposed sanctions on the Faroes in 2013 in a dispute over fishing quotas. The next year, Norway cut its fish exports to Russia due to sanctions against Russia for its seizure of the Ukrainian territory of Crimea. During this time, Russia reached a deal with the Faroes (not a member of the EU) to increase imports from the islands.

Fascinating Features of the Faroes

The capital of the Faroe Islands is Torshavn. A site on the city's harbor called Tinganes was where early Norse settlers gathered to trade, to settle any disputes, and to decide on any policies they wanted to follow in the meeting they called the "thing." Since the Faroese parliament meets there now, the Tinganes "thing" in the form of its modern successor is considered the oldest parliament in the world. Obviously it has worked for the Faroese.

Many property owners in the Faroes find that grassy sod makes a good roof for their structures. Such roofs help the homes and barns so equipped handle the 300 days with rain that they enjoy each year.

Christianity has a strong presence in the islands. The gospel came to the islands about 1000. Today about 85% of the population are Evangelical Lutherans. Ten percent are Christian Brethren, and other groups have smaller percentages.

St. Olav's Church in Kirkjubour was built prior to 1200, making it the oldest church still in use in the islands. It used to have a churchyard in front, but erosion by the sea has caused the building to sit right on the seacoast.

The Roykstovan might be the oldest wooden house in the world that is still inhabited. Legend has it that the wood was driftwood that came from Norway. The islands have no native forests, so the wood had to come from somewhere. The house was built in the 1000s. The same extended family has lived in the house for seventeen generations, dating back to the mid-1500s.

An activity that used to take place in several European countries but continues in the Faroe Islands is the chain dance. People perform this dance at many special, festive occasions. Participants form a circle and hold hands. They move from side to side and progress around the circle. The leader begins verses about Faroese or Norse history, love and warfare, or one of many other subjects, and the other dancers join in. The verses can run into the hundreds.

Besides the ferries and bridges that connect the islands, the people of the Faroes use 19 tunnels, most dug through mountains to connect different parts of an island but three of which run beneath the coastal waters. Faroese drivers appreciate the shortened driving times.

Faroese coastal areas present many beautiful and stunning scenes. Included in these are sheer rock cliffs that rise over 2,000 feet above the sea, Lake Sorvagsvatn that at a certain angle of sight appears to float above the sea, and Bosdalafossur waterfall that descends from the lake and plummets about a hundred feet into the ocean. The islands have the world's largest colony of storm petrel birds, which are active primarily at night.

In 1986 Faroese Ove Joensen, on his third attempt, rowed the 900 miles from the Faroes to Copenhagen, with only a stopover on the Shetland Islands in northern Scotland. Conquering geography in this way was the route he chose to take.

And then there's the language of the islands. It's not Danish, and it's not Norwegian. It's related to those tongues, but it is distinctly Faroese. It works for them.

Paul had a single-minded focus on the course of life he knew he had to take:

Brethren, I do not regard myself as having laid hold of it yet; but one thing I do: forgetting what lies behind and reaching forward to what lies ahead, I press on toward the goal for the prize of the upward call of God in Christ Jesus.
Philippians 3:13-14

Assignments for Lesson 58

Gazetteer Read the entries for Denmark, the Faroe Islands, Norway, and Sweden (pages 117, 119, 124, and 125).
Read "61 Mountains—In One Year!" (pages 277-278).

Worldview Copy this question in your notebook and write your answer: How do you explain the origin of the conscience, the capacity to know right and wrong?

Project Continue working on your project for this unit.

Literature Continue reading *Kidnapped*.

Student Review Answer the questions for Lesson 58.

Surveying in Sweden (2017)

59 Surveying the Matter: The Struve Geodetic Arc

Imagine the angry confrontation:

"That's my property!"

"No, this is my property. Your property is over there!"

On a larger scale:

"The border between our countries is here!"

"No, the border is a mile further east! You are taking our land illegally! We will fight you for it!"

What difference does a mile make? Probably not much—unless someone strikes oil or discovers coal on the disputed territory, or unless the people who live in that area thought they were citizens of one country but another country claims them. Then it makes a big difference.

Other questions beg for accurate answers also, such as:

Where is the new road supposed to go?

What exactly is the land that is for sale?

What are the boundaries of the land I just bought?

On a much larger scale, how big is the earth? Does that make any difference?

Christopher Columbus miscalculated it. That made a difference.

What if you wanted to send a man into orbit around the earth? You would need to know the size of the earth to know how much fuel you would have to have to get him into orbit, keep him there, and get him back. Would it be wise to send a man into orbit around the earth with only a guess as to how big the earth is?

The way to resolve these issues and to find the answers to many other similar questions is by means of surveying.

Surveying is a branch of civil engineering. It involves measuring the position of an object on the surface of the earth in relation to other objects by determining horizontal distances, elevations, angles, and directions. Surveying gives people the information about the earth that they need in order to do what they want to do.

Surveying has a long and illustrious history. The construction of the Egyptian pyramids apparently involved surveying. The ancient Sumerians apparently knew some techniques of surveying. The young George Washington was a surveyor in the western lands of the American colonies.

Friedrich Georg Wilhelm von Struve

Jacob Struve was born in 1755 in the German state of Holstein. He went to school in Altona, Denmark, studied at the university in the German

Lesson 59 - Surveying the Matter: The Struve Geodetic Arc

city of Gottingen with an emphasis in mathematics, and then entered an academic career.

Struve and his wife had seven children. Our interest is in his son Friedrich Georg Wilhelm von Struve, who was born in 1793 (some sources include the von in his name, while others do not). We will return to the story of the Struve family later.

When Napoleon's armies invaded the area where they were living, Jacob moved his family to Estonia, which was then in the Russian Empire, so that his son would not have to join Napoleon's army. Friedrich Georg Wilhelm von Struve eventually became a professor of mathematics and astronomy at the University of Dorpat (now called Tartu) in Estonia. He later became director of the observatory at the university, which housed the best refracting telescope yet made.

Von Struve had an interest in double or binary stars. He surveyed 120,000 stars and found 3,112

"Hogland, Z" Geoderic Arc Site in Russia

binaries, about three-fourths of which had not been identified. At the request of the tsar, von Struve moved to St. Petersburg, Russia, to supervise the building of a new observatory and later became its director. He received international recognition for his work on binary stars. In 1843 von Struve became a Russian citizen.

The Struve Geodetic Arc

In 1816 von Struve undertook the leadership of a project to determine the accurate measure of a meridian of longitude, as a step toward determining the size and shape of the earth. In Lesson 2 we discussed the work of Eratosthenes in attempting to determine the size of the earth. In the 1800s, the question still awaited an accurate answer. Von Struve decided to measure the meridian that passed through the observatory at the university where he worked. He built on surveying work that people had done a few years earlier, and he obtained the participation of many surveyors and scientists.

The project utilized the technique of triangulation, which surveyors had developed in the 1500s. In this method, the surveyor begins with a given known point A. He then establishes a second point B and measures the distance between them. Line AB becomes his baseline. He then establishes a third point C and measures the size of the interior angles of triangle ABC.

"Fuglenaes" Geodetic Arc Site in Norway

Using trigonometry, the surveyor can determine the length of sides AC and BC to establish the exact location of point C. Then side BC becomes the baseline for another triangle. The surveyor repeats the process to find point D. In this way the surveyor determines the exact location of a series of fixed points and the exact length of the sides. This process measures a series of short distances accurately in order to derive an accurate long-distance measurement using the connected triangles.

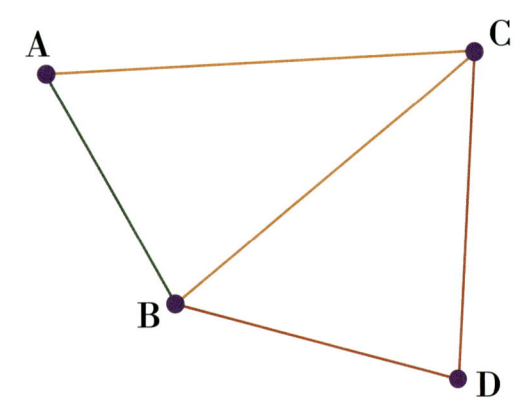

The measurement of the meridian von Struve decided to measure extended from the Arctic coast of Norway and continued 1,752 miles to the coast of the Black Sea in Ukraine. The arc included 258 main triangles and 265 main station points. At the time, the line only passed through two countries, the Union of Sweden and Norway and the Russian Empire. With subsequent political reorganization, the line now goes through ten countries: Norway, Sweden, Finland, Russia, Estonia, Latvia, Lithuania, Belarus, Ukraine, and Moldova.

A series of markers along the meridian memorializes the work of von Struve and others, as shown on the previous page. These markers take different forms, such as a hole filled with lead in a rock surface on the ground, engraved cross-shaped marks on rock surfaces, stone markers, and tall elaborate monuments. This portion of that meridian of longitude is now called the Struve Geodetic Arc. Geodetic means related to the science of measuring the earth, which is geodesy. Von Struve was involved in the project for 39 years, from 1816 until 1855.

The Struve Family

We benefit from the dedication of people who devote their lives to scientific inquiry. We want to know how the world functions, and scientists do the hard work that gives us that information so we can

Key Sites in the Surveying of the Struve Geodetic Arc

understand the world, preserve it, and help make it better. We should be thankful for the work of a scientist. An extended family that includes many scientists over several generations deserves, well, to be surveyed.

Jacob Struve (1755-1841) was the father of Friedrich Georg Wilhelm von Struve (1793-1864), who became an astronomer.

Friedrich Georg Wilhelm von Struve was the father of Otto Wilhelm von Struve (1819-1905), who became an astronomer.

Otto Wilhelm von Struve was the father of Karl Hermann von Struve (1854-1920), who became an astronomer. Karl's brother, Gustav Ludwig von Struve (1858-1920), also became an astronomer.

Karl Hermann von Struve was the father of Georg Otto Hermann Struve (1886-1933), who became an astronomer. Georg Otto Hermann Struve had two sons, neither of whom became astronomers.

Gustav Ludwig von Struve had a son Otto Struve (1897-1963), who was an astronomer. Otto and his wife did not have any children.

In addition, other members of the Struve family were educators, writers, political activists, scientists, government officials, and an ambassador.

The Value of Work

Jacob Struve once wrote in a letter to his son Friedrich, "We Struve can not live happily without continuous work, because from youth we learn that it is the most useful and best virtue of human life." The descendants of Jacob Struve learned this lesson well and made a difference because they followed it.

Solomon spoke of the value of work several times in Proverbs, including this verse:

In all labor there is profit,
But mere talk leads only to poverty.
Proverbs 14:23

Assignments for Lesson 59

Geography   Complete the map skills assignment for Unit 12 in the *Student Review Book*.

Project   Continue working on your project for this unit.

Literature   Continue reading *Kidnapped*.

Student Review   Answer the questions for Lesson 59.

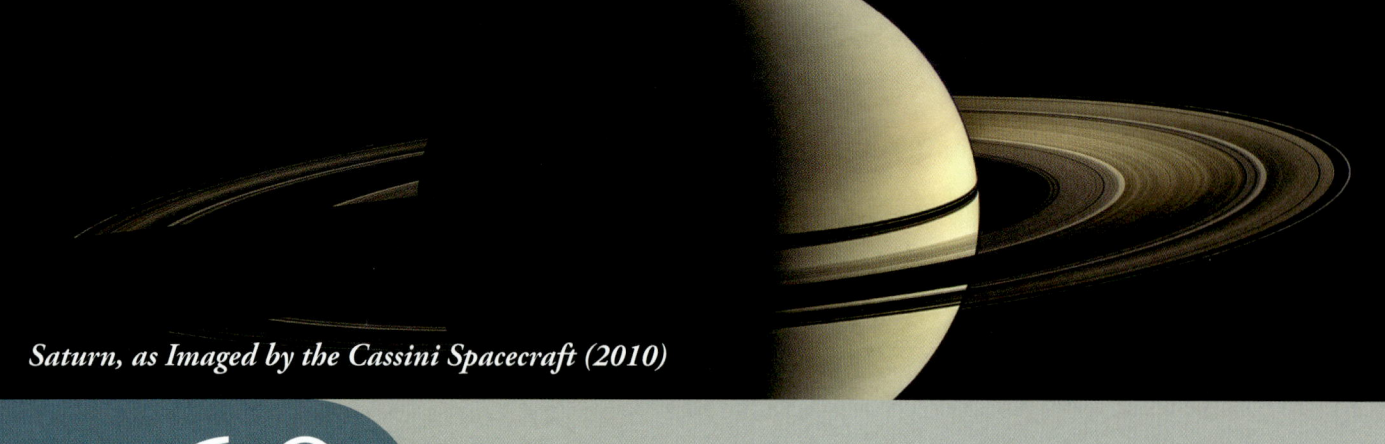

Saturn, as Imaged by the Cassini Spacecraft (2010)

60 The Existence of God

A worldview is a person's view of the world. The first and most important element of a person's worldview is what that person believes about God, the ultimate reality. What a person believes about God influences how that person sees the world.

Evidence for God

Over the centuries, people have established several lines of logical evidence that support belief in the existence of God. Here are four.

God as First Cause. Thomas Aquinas observed that everything in our world has a cause. A chicken lays an egg; an egg becomes a chicken. An acorn falls from an oak tree and grows to become another oak tree. Day and night occur because of the earth's rotation as it orbits the sun. Events happen in history because previous events happened in history. However, Aquinas said, as we proceed back through the causes and effects in our world, we must come to something that did not have a cause, a first cause that was not itself caused by anything else. Aquinas said that this First Cause is God.

In other words, our world had to begin in some way. We do not know of any physical matter in our world that is eternal or that creates itself. Something or Someone must have existed before everything else and must have begun the world in which we live. That Someone is God.

The Evidence of Design. The seasons follow one another regularly, always in the same order. Leaves do not sometimes fall up and never come down. The moon has not gone spinning off on its own but instead orbits the earth regularly. The human body has intricate systems that work together and enable human life. In other words, our world has order that random chance cannot explain. These patterns of order are evidence of design, which implies a Designer behind it all. That Designer is God.

The complexity and rhythm of our world, from the majesty of the universe to the intricacies of our human bodies, is so great that the probability that our universe came into existence and continues to exist as it does merely by chance is so small that it is essentially zero. It is not reasonable to think that the order we see is the result of an irrational process of chance. The most reasonable explanation for how the world operates, like the most reasonable explanation for a watch, a house, or a child, is that an Intelligence created it and sustains it.

The Moral Argument. Everyone has a moral oughtness within, a conscience that tells him or her that some actions are right and other actions are

Lesson 60 - The Existence of God

wrong. This capacity to determine right and wrong came from somewhere beyond ourselves and our society. It could not have evolved into existence. The only reasonable conclusion is that it comes from God.

The Evidence of Jesus. Jesus is either trustworthy or He is not. Those are our only two options. Jesus believed in God. If Jesus were mistaken about this, how could we trust any part of His teaching? The testimony of Jesus about God is pivotal.

What Difference Does It Make?

The existence of God makes a profound difference in how people live. In the novel *The Brothers Karamazov*, the author Fyodor Dostoevski has one character say that, without God, "All things are lawful then." In other words, without God—without a Creator, an Author of truth, and a Judge at the end of time—we would have no absolute laws or standards. We would only have standards that people and societies make up, and those can change depending on changing conditions and varying worldviews. Where would laws and standards come from otherwise? Without a source outside of human beings, there could be no permanent, eternal truth; no right and wrong that apply everywhere in every age. There would be no logical reason to obey any laws or to treat another person with respect. We see this in the lives of people who do not believe in God or who live as though God does not exist. Their worldview leads to a certain lifestyle.

Beyond questioning social standards and expectations, if God did not exist we would have to question logic and reason. As C. S. Lewis asked in *Miracles*, how could we trust what we call logic and reason if they resulted from minds that were merely the result of an illogical, irrational process that had no designer behind it?

A Universe Without Purpose?

If God does not exist, the universe has no purpose. The vast reaches of space, the planet on which we live, our individual selves—all would be pointless because there would be no point to, no purpose for, no mind behind, their coming into existence. All this would have just happened, and happenstance has no purpose. People might propose various ideas about a purpose for the universe or parts of it, but we could not know which idea is correct. And again, why should we trust the thoughts of human minds that are merely the result of an irrational process?

St. Thomas Aquinas
by Antonio del Castillo y Saavedra (Spanish, c. 1649)

Everything in our world has a purpose. Air, leaves, sand, water, the moon, the planets, food, automobiles—you name it, and it has a purpose. Does it not logically follow that the universe itself has a purpose, namely to glorify God?

Theism

We call belief in the God of the Bible theism. The word comes from *theos*, the Greek word for god (the same Greek word can also mean God, depending on the context).

One form of theism is traditional Judaism, which accepts the existence of the God whom the Old Testament, the Jewish Scripture, reveals.

Another form of theism is Christianity. Christianity holds to the belief in the God of the Bible and accepts Jesus as the Messiah or Christ. Christianity has multiple subgroups within it. These different subgroups (for instance, Roman Catholic, Orthodox, Anglican, Methodist, Presbyterian, Baptist, congregational, evangelical, liberal, Pentecostal, and so forth) hold some beliefs in common but have many variations in their doctrines about God and His will. These different doctrines impact how adherents of the different groups see the world, God's involvement with it, and how people should live. However, the differences in worldview among the traditional Christian groups are less than the differences between the worldview of Christian groups as a whole and the worldviews of unbelievers.

Variations of Theism

We can mention here two other worldviews that hold to the existence of one God but whose ideas about God are very different from the way the Bible portrays God.

Deism (from *deus*, the Latin word for god or God) is the belief that God exists and that He created the world but that He is not generally involved in the world and the lives of people. Deism holds that the world operates on the basis of natural law, not the will or actions of God. Deists generally believe that prayer (asking God to intervene in the world) and the belief that "God did" this or that specific act, are illogical. However, deists can be inconsistent on this point. For instance, Benjamin Franklin, who described himself as a deist, urged the Constitutional Convention of 1787 to begin each session with prayer.

A fairly widespread modern view of God has been called moralistic therapeutic deism. This is the idea that God simply wants us to be happy and nice to others and that He only becomes involved in our lives when we ask Him to, usually in desperation. This worldview gives a passing recognition to the existence of God but rejects the idea that His existence makes much difference in the world or in our lives. In other words, people who hold this view try to have it both ways: yes, God exists; but no, He doesn't make much difference.

Another variation of theism is the belief of some religious groups in a supreme being that they call God, but their ideas about that being are so different from those of Biblical theism that they apparently believe in a different deity altogether. Each group has its authoritative book and claims to have its prophets.

Here is one way to look at these alternative views of God. Suppose someone tells you that he knows Ray Notgrass. He describes Ray as deceptive, malicious, conniving, revengeful, and cruel. That

---

This image shows the beginning of John 1 in a Greek New Testament, including the word Θεός (theos).

Lesson 60 - The Existence of God

person's claimed knowledge of Ray is so different from the way Ray really is that you have to conclude that that person does not really know Ray at all. He might have the name right, but his concept of Ray is incorrect.

Worldviews About Worldviews

Some worldviews, while not denying the existence of God, equivocate about the significance of His existence.

Postmodernism is the idea that no single worldview is a valid basis of truth for all people. Postmodernists believe that individuals or groups can express their own sense of meaning, that such an expression is valid for them, and that this is all that matters. In other words, for them the absolute is the denial of an absolute. This, of course, is logically contradictory.

Postmodernism is a response to two previous views. As they see it, religious authority, such as the Bible or the Roman Catholic Church, attempted to establish meaning for all people. This gave way to modernism, expressed in the Enlightenment and rationalism, which said that man's reason is the authority for finding meaning in life. The rejection of all authority, both religious and rational, is postmodernism. One has to wonder what will come after postmodernism; post-postmodernism?

Postmodernism is an expression of pluralism. Pluralism is a worldview about worldviews. It says that there are many valid paths, a plurality of paths, to truth/wholeness/salvation/heaven—however each path might describe man's ultimate goal. Pluralism says that we cannot know which one is right, so we must accept them all as equally valid. Pluralists say that we cannot and therefore should not decide whether any path that someone says he is following is right or wrong.

Agnosticism is the belief that we cannot know whether God exists or not. It's a shrug of the shoulders, and agnostics are happy to leave it at that. They say we should leave it at that. Agnostics say that, since this is the case, whether He exists or not does not matter in our worldview or in our actions.

We will look at atheism, the flat-out denial of the existence of God, in a later lesson.

Deciding to Believe

At the feast described in John 7, the Jews were astonished at Jesus' teaching. Jesus told them, "My teaching is not Mine, but His who sent Me. If anyone is willing to do His will, he will know of the teaching, whether it is of God or whether I speak from Myself" (John 7:17).

The existence of God is the ultimate worldview issue. It is basic to a correct understanding of our world. We have evidence and logical arguments that demonstrate His existence. However, God does not force anyone to believe. We have evidence about God and Christ from Scripture and from Creation, but a person can reject that evidence and refuse to believe. Every person must consider the evidence and decide whether or not he or she is going to believe in God. If a person decides to believe—if, in Jesus' words, he is willing to do God's will—then he will know whether the teachings of God and Christ are true and reliable.

A large part of today's world conveys the idea that believing in God is old-fashioned, superstitious, and irrational. But many intelligent people both today and in the past have believed in God. The difference between belief and unbelief is not a matter of intelligence. Nor is believing or not believing something you should do because it is the socially acceptable thing to do among your group. Belief—faith, trust—is a matter of the will. Are you willing to do God's will? If you are, Jesus says that you will know whether the teaching of Christ is true. This is not to say that belief involves shutting your mind off and taking a leap of faith. It says that the evidence is there, but you have to decide what you are going to do with it.

Belief in God is not merely intellectual assent to a set of doctrinal truths. As we saw in Lesson 55, belief or faith in the Biblical sense is trust, a commitment of one's life to God's truth and its consequences, a willingness to undergo imprisonment, torture, and even death because of your commitment to His way.

Thomas Howard, in his 1969 book *Chance or the Dance?*, says that belief in God means that everything means everything, while the denial of God means that nothing means anything. He notes the irony of the fact that this age, when many people believe that nothing lies behind everything, is called enlightened. For Howard, the choice of worldviews comes down to this: Is it all chance, with no direction, or is it the Dance: a marvelous, beautiful, intricate dance that is planned and created, with right and wrong steps that require right choices and which results in joy and in arriving at a definite end?

This curriculum teaches that the true basis for living in our world and for a correct worldview is the God whom the Bible reveals to us: the holy, loving, patient, purposeful, redemptive, one true God. He is also the Judge who will determine the eternal destiny of every person. This paragraph is not a complete description of God, but it is an attempt to say that we believe in the existence of God as He is revealed in Scripture.

The Bible uses pointed words to describe the need for faith and the consequences of both belief and unbelief.

And without faith it is impossible to please Him, for he who comes to God must believe that He is and that He is a rewarder of those who seek Him.
Hebrews 11:6

The fool has said in his heart, "There is no God."
They are corrupt, they have committed abominable deeds.
Psalms 14:1

Assignments for Lesson 60

Worldview   Recite or write the memory verse for this unit.

Project   Finish your project for this unit.

Literature   Continue reading *Kidnapped*.

Student Review   Answer the questions for Lesson 60. Take the quiz for Unit 12.

Autumn in Moldova

13 — Eastern Europe

The history of the Jews in Eastern Europe demonstrates the significance of where people live. The Hungarian (or Magyar) people have developed a defiant, independent spirit, influenced to a great degree by the location of their homeland. Bohemian composer Bedrich Smetana expressed his love for his country by focusing on a geographic feature in his work titled "The Vltava," better known as "The Moldau." Because of the geographic location of Ukraine, people have sought it and fought over it for centuries, just as is happening today. In our worldview study, having discussed the existence of God, in this unit we consider the nature of the God in whom we believe.

Lesson 61 - The Jews of Eastern Europe
Lesson 62 - Defiant Hungarians
Lesson 63 - The Vltava (The Moldau)
Lesson 64 - A War They Didn't Want: Ukraine
Lesson 65 - The Nature of God

Memory Verse — Memorize Galatians 5:1 by the end of this unit.

Books Used — The Bible
Exploring World Geography Gazetteer
Kidnapped

Project (Choose One)

1) Write a 250-300 word essay on one of the following topics:
   - What is your perspective on the nature of God? This is sometimes called the attributes of God or the characteristics of God. Include Scripture references in your essay. See Lesson 65.
   - Research and write about the life of Joseph Cardinal Mindszenty. See Lesson 62.
2) Watch an interview of or speech by a Jewish Holocaust survivor. Afterwards, write down your thoughts about the presentation. Tell why you believe it is important never to forget what happened in the Holocaust. See Lesson 61.
3) Write a story about two friends who live in a land that a foreign power invades. In the story, work in why the invasion took place and the prospects for the future. See Lesson 64.

Nozyk Synagogue in Warsaw, Poland

61 The Jews of Eastern Europe

Where we live affects how we live. The lives and worldview of South Sea islanders, for instance, are different from the lives and worldview of Arab Muslims. The worldview of someone who lives in an Italian port city will probably be different from that of someone who lives on the high plains in the United States.

The groups of which we are a part also affect how we live. A Hispanic child in an American inner city whose father does not live with him will have a different life from the daughter of a physician from India who lives in an American suburb with both of her parents.

The experience of millions of Jews in Europe bears this out. Jews who lived in the United States during the 1920s through the 1940s had a different life experience from that of Jews who lived in Germany and Eastern Europe during the same period. You probably know something about the experience of European Jews during the Holocaust. This lesson provides information about the Jewish experience in Eastern Europe leading up to that horrible time.

For the purposes of this curriculum, we consider Eastern Europe to include the countries of Belarus, Bulgaria, Czechia, Hungary, Moldava, Poland, Romania, Russia, Slovakia, and Ukraine. In this lesson we especially look at the experience of Jews in Poland. We also include some information about Jews in Germany.

Jews Come to Europe

We don't know exactly when, why, or how Jews began arriving on the European continent. Some might have gone there after the Babylonian captivity during the time of the Persian Empire. Some evidence indicates that Jews might have been living in Central Europe even before the Roman Empire spread into that region.

Historical records show that Jews lived in Greece by the 200s BC. One historian estimated that seven thousand Jews lived in Rome during the time of Augustus (27 BC-14 AD). The first century AD Jewish historian Josephus stated that Jews were living in Europe during his lifetime. Apparently a settlement of Jews existed in Croatia in the 100s AD. Records from the 400s-500s indicate the presence of "Syrians" in Europe (probably a reference to Jews). It is a fairly safe assumption that Jews in significant numbers moved to Europe after the Muslim conquest of Palestine in the 600s AD.

During the 800s AD, kings in Western Europe of the Carolingian dynasty (Charlemagne and his

349

heirs) offered protection to Jews in return for their help in trade. During and after this period, many Jews emigrated to central and western Europe from southern Europe. Historians estimate that almost 20,000 Jews lived in German lands by the time of the first Crusade at the end of the 1000s.

Culture

Although the Jews of the Diaspora were scattered geographically, they were largely united in their religion and the culture they shared. Because they were scattered, they had little political and economic power. Their strength lay in the unity of their faith and culture.

We can define culture as the way a certain group of people think and the way they do things. The Jews of Europe developed many cultural traditions based on their faith and their interpretation of Scripture. They observed the Jewish festivals commanded in the Law. They avoided work on the Sabbath. They did not eat food that the Scriptures identified as unclean. They kept many traditions involving wedding ceremonies and other aspects of life that Jews had observed for centuries.

Family life was very important to Jews. They scrupulously maintained the accepted roles of husbands, wives, and children. They had great loyalty to their families and respected the authority of the fathers. Parents generally arranged the marriages of their children. The social rank and financial stability of the families involved influenced the arrangement much more than any romantic feelings the young man and woman might have had for each other. Among poor families, a man had to wait to be married until he had an occupation that would support his own wife and children.

A few Jewish men devoted themselves completely to the study of the Tanakh (the complete Hebrew Scriptures, what Christians call the Old Testament) and the Torah (the first five books of the Tanakh). While most devout Jewish scholars supported themselves in a profession, the wives of some scholars worked so that their husbands could devote their full time to the study of the Scriptures.

Polish Jew (c. 1700)

The traditions that Jews followed not only were the right things to do in their eyes, but they also provided stability and order in the midst of persecution and change and as they lived in a world whose ways were very different from their own. Jews believed that God called them to be distinct, to be God's holy or set apart people. Some Jews compromised their traditions in order to get along in the world, but this caused them to lose their grounding. When they compromised, they risked being outsiders to both the Jewish and Gentile worlds. The influence of the Gentile world was a major reason for the rise of Hasidic Judaism in Eastern Europe in the 1700s with its even stricter rules and traditions as a way to hold on to their spiritual roots.

Lesson 61 - The Jews of Eastern Europe

Most Jews who lived in Eastern Europe and western Russia lived in small towns called shtetls (three syllables: shh-TET-ull). A shtetl usually had a population of several hundred to several thousand. They typically had cobbled streets, contained a jumble of homes and shops, and were surrounded by fields and forests. Most of the population in a typical shtetl was non-Jewish, so Jews and non-Jews had daily contact. Most of the Jews in a shtetl ranged between poor and desperately poor. Beginning in the 1880s, many Jews moved from the shtetls to larger cities, both for work and for a greater measure of security.

Even though they had a rich spiritual and cultural life, many Jews were materially impoverished. In the 1890s it was common for Jews in Russia to work 14-16 hours per day in factories and earn perhaps two to three rubles per week, barely enough to survive.

One important aspect of Jewish culture in Europe was the Yiddish language. Yiddish developed from the Jews' widespread use of Hebrew, German, and other languages. It utilized the Hebrew alphabet. Hebrew was the holy language in which the Scriptures were written. Yiddish was a level beneath Hebrew; it was the everyday language that the people used. Almost all Jews who lived in Eastern Europe spoke Yiddish. Yiddish maintained the Jews' cultural identity wherever they lived. The language provided cultural continuity especially for those who emigrated to other countries such as the United States.

Victims of Prejudice

People have hated and mistrusted Jews for millennia. In the ancient world, people regarded Jews as strange because they didn't worship like the pagans did, they didn't eat what everybody else ate, and they had strict rules about what they could touch and where they could go. People didn't like Jews because they were different.

In the Middle Ages Jews were often the victims of violence and rumor, as we discussed in Lesson 15. For instance, as Crusaders advanced toward Palestine, sometimes for absolutely no reason they assaulted Jews in the cities through which they passed. One of the vilest rumors about Jews was the blood libel, which was the belief that Jews kidnapped and killed the babies of Christians and mixed their blood in their matzoh bread at Passover.

Jews often lived as marked individuals. In some ways they marked themselves. For instance, Jews commonly placed a mezuzah, a small box containing a portion of Scripture, on their doorpost in obedience to Deuteronomy 6:9. This marked the Jews' dwelling places. In addition, however, in some places the law marked Jews by forcing them to wear an emblem such as a yellow circle on their clothing, so others would know that they were Jews.

From time to time the ruling authorities of a city or country ordered all Jews to leave. We read in Acts 18:2 of one instance of this in the first century AD, when the Roman emperor Claudius ordered all Jews to leave Rome. This occurred in some European countries during the Middle Ages. When the government forced Jews to leave their homes, authorities usually confiscated their property. In many instances, later rulers allowed Jews to return to their homes, but they had to pay a ransom or tax to the authority to be able to return.

This illustration from the 15th-century Nuremberg Chronicle *depicts Jews being burned alive for allegedly dishonoring the Eucharist.*

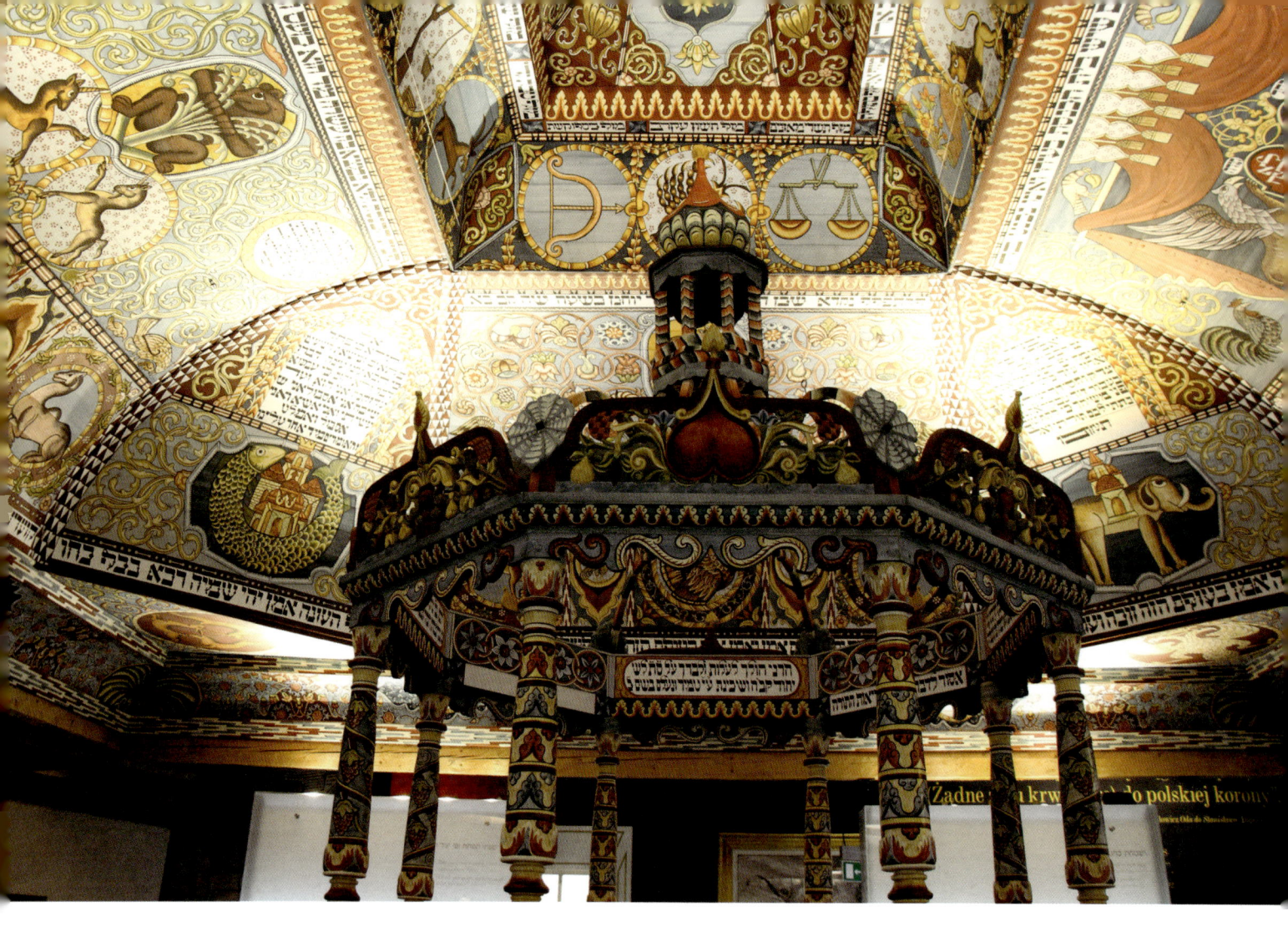

Jews established a synagogue in Gwoździec, Poland, around 1640. A humble exterior held an elaborate interior. The building was damaged during World War I and completely destroyed during World War II. The Museum of the History of Polish Jews in Warsaw features this reconstruction of part of the original building.

Pogroms

As we saw in Lesson 15, mistreatment of Jews took many forms in different times and places. In some times and places the persecution included physical attacks. These government or government-endorsed attacks were called pogroms, from a Russian word that means to wreak havoc or to destroy.

Czar Nicholas I, who ruled Russia from 1825 until 1855, was especially harsh toward Jews. His reign saw over six hundred anti-Jewish decrees, including such actions as the banning of Hebrew and Yiddish language books and expulsions of Jews from certain villages. Many Russians wanted to destroy or eliminate Judaism by demanding that Jews convert to Christianity upon pain of banishment or death. These attacks came from simple prejudice or as a result of blaming Jews for economic problems or political instability.

Czar Alexander II, who assumed the throne in 1855, was more lenient toward the Jews; but political revolutionaries assassinated him in 1881. His successor, Alexander III, returned to a more anti-Jewish policy. During the 1880s, Russia enacted many laws that negatively affected Jews, such as ones that limited the number of Jews who could attend secondary schools and universities. Jews who graduated from law school were forbidden to join the Russian bar and practice their profession (which severely limited the number of Jews who attended law school). Laws prevented Jews from engaging in

Lesson 61 - The Jews of Eastern Europe

business on Sundays and from obtaining a mortgage from a non-Jewish lender.

Such laws encouraged rowdies and terrorists to attack Jews without fear of legal punishment. A pogrom might involve burning a village, destroying the homes of Jews, or a physical attack by a mob on Jews in a village, town, or city. Sometimes the violence included sexual assaults on Jewish women. If authorities were not directly involved in an attack, they would often be slow or half-hearted in responding to quell the violence. Many Jews lost their property or their lives in these pogroms.

In the face of such restrictions and attacks, beginning in the late 1800s many Jews found comfort and hope in the Zionist movement, which encouraged the creation of a safe homeland for Jews, ideally in Palestine.

Poland

Jews often faced inconsistent treatment from governments over the passage of time. For instance, although Jews received protection from kings in the Carolingian dynasty, in succeeding years governments in western Europe accused and targeted Jews, so many decided to go elsewhere. Poland became a refuge for many Jews during the Middle Ages. In the time of the Crusades, many Jews sought safety in Poland. Rulers of Poland welcomed Jews and hoped that they could help their country increase its international trade. In the 1200s, Poland enacted laws to protect Jews.

By the end of the 1500s, Polish Jews had their own parliament. In some ways Poland viewed Jews as a fifth class of citizens: nobility, clergy, merchants, peasants, and Jews. However, their success sometimes resulted in attacks by those who were fearful or jealous. Some Polish cities during this period forbade Jews from living there. Nevertheless, scholars believe that by the end of the 1500s some 80% of all Jews in the world lived in Poland.

In the late 1700s Prussia, Russia, and Austria undertook three partitions of Poland that ended in the disappearance of Poland as a country from the map of Europe for 123 years. Most Polish Jews ended up in territory that Russia controlled. The former area of Poland that became part of Russia, combined with parts of Lithuania, Belarus, Ukraine, and Moldova, was known as the Pale of Settlement. The Pale was an area 386,100 square miles that stretched from the Baltic Sea to the Black Sea.

Jews who lived in Russia were legally allowed to live only there. The 1897 Russian census revealed that 4.9 million Jews lived in this area. They constituted 94% of the Jews living in Russia at the time and 11.6% of the total population in the Pale. This limitation on where Jews could live was part prejudice and part an attempt to limit competition from Jews for jobs. We must also note, however, that at the same time serfs and other groups in Russia faced similar restrictions on where they could live.

This map of the Pale of Settlement is from the 1905 Jewish Encyclopedia. Darker shading indicates a larger percentage of Jews in a given region.

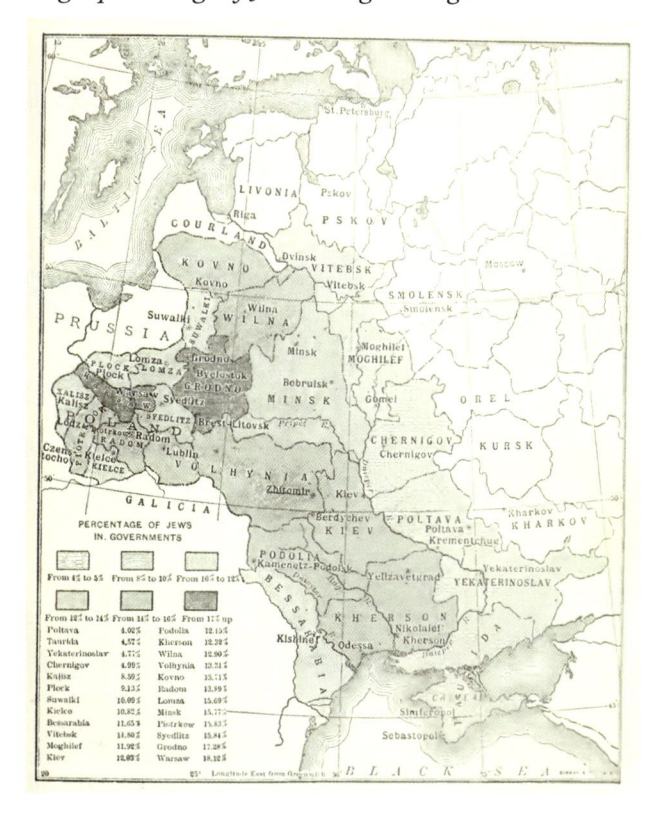

As part of the outcome of World War I, Poland was reconstituted as an independent country. The growth of Polish nationalism in the minds and hearts of many Poles leading up to this period resulted in greater hostility toward Jews. During the prewar period, most Jews lived in small towns. However, as industry developed in the 1920s, many Jews moved to the cities to work in factories. By the end of the 1920s, one-fourth to one-half of the population in larger Polish cities was Jewish. Jews became contributors in all walks of life and developed a thriving culture. Some 160 newspapers in Poland were published for Jewish audiences. These newspapers were published in Polish, Hebrew, and Yiddish.

The 1930s brought economic hardship as a result of the worldwide Great Depression. Many people blamed the Jews for these economic problems, and discriminatory treatment against Jews rose to new levels. For instance, some Polish universities set quotas for the number of Jews they admitted, and they segregated Jewish students on campus.

On the eve of World War II, about 3.5 million Jews lived in Poland. They constituted the largest minority in the country. They were 10% of the total population of Poland, the highest percentage of Jews in any country in Europe. However, less than 10% of that total survived the war in Poland. Many fled the country, some using fake identification papers. Many went to the Soviet Union since it was the nearest country that opposed Germany. Few Polish Jews survived the concentration camps. After the war, anti-Jewish activity continued in Communist Poland and the Soviet Union.

Emigration

In the face of legal discrimination and physical violence, what were the Jews to do? It was extremely difficult for the typical Jew in Eastern Europe to improve his lot. Many doors were closed to him. The Jews could not elect representatives who shared their goals and who could change discriminatory laws. Factories where Jews worked in large numbers witnessed occasional strikes, but the labor union movement was not strong so violent repression of strikers was common. What many Jews did was leave, and America gave them somewhere to go. Those who were the first to move to America often wrote to their relatives and friends in the old country encouraging them to join them in America.

Leaving home, whether for America or for some other country, was not easy for Jews in Eastern Europe. First, they had to save money for tickets. This could take several years. When they did leave, they had to keep up with the possessions they could carry or risk having them stolen. Many times entire families emigrated, which meant that they understood this to be a permanent move. They endured painful separations from extended family. There was no direct route, so they often had to pass through several countries or cities to get to a port. Between 1905 and 1914, 700,000 Jews from Eastern Europe passed through Germany. On the journey, they might have to confront anti-semitic (anti-Jewish) border guards who might turn them back or demand a bribe. For the most part, Jews in Western Europe did what they could to help their Eastern European brethren on their journey.

Once in the United States, Jews had to receive the approval of immigration service workers, find a place to live, and find jobs without being cheated by swindlers in a land where few of them could speak the language. As the number of Jewish immigrants from Eastern Europe grew, many Jews already in the United States were hostile to them. Most of these American Jews (or their parents) had come from Germany a generation or two earlier. The peak of Jewish immigration from Germany was in the 1840s. These German Jews were, as a rule, better educated than Eastern European Jews. Many were doctors and university professors. A good number started out in America as peddlers, but German Jews made significant headway in banking, the law, and as artists.

Jews Visiting the Auschwitz-Birkenau Concentration Camp in Poland (2017)

Jews from Eastern European, by contrast, were farmers, small shopkeepers, and factory workers. The Jews already in America feared that they might lose their jobs to the Eastern European Jews, who would be willing to work for lower wages. Eventually, however, Jews in America did provide help for the later-arriving Jews. The Eastern European Jews tended to settle in poorer neighborhoods of major American cities. Many of them found work in the garment industry, cigar manufacturing, food processing, and construction. Cincinnati became a popular place for Jewish immigrants to settle.

The Jews who left their homeland survived physically. They also preserved something of their culture, at least for a time. The danger they faced was that they and their faith and their culture might be swallowed up in the dominant American way of life. In addition, many white Americans disliked and distrusted the huge numbers of largely poor and poorly educated Eastern Europeans who spoke a different language and followed a different religion from theirs. Some Americans saw the Jewish immigrants as a threat to the American way of life, a threat to their jobs, and promoters of socialist revolution. Some of the Jewish immigrants were political and economic radicals, but that did not justify discrimination or persecution.

The Changing Population of Jews in Europe

Between the assassination of Czar Alexander II and the outbreak of World War I (in other words, from 1881 to 1914), about one third of the Jewish population in Eastern Europe—some 2.5 million people—left their homeland. Some two million came to the United States, while the rest went to other countries, including Western Europe and Palestine.

In 1880 the Jewish population of the United States was about 250,000, and only one of six were from Eastern Europe. By 1920 the United States was home to four million Jews, about 80% of whom were from Eastern Europe.

Emigration and the Holocaust radically altered the Jewish population of Eastern Europe. The best estimates are that, in 1939, the world population of Jews was about 16.6 million. Of that number, 9.5 million lived in Europe. In 1945, the Jewish population of Europe was 3.8 million.

In 1939 about 3.4 million Jews lived in the European portion of the Soviet Union. Today the number of Jews who live in what was the Soviet Union is 310,000.

In Eastern Europe outside of Russia (Poland, Hungary, Romania, etc.), there were about 4.9 million Jews in 1939; today that number is less than 100,000.

Worldview Lessons

The story of the Jews in Eastern Europe provides several lessons about worldview. The Jews by and large saw themselves as God's chosen people, but as such they also saw themselves as God's suffering servant that Isaiah described. They saw themselves as strangers and aliens in foreign lands who longed for a secure homeland. They awaited the coming messiah that they believed God would send. For the most part Jews struggled in their relationship to the Gentile world. Some compromised their Jewish traditions to get along more with Gentiles, while others withdrew even more from the Gentile world and became more devoted to strict traditions. The majority of Jews continued to struggle but made a life for themselves within their Jewishness in a world that often treated them harshly. Most put their faith in God to help them through.

We must also consider the worldview of non-Jews regarding the Jews. Many Gentiles treated Jews with varying levels of suspicion, prejudice, and in some cases even contempt and hatred. These Gentiles believed that Jews were beneath being treated as equals but also feared what Jews might do. Many Gentiles (perhaps unconsciously) believed in a limited economic pie, and thus they saw any Jewish financial success as a threat to their own success. A Jew holding a job meant to them that a Gentile would not have a job. Some Gentiles blamed Jews for what they saw as wrong in the world, from plagues to economic hard times. On the other hand, some Gentiles, such as some in Poland, saw Jews as simply fellow human beings who could contribute to society.

People have been oppressing Israelites for a long time. The Egyptians' oppression of the Israelites was the start of God's redemption of Israel from slavery.

Now, behold, the cry of the sons of Israel has come to Me; furthermore, I have seen the oppression with which the Egyptians are oppressing them.
Exodus 3:9

Assignments for Lesson 61

Geography   Study the map of Eastern Europe and read the entries for Belarus and Poland (pages 126, 127, and 132).

Worldview   Copy this statement in your notebook and write your answer: Write a paragraph describing God.

Project   Choose your project for this unit and start working on it. Plan to finish it by the end of this unit.

Literature   Continue reading *Kidnapped*. Plan to finish it by the end of this unit.

Student Review   Answer the questions for Lesson 61.

Parade in Budapest, October 25, 1956

62 Defiant Hungarians

On October 23, 1956, tens of thousands of demonstrators took to the streets of Budapest, Hungary, to protest the Communist oppression that was strangling their country. Their national government was a puppet whose strings the Soviet Union controlled, and the protesting Hungarians wanted to be free.

The demonstrators, largely young factory workers and university students, tore down a statue of Joseph Stalin and dragged it through the streets. They focused their wrath on the government-run radio station. The station was the mouthpiece of the government, and the demonstrators hoped that if they seized the station they could get their call for reforms into the ears of many more Hungarians. The protesters wanted free elections, an impartial legal system, private ownership of farms, the withdrawal of Soviet troops from their country, the right to leave the Warsaw Pact (the Soviet-imposed organization of Communist Eastern European nations begun in response to NATO), and the freedom to be neutral (neither pro-Soviet Union nor pro-United States) among the nations of the world.

The evening's event quickly became more than a protest march. Many of the demonstrators had brought guns, and the Soviet troops and Hungarian government security forces on the scene were willing to use their arms as well. Before the night was over, scores had been killed or wounded. Before two weeks had passed, about 2,500 people had lost their lives. Before the crisis ended, around 200,000 Hungarians had fled to neighboring Austria or elsewhere.

Hungary is a beautiful land that occupies a strategic location in Eastern Europe. The unfortunate aspect about its location is that it once shared a border with the Soviet Union. Therein lay its misery.

The Land and Its People

About the size of Indiana, Hungary occupies a place on the main route between Europe and the Balkan Peninsula and between Ukraine and the Mediterranean region. In its west are hills and low mountains (the highest point in Hungary is about 3,300 feet elevation); to the east are plains. Hungary is landlocked; but the beautiful Danube River flows through its center, connecting it with other parts of Europe. The Danube and Tisza Rivers divide the country into three parts.

The ethnic group known as Magyars migrated from the Carpathian Mountains to the east and in 896 settled the region that we are discussing. Stephen, son of the Magyar chieftain Geza, was born about 970. Stephen was born a pagan but was baptized and taught as a Christian. When Geza died in 997, Stephen fought off a challenge by his pagan cousin and claimed the leadership position of his nation. The Byzantines called this people the Oungroi, which means the people of the alliance of ten tribes. The term Oungroi eventually became Hungarians and their land became Hungary. On Christmas Day in the year 1000, Stephen accepted a crown from Pope Sylvester II and became the first king of Hungary.

A period of inept rule in the early 1200s led Hungarian nobles to demand rights for themselves and a limitation on the powers of the king. The nobles crafted a document, the Golden Bull, that every succeeding king had to accept. The nobles had the right to approve a person before he became king. This created an early constitutional monarchy.

Mongols invaded Hungary in 1241. They devastated the country, and half of the population either died or left. However, the Hungarians recovered. The last of Stephen's descendants died in 1301. Foreigners ruled Hungary from 1308 with only two brief exceptions.

Hungary served as a wall of resistance that delayed (but did not end) the advance of Ottoman Turks into Europe. In 1389 the Turks defeated the Serbs at the Battle of Kosovo in the Balkans, and thus the Turkish threat came closer. In the mid-1400s, the Hungarian political and military leader Janos Hunyadi defeated invading Turks on several occasions, most notably at the Battle of Belgrade in 1456.

But then in 1526, the Ottoman Turks defeated Hungary at the Battle of Mohacs. As a result, three powers claimed portions of Hungary: the Ottomans in the central part, the Habsburgs of Austria to the west, and Transylvania (now part of Romania) to the east. The Habsburgs declared themselves to be the only true rulers of Hungary, and they and the Turks engaged in a long-running conflict for control. The Habsburgs finally expelled the Turks in 1686 and ruled practically all of Hungary until 1867.

During this period of Habsburg rule, in 1848 a series of liberalizing revolutions took place or were attempted in several countries throughout Europe. Lajos Kossuth led such a revolt in Hungary. Kossuth succeeded in establishing a reform government, but it only lasted until the following year. When it fell apart, Kossuth left the country never to return. Many Hungarians see Kossuth as a hero of independence.

Lesson 62 - Defiant Hungarians

The Habsburg dynasty reasserted itself over Hungary, but in 1866 Prussia soundly defeated the Habsburgs in war. Now weakened, the Austrian Habsburgs had to compromise with Hungarian leaders to regain some of their stature and strength. The Compromise of 1867 created a dual monarchy over the two largely independent countries of Austria and Hungary. The two nations had separate governments and operated independently in domestic matters, but they shared one monarch and one foreign policy. This created the Austro-Hungarian Empire. Hungary is slightly larger than Austria in terms of land area.

Austria-Hungary annexed Bosnia and Herzegovina in 1908 to limit Ottoman influence in the Balkans. Because of Austria's close connection with Germany, Hungary got drawn into World War I on the side of Germany. The defeat of Germany and Austria-Hungary brought an end to the Habsburg dynasty. At the end of the war, Austria and Hungary became two completely separate nations.

After a period of instability, the last commander of the Austro-Hungarian navy, Miklos Horthy de Nagybanya led a group of arch-conservative military officers in ruling Hungary. He hated both Communism and Naziism, but since Germany supported Hungary's desire to reclaim land that it had lost in the settlement following World War I, Hungary's leaders sided with Nazi Germany. Horthy tried to negotiate a secret separate peace between Hungary and the Allies, but Germany found out and imprisoned him.

During the war, Hungary was a victim of the Nazi Holocaust. About 825,000 Jews lived in Hungary before the war. While Raoul Wallenberg was on the staff of the Swedish diplomatic mission in Budapest, he and his colleagues arranged for thousands of Jews to receive diplomatic protection from the Nazis. Over a half-million Hungarian Jews perished in the Holocaust, but Wallenberg and others were able to save the lives of untold thousands of Hungarian Jews.

The Romans established the settlement they called Aquincum on the west bank of the Danube, in the century before Christ was born. This developed into the city of Obuda or Old Buda. Another town, Buda, was built further west on a hill. It became the seat of Hungarian monarchs. (Note Buda Castle in the foreground on the left.) The city of Pest (seen on the right) was established on the plains of the eastern bank of the Danube, and it became a commercial and industrial city. In 1872 the three cities merged and became Budapest.

Budapest was heavily damaged by fighting during the war. Russian forces took control of Hungary in April of 1945, even before the surrender of Germany. In 1948 the Soviet Union installed a puppet government in Hungary. Hungary shares a border with Ukraine, which at that time was part of the Soviet Union. Thus the Russians wanted Hungary as part of their buffer of protection from invasion and as a place where they could exercise Communist control over the hearts and minds of more people. On the national flag, the Soviet crest replaced the Holy Crown of St. Stephen.

The 1956 Revolt

The Soviets' hardline rule appeared to be changing as 1956 unfolded. In February, new Soviet leader Nikita Khrushchev delivered a secret speech before a gathering of high Soviet officials. In the speech he reported some of the crimes and inhuman actions of former leader Joseph Stalin, who had died in 1953. Not many people outside of the gathering knew about the speech, but those who did wondered if the speech signaled a new day in the Soviet Union.

In June 30,000 workers demonstrated for better working conditions in Poznan, Poland, another Communist-controlled country. The demonstrations turned into a riot that the Soviet-backed authorities put down harshly, but the event raised the possibility that the Communist grip on Eastern Europe might not be as firm as once thought. Some wondered whether open political debate and even criticism of Communist policies might be possible.

This message on a bookstore window in Budapest in 1956 says, "Russians, go home!"

Then came the night of October 23, 1956, in Budapest, as described at the start of this lesson. The street demonstrations and the violence continued for a few days as the revolutionaries tried to gain control and the authorities tried to regain control. Then for about five days, Budapest experienced the sunshine of peace and freedom as the Soviet government determined its next move.

On November 1, 1956, Soviet leaders put Imre Nagy at the head of the Hungarian government to try to insure order. Nagy had been the head of government once before. In 1953 he had promised that reforms would take place, but two years later he was dismissed and driven from the Communist Party. Now Nagy asked the Soviets to withdraw their troops from Hungary, which they did. But he also announced bold steps that he planned for Hungary to take. Nagy announced that Hungary would hold multiparty elections and that it would withdraw from the Warsaw Pact.

In announcing these changes, Nagy again had gone too far. Soviet agents abducted him. He stood trial, was found guilty of treason, and was executed. The Soviets replaced him with Janos Kadar, someone who would more closely hew the party line.

The situation was at a standoff. Western countries feared that intervening on behalf of the reformers would trigger a stronger Soviet response and perhaps even a wider war. The Soviets feared what the United States might do if Russia took a stronger stance. Although the United States had encouraged people in Eastern Europe to stand up against their Communist rulers, when it actually happened in Hungary, the U.S. response was muted. America offered no direct military aid to the rebels, although the U.S. did send assistance to Austria to help refugees. The unfolding Suez Canal crisis in Egypt diverted world attention from what was happening in Hungary (see Lesson 22).

Encouraged by believing that the Americans would not respond, on November 4 the Soviets re-entered Hungary in force and crushed the rebellion. Russian tanks destroyed much of Budapest. The

Lesson 62 - Defiant Hungarians

ad hoc army of protesters was no match for Soviet military might. Scattered resistance continued in a few places for a few days; but the revolution was effectively over, and the Hungarians had lost.

Janos Kadar remained in power for most of the duration of the Communist regime in Hungary. Over time he gradually oversaw a few moderate changes in Hungarian life, government, and economics. Hungary experimented with what some called goulash Communism, a mixture of Communist and capitalist activities. Through it all, however, Communist Hungary was a loyal ally of the Soviet Union in world affairs. Kadar was removed from power in 1988, when matters were unraveling in Hungary and the rest of the Soviet bloc.

Communism in Hungary went the way of Communism in all of eastern Europe in 1989 and 1990. In 1990 Hungary held its first multiparty election and announced its intention to develop a free market economy. The country joined NATO in 1999 and the European Union in 2004. Every year, October 23 is a day of national celebration. Flags that flew during the period of Soviet domination are brought out and flown again—but now with the Soviet crest cut out of the middle.

The population of Hungary is about ten million, and that number has been shrinking. The country has a low birth rate. It has an aging population, and it has also lost many people who have emigrated to other countries. The current Hungarian government has begun offering significant financial incentives to couples who have children, such as grants to help them buy new homes and income tax deductions for value-added (sales) taxes that they pay.

Almost everyone in Hungary is ethnically Hungarian. Hungarians are about one-third Roman Catholic and 11-12% Calvinist. About 45% of the population indicate no religious preference or do not give a response. The country's troubled history under an atheist Communist government during much of the twentieth century is probably a major reason for this large number of those who do not profess faith.

Hungarian Catholic leader Joseph Mindszenty (born 1892) was an outspoken and lifelong opponent of totalitarian government. He was arrested in 1919 during a period of instability after World War I, and again in 1944. In 1948 the Communist government seized the lands belonging to the Lutheran and Calvinist churches, which agreed to financial arrangements with the state to maintain their schools (which would have to become secularized). Cardinal Mindszenty refused to secularize the Catholic schools, and the Communist government arrested him, found him guilty of treason, and sentenced him to life in prison in Hungary.

Mindszenty was released during the 1956 revolution. When the Communists regained control, Mindszenty sought asylum in the U.S. embassy and spent the next fifteen years there. He left Hungary in 1971 and visited the Vatican, then settled in Vienna until his death in 1975. After the Communist government of Hungary fell, Mindszenty's body was reinterred in Esztergom, Hungary, where he had served as archbishop.

The Holy Crown of St. Stephen is the symbol of Hungary and remains an object of Hungarian pride. During World War II, the Hungarian army entrusted the crown to American soldiers so that it would not fall into German or Russian hands. The U.S. stored the crown at Ft. Knox, Kentucky, until 1978. President Jimmy Carter decided to return the crown to Hungary. Hungary was still a Communist nation at the time, but it had shown progress in protecting human rights and Carter wanted to recognize this. The return of the crown was a step that helped pull Hungary toward the West and away from the Soviet Union. The crown is shown here on display with other royal regalia at the Hungarian Parliament building in 2017.

Hungary is a country most Americans know little about. Very few Americans will ever visit it. Why do we include a lesson about Hungary in this curriculum? For several reasons:

- It shows how every person matters.

- It reminds us what bravery, sacrifice, and other noble traits look like in real life.

- It is a lesson in human geography: how people respond to and are affected by the geographic setting in which they live.

- It is a vivid lesson in the value of freedom. We must never give up freedom. Freedom is the soil that encourages the growth of great lives and great accomplishments. Such greatness might not happen every day, but it will eventually come if people stand for freedom, including freedom for those with whom they disagree.

Paul encouraged the Christians in Galatia to stand firm in the spiritual freedom that Christ gave and not to give in to control by others.

Lesson 62 - Defiant Hungarians

It was for freedom that Christ set us free; therefore keep standing firm and do not be subject again to a yoke of slavery.
Galatians 5:1

Assignments for Lesson 62

Gazetteer — Read the entries for Bulgaria, Hungary, and Romania (pages 128, 130, and 133).

Worldview — Write a paragraph in your notebook describing God's love.

Project — Continue working on your project for this unit.

Literature — Continue reading *Kidnapped*.

Student Review — Answer the questions for Lesson 62.

Vltava River Near Teletín, Czechia

63

The Vltava (The Moldau)

How would you express your appreciation of the beauty of a particular feature of geography?

Some people might describe it in words. Some might write a poem about it. Others might take photographs or execute a painting.

Bedrich Smetana composed a piece of music that has thrilled concert-goers and inspired the people of his country for a century and a half.

Bohemians, Moravians, and Slovaks

The ethnic groups of Eastern and Central Europe have histories and cultures that go back for centuries, if not for millennia. Some of these groups are related, while others have battled each other repeatedly. Most of them have endured domination by foreign powers at various times.

In ancient times, the Boii tribe of Celtic people lived in what we now call Bohemia. The Czechy Slavonic tribe lived in that region from around the 500s. The homeland of the Moravians lies just to the east of Bohemia. The terms Bohemians and Czechs generally refer to the same people. In fact, Bohemians usually call themselves Czechs. Sometimes people who use these terms mean to include the Moravians while at other times Moravians are considered to be distinct from Bohemians. The Slovaks, who lived to the southeast, are related but are never included as Bohemians or Czechs.

The district called the Sudetenland includes some of western and northern Bohemia and northern Moravia. Its people are predominantly German. A portion of the region of Silesia, which lies mostly in Poland, is in northeastern Moravia.

Austro-Hungarian Domination

The Austro-Hungarian Empire came into existence in 1867 and claimed control over the Czech and Slovak lands. German became the second language of the region. As you can imagine, the Bohemians resented this takeover and longed to be independent once again. This desire found expression in many ways. One that has endured is in the form of an orchestral suite by the Bohemian composer Bedrich Smetana.

Smetana wanted to express his love for his country and to stir the hearts of his countrymen with the same passion. He titled the suite *Ma Vlast (My Country)*, and his purpose was to express what he called "musical pictures of Czech glories and defeats." The six movements included tributes to

Lesson 63 - The Vltava (The Moldau)

chivalry at a medieval castle, Czech legends and warriors, the area surrounding the Elbe River, and his hope for eventual Czech victory and freedom.

The Vltava (The Moldau)

The most famous part of *Ma Vlast* is the second movement, titled "The Vltava," the name of what is called the Czech national river. The piece is better known by the German name of the river, "The Moldau."

At 267 miles, the Vltava is the longest river in Czechia. It begins in the Bohemian Forest in the southwest portion of the country. At first the river flows to the southeast but then turns north. It continues through Prague, the capital, then flows into the Elbe north of the city.

Smetana had taken walks along the Vltava and had been struck by its beauty. His piece musically portrays a journey on the river from its headwaters to its flow beyond Prague. Smetana composed this movement in 1874, just a few years after the Austro-Hungarian takeover. It was first performed the next year. The first performance of the complete *Ma Vlast* took place in 1882.

The History of the Czechs Before World War II

The ethnic conflict and resentment that grew during Austria-Hungary's control of this region was part of the strife that sparked the beginning of the First World War. One of the principles that U.S. President Woodrow Wilson emphasized when America entered the war was the need for the autonomy (self-rule) by national groups as opposed to their being dominated by foreign powers. Near the end of the war, in October 1918, Bohemia, Moravia, and Slovakia formed a new nation called Czechoslovakia. The new country became prosperous and enjoyed a rich culture.

The Sudetenland also became part of Czechoslovakia, but the German-speaking majority there resented this. As Adolf Hitler's power grew in Germany, the German Sudeten people expressed their support for him. In the late 1930s, Hitler set his sights on taking over the Sudetenland and began to demand that Germany be allowed to do so.

In September of 1938, British prime minister Neville Chamberlain met with Hitler in Munich, Germany, and agreed for Hitler to take over the Sudetenland without opposition from other countries in Europe. Historians call Chamberlain's policy appeasement, because the British prime minister hoped that giving Hitler the Sudetenland would appease his further territorial desires. It did not. Germany took over the rest of Czechoslovakia in March of 1939.

Statue of Bedrich Smetana in Prague

In the 1968 photo above, protesters in Prague carry the Czechoslovakian flag past a burning Soviet tank. In the 1989 photo below, mourners commemorate students killed by the Communist leadership during the Velvet Revolution. In the center is Václav Havel, who served as president of Czechoslovakia (1989-1992) and then of the Czech Republic (1993-2003).

Lesson 63 - The Vltava (The Moldau)

After World War II

As World War II was drawing to a close, Soviet forces seized Czechoslovakia and installed a Communist government. The new government expelled ethnic Germans from the Sudetenland and repopulated the area with Czechs. In 1968 thousands of Czechs gathered in the streets of Prague asking for greater freedoms and democracy. The Soviet Union responded by sending troops and tanks into Prague to quell the protests.

Communists ruled the country until 1989, when Communist governments were falling throughout Eastern Europe. The peaceful transition to democracy in Czechoslovakia came to be called the Velvet Revolution. However, the changes were not over. On January 1, 1993, Czechoslovakia ceased to exist. Bohemia and Moravia became the Czech Republic (they later also accepted a shorter version of their name, Czechia). The remaining portion became the independent country of Slovakia. Czechia and Slovakia have become members of both NATO and the European Union.

The Czech people have often received harsh treatment from foreign powers. The way of Jesus is the way of kindness and compassion. When the disciples of John the Baptist came to Jesus to ask if He were the Messiah or if they should look for another, Jesus told them that they could perceive who He was not by His strident military actions but by these miracles of compassion, which fulfilled a prophecy of the Old Testament.

. . . the blind receive sight and the lame walk,
the lepers are cleansed and the deaf hear,
the dead are raised up, and the poor have the gospel preached to them.
Matthew 11:5

Assignments for Lesson 63

Gazetteer   Read the entries for Czechia, Moldova, and Slovakia (pages 129, 131, and 135). If you can, listen to a performance of "The Vltava." Visit NOTGRASS.COM/EWGLINKS for one recording. As you listen, follow the description of the music in the *Gazetteer* on pages 279-280.

Worldview   Copy this question in your notebook and write your answer: What does it mean to you that God's thoughts and ways are higher than man's (Isaiah 55:9)? To what extent can we know and understand Him?

Project   Continue working on your project for this unit.

Literature   Continue reading *Kidnapped*.

Student Review   Answer the questions for Lesson 63.

Yalta, Crimea, on the Black Sea

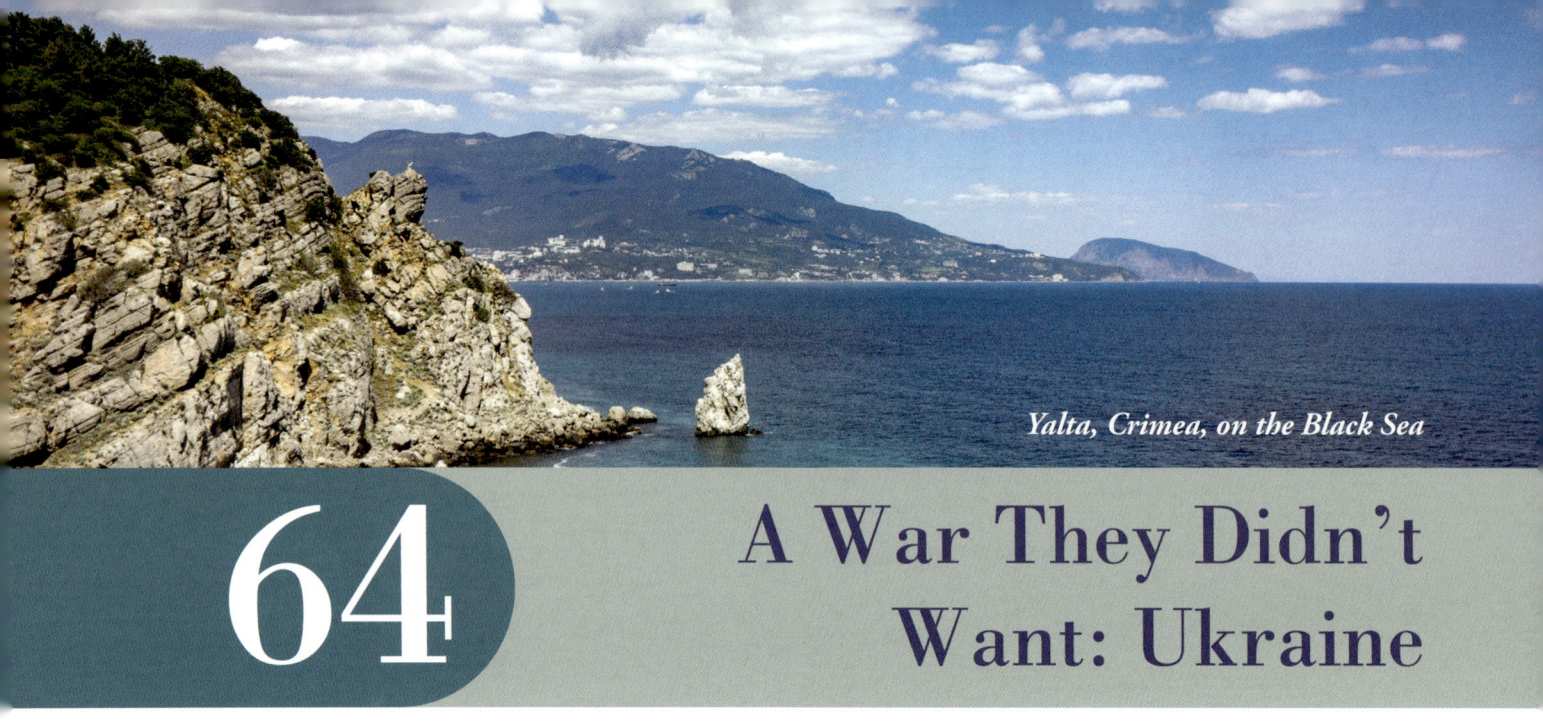

64 A War They Didn't Want: Ukraine

Igor Kozlovsky is a minister and professor of theology. He spent 700 days in captivity, enduring inhumane treatment and torture, in his homeland. His crime was supporting his country's government.

Kozlovsky lives in the eastern part of Ukraine, where an unofficial war has cost thousands of lives. The roots of this conflict go back hundreds of years. The central issues in this tangled and complicated conflict involve geography.

Background

Ukraine lies along the northern shore of the Black Sea. The country is south of Russia and east of other Eastern European countries, such as Poland, Slovakia, Hungary, Moldova, and Romania. A little smaller than Texas, Ukraine is rich in farmland and natural resources. Located on the East European Plain, most of its surface area is plains and steppe lands. The Carpathian Mountains are in the west.

Dangling from Ukraine like a bell in the Black Sea is Crimea or the Crimean Peninsula. Most of the peninsula continues the East European Plain, but the Crimean Mountains lie near the southern coast and are parallel to it. The Crimea is home to the major port of Sevastopol.

Ukraine's geographic location and rich resources have made it a target of foreign invaders many times. Around 1000 AD, Ukraine and its capital Kiev were home to the Kievan Rus people. For many years the Kievan Rus made up the largest and most powerful state in Europe. Even today, Ukraine is the second largest country in Europe behind Russia in terms of land area. Over the years, and partly because of invading Mongols, the Kievan Rus spread north to the region around Moscow. They became who we know today as the Russians. In addition to the Mongols, Poles, Lithuanians, and the Ottoman Turks at different times also have invaded and either controlled or tried to control Ukraine and Crimea.

Despite (or because of) these repeated invasions, Ukrainians developed a strong cultural and religious identity. About two-thirds of Ukrainians are Orthodox Christians. Ukraine knew a brief period of independence from the mid-1600s until the late 1700s, but eventually the Russian Empire absorbed it. Many Russians settled in eastern Ukraine and in Crimea, and Russia imposed its language and culture on Ukraine.

The Crimean War, fought between 1853 and 1856, involved the major world powers Russia and the Ottoman Empire who fought over control of Crimea and other issues. Both countries wanted to

Lesson 64 - A War They Didn't Want: Ukraine

increase their power in the Black Sea region. England and France sent troops to support the Ottomans. Fighting took place for three years, much of it in the Crimea, until Russia surrendered. In the treaty that ended the war, Russia continued to control Ukraine, including the Crimea.

Communism and After

When the Communist Revolution took place in Russia in 1917, Ukraine declared its independence from Russia. However, the Communist government in Russia took it back in 1920. Under Communist rule, Ukraine suffered greatly. The Soviets created two famines by manipulating the food supply which cost the lives of about eight million Ukrainians. During World War II, another 7-8 million died. The Communist government of the Soviet Union considered Crimea a separate republic until 1954, when the central government transferred control of Crimea to Ukraine.

After the collapse of the Soviet Union, Ukraine became a separate country in 1991 and continued to control Crimea. However, many ethnic Russians in Ukraine have wanted to remain allied with Russia. On the other hand, in reaction to the years of suffering under Russian rule, many other Ukrainians have wanted closer ties to Europe, including the possibility of joining NATO and the European Union. In the 2000s, Ukraine struggled with pro-Russian and pro-European political factions battling each other, sometimes in literal battles. In addition to this basic conflict, corruption and the struggles of changing from Communism to freedom have weakened the country.

Street Scene in Kharkov, Ukraine (then part of the USSSR, c. 1981)

These people gathered in Kharkov, Ukraine, in March 2014 to protest Russian aggression.

Russia's Interest

Russia has an interest in controlling Ukraine for two main reasons: the ethnic Russians who live there and the economic and military advantages that would come with controlling Ukraine.

Ethnic Russians make up about 17% of the population in Ukraine, most of whom live in the eastern part of the country. A majority of people who live on the Crimean Peninsula are ethnic Russians. These people speak Russian as their first language. A Russian law guarantees that Russia will protect its "citizens" (however the Russian government defines that term). Moreover, Russian President Vladimir Putin has said that "Ukraine is a made-up country," which reveals his lack of respect for Ukrainian sovereignty.

Concerning economic and military issues, Russia faces limitations to being the trade and military sea power it would like to be because of its geography. Its northern coast and its Siberian coast are frozen for much of the year. Russian ships can get to the Atlantic Ocean from the Baltic Sea, but to do that they must pass through the Skagerrak Strait, which NATO members Norway and Denmark control. That is not a problem in peacetime, but in a war the Russian Baltic fleet would be bottled up.

Russia faces a similar situation to its south. Russia has no good warm-water port on the Black Sea. For many years Russia leased a port from Ukraine at Sevastopol in Crimea. However, in recent years Russia has not been sure it can count on Ukrainian friendship.

Access from the Black Sea into the Mediterranean is through the Bosphorus and the Dardanelles, which are controlled by Turkey, now a member of NATO. Russian ships cannot get to the Atlantic Ocean except through the Strait of Gibraltar or to the Indian Ocean except through the Suez Canal, both of which are controlled by other countries.

Lesson 64 - A War They Didn't Want: Ukraine

Complicating Factors

But the complexity of the situation does not end there. European countries depend heavily on imports of Russian energy. At times, Russia has limited the energy supplies it has been willing to send to Europe when the European community took positions that Russia did not like. Moreover, much of the gas and oil that Russia sells to European countries passes through pipelines that run through Ukraine. Russia is seeking alternative routes.

When Ukraine's pro-Europe party has been in power, the country has made significant moves closer to Europe. For instance, in 2014 Ukraine purchased all of the nuclear fuel it needed from Russia. By 2016 that had fallen to 55%. In 2014 Ukraine purchased all of its natural gas from Russia. Two years later, Ukraine purchased all of its natural gas from Europe.

Russia's Actions

In 2014 Russia seized control of the Crimea. A "referendum" (perhaps fraudulent) held almost immediately overwhelmingly endorsed annexation, which Russia promptly announced. The international community applied weak sanctions against Russia for this move, but Putin's government thought the annexation was worth the slight punishment it received. Control of Crimea doesn't get Russia through the Bosphorus and the Dardanelles, but it does secure Russia's presence in the Black Sea region.

Russia has also built a 12-mile-long bridge between Russia and the Crimean Peninsula, to make trade and traffic between Russia and the Crimea easier and to secure its hold on the peninsula. Putin helped open the bridge in late 2019 by driving a truck across its length from Russia to Crimea. The world community condemned the bridge as a violation of Ukraininan sovereignty.

However, Russia is not only interested in Crimea. Since 2014, about 35,000 pro-Russian Ukrainians, foreign mercenaries, and regular Russian army troops (all backed by Russia) have been fighting about 60,000 Ukrainian loyalists in the Donbas region of southeastern Ukraine. As of late 2017, about 10,300 Ukrainians had lost their lives in the fighting, including some 2,500 civilians. An estimated 1.7 million people have been displaced because of the fighting. In addition, land mines are a continuing problem and have caused many casualties.

Much of the coal that Ukraine usually produces is mined in the region in which the fighting is taking place, but production and distribution have been difficult because of the fighting. Ukraine and Russia continue trading, although at lower levels than before the fighting began. Ukraine's trade with Europe has increased during the same time period.

Other Countries Get involved

In 2017 the United States government agreed to sell small arms and coal to Ukraine. This greatly assisted the Ukrainian people in their struggle for independence.

Volodymyr Zelensky was elected president of the Ukraine in 2019. He is shown here speaking at a National Flag Day event in August 2020. He spoke of the blue-yellow flag "under which our Armed Forces courageously defend the sovereignty and territorial integrity of the state."

Sunflower Field in Ukraine

Nor are the nations we've mentioned so far the only players in Ukraine. China has recently taken a great interest in the region. China wants to increase its trade with Europe. It has undertaken what it calls its Belt and Road Initiative (see Lesson 93). This involves China's investment of billions of dollars in road, railroad, and seaport facilities in Central Asia, Pakistan, eastern Africa, and elsewhere—including $7 billion in Ukraine, a key part of China's proposed trade route. Through these projects, China will get secure trade infrastructure (and perhaps secure military transport infrastructure also if the need ever arises), and the host countries get large, modern transport facilities.

Ukraine has significantly increased its trade with China. China once got 97% of its imported corn from the United States; today it gets 95% from Ukraine. China is the Ukraine's best customer for military equipment that Ukraine produces, with purchases totaling some $90 million in 2016. In addition, thus far China has respected Ukraine's territorial integrity and sovereignty; Russia has not.

What Difference Does This Make to Us?

What difference does this make to people living in other parts of the world? Why should we care what China buys from Ukraine and whether Russia builds a bridge to Crimea?

Russia and China both want to increase their trade and influence around the world. Russia's Putin sees the breakup of the Soviet Union as a "geopolitical disaster" and wants his country to regain the strength and status in the world it once had. And for all its movement toward capitalism, China is still a Communist country. In the eyes of the Chinese government, the fall of the Soviet Union was a tragedy. That event said that Communist states can collapse, and China's government does not want to be next.

Russia and China are expanding their influence, not just in Ukraine but in other parts of the world. Thus far China is expanding its influence economically, but Russia has been willing to move troops and rattle its saber. These trends might well impact the influence of the United States, both in Eastern Europe and elsewhere.

The situation in Ukraine and Crimea is unstable. How far will this instability spread? How long will the conflict in eastern Ukraine last? What will be the outcome if Ukrainian forces collapse? Will the United States make a military commitment somewhere, at some point? Will Europe? How will this affect Ukraine's movement toward closer ties with Europe, perhaps even membership in NATO and the EU, moves which Russia strongly opposes?

Lesson 64 - A War They Didn't Want: Ukraine

In addition, the continuing conflict is costing money that could be spent improving people's lives in Ukraine and elsewhere.

The Big Issue Becomes Personal

Igor Kozlovsky spoke out in support of the Ukrainian government when pro-Russian forces declared the Donetsk People's Republic. He was seized outside of his home on January 27, 2016. Kizlovsky was beaten to the point that his arms and legs swelled to twice their normal size.

In other incidents, in the Ukrainian city of Gorlovka, pro-Russian separatists entered a church building during a Sunday assembly and told the attendees to leave. In Donetsk, militants seized another church building and used it as a barracks for their fighters.

Igor Kozlovsky and his wife care for a son in his 30s who has Down syndrome and is confined to a wheelchair because of partial paralysis. Kozlovsky lost his income because he was not able to work, and his wife quit her job to care for their son. Kozlovsky was released on December 27, 2017, after 23 months in captivity.

"We must be disciples of love," Kozlovsky said after his release. "We may disagree, but we must never turn to guns. It is important to talk, not fight." Through his ordeal, we see once again why geography has practical importance in the everyday lives of everyday people.

Jesus taught His followers how to respond to evil attacks.

You have heard that it was said, "An eye for an eye, and a tooth for a tooth." But I say to you, do not resist an evil person; but whoever slaps you on your right cheek, turn the other to him also.
Matthew 5:38-39

Assignments for Lesson 64

Gazetteer   Read the entry for Ukraine (page 136).

Geography   Complete the map skills assignment for Unit 13 in the *Student Review Book*.

Project   Continue working on your project for this unit.

Literature   Continue reading *Kidnapped*.

Student Review   Answer the questions for Lesson 64.

Ciucaș Mountains, Romania

65 The Nature of God

You say you believe in God. What is He like? How would you describe Him? What is the character, the nature of the God you say you believe in?

Is He a cruel overlord that demands human sacrifices to be thrown into a volcano to appease His wrath? Is He an angry God who holds us over the fires of hell on a spider's thread just waiting to cast us into the abyss? Is He a jolly grandfather sitting in heaven who wants His children to be happy but who for some reason can't do anything about the suffering that exists in the world?

What a person believes about the nature of God is as important to his worldview as is the question of whether or not he believes in God. His understanding of the nature of God determines whether that person cowers in fear at the thought of a vengeful God or lives in trust of a loving God who is like a good shepherd.

Attributes of God

The Bible speaks often about the nature of God. Here are a few passages.

This is God's description of Himself:

The Lord is slow to anger and abundant in lovingkindness, forgiving iniquity and transgression; but He will by no means clear the guilty, visiting the iniquity of the fathers on the children to the third and the fourth generations (Numbers 14:18).

Moses says:

*The Rock! His work is perfect,
For all His ways are just;
A God of faithfulness and without injustice,
Righteous and upright is He.
(Deuteronomy 32:4)*

The psalmist says:

*Once God has spoken;
Twice I have heard this:
That power belongs to God;
And lovingkindness is Yours, O Lord,
For You recompense a man according to his work.
(Psalm 62:11-12)*

Lesson 65 - The Nature of God

The prophet Isaiah says:

Seek the Lord while He may be found;
Call upon Him while He is near.
Let the wicked forsake his way
And the unrighteous man his thoughts;
And let him return to the Lord,
And He will have compassion on him,
And to our God,
For He will abundantly pardon.
"For My thoughts are not your thoughts,
Nor are your ways My ways," declares the Lord.
"For as the heavens are higher than the earth,
So are My ways higher than your ways
And My thoughts than your thoughts."
(Isaiah 55:6-9)

The apostle Paul refers to ". . . . the hope of eternal life, which God, who cannot lie, promised long ages ago. . . ." (Titus 1:2)

Red Deer, Belarus

Learning About God Through Creation

We can learn something about the nature of God from geography. In Romans, Paul gives us the big picture of reality. In that letter he deals with God, sin, Christ, justification, redemption, the meaning of Israel, and how Christians should live and how they should get along with one another. That's a lot of ground to cover. Near the beginning of the letter, Paul refers to geography.

In the first chapter of Romans, Paul points out the reality of sin in the world. He declares that people are in rebellion against the God that they should know and worship. Paul says:

"[T]hat which is known about God is evident within them; for God made it evident to them. For since the creation of the world His invisible attributes, His eternal power and divine nature, have been clearly seen, being understood through what has been made, so that they are without excuse" (Romans 1:19-20).

The created world, Paul says, demonstrates God's divinity and power. Because the universe portrays these characteristics so clearly, people have no excuse for their rebellion against Him. Unbelief is not the result of a lack of evidence about God. Unbelievers simply choose not to acknowledge God in their thinking, even though evidence for Him is everywhere. As a result, God gives those who reject Him over to a depraved lifestyle. Not only do they not acknowledge God, but they suppress the truth by their unrighteous lives. They prevent themselves and others from knowing the truth (Romans 1:18).

The evidence that the physical world provides gives us the responsibility to believe in God, but it doesn't give us salvation. We can't know how to be saved by admiring trees and waterfalls. Creation tells us that everyone should acknowledge and worship God as the Creator, but it is in the Bible that we learn of our need for salvation and of God's provision of salvation through Jesus Christ.

Medieval Paintings in the Boyana Church, Sofia, Bulgaria

The Nature of God

Paul teaches us many attributes of God in his letter to the Romans.

(1) God has a plan. This world is not a matter of chance. God made promises before Christ came regarding His Son (Romans 1:2-3). God had a plan for Israel that He was working out in Paul's day (see chapters 9-11).

(2) God is a God of grace and love. He gave Jesus to be our righteousness before Him. God made the gospel of salvation known because mankind needed it as a result of our sin (1:17-18). God provided for our reconciliation to God even while we were His enemies (Romans 5:10). God is good and His purpose is good; therefore we should never blame God for our sin or for the suffering we endure.

(3) God is a God of eternal power who has a divine nature (1:20). His power operates on an eternal plane, not within the limitations of the mind and power of a mere human. God's nature is divine, which means He does not think or act the way humans do. Not only does He have a plan, but He is able to accomplish it. He has power, and He uses that power in love and for our good.

No doubt the Gentile Christians who read or heard Paul's letter to the Romans had in their background the typical Roman thinking regarding creation. Roman ideas about creation were much less clear and straightforward than the Genesis account. They believed that there were gods and there was chaos, and somehow one or more of the gods brought forth the world out of the chaos. Paul says that the world provides evidence of God's nature. It does not provide evidence of many deities overseeing a chaotic world.

(4) Because man chooses rebellion, God chooses to let people endure the consequences of

Lesson 65 - The Nature of God

their choices. Although the evidence for God is clear, people choose to reject Him. They neither honor God nor thank Him. They engage in futile speculations (in other words, the brain candy we call mythology and idle talk). This darkens their already foolish hearts. They claim to be wise, but in God's eyes they are fools. They exchange the worship of the imageless Creator for representations of what He has created, the true God for what is no god, the truth of God for a lie. This is not a smart trade!

You become like what you worship. If you worship an animal, you will likely live like an animal. If you worship yourself, you can go no higher than you. If you worship the One who is high and holy, you lift yourself up toward Him and become more like Him.

Because humankind rejected God, God gave them over to the consequences of their corrupt worldview. Mankind developed impure hearts that resulted in the dishonoring of their bodies. They developed degrading passions. God gave them over to a depraved mind that led them to practice, accept, and approve the list of sins that Paul gives at the end of the first chapter of Romans.

God is willing to let people face the consequences of their choices. We need to be careful what we think. When we accept Christ, we are saying to God, "Your will be done." On the other hand, when we reject God and want nothing to do with Him, God says to us, "Your will be done." If we reject God and move away from Him, He is willing to leave us there. He calls to us, and He wants us to come to Him; but He will not beg or change His terms. This does not make God hard or unfair. It means that God is righteous and that He gives us the freedom to choose the course of our lives. He will bless us and guide us if we choose to follow Him, and He will deal with us justly if we don't.

In 1 Peter 2:23, Peter says that Jesus entrusted Himself "to Him who judges righteously." God is going to get judgment right. He is not going to let anyone in who shouldn't be in, and He is not going to leave out anyone who should be in. We can trust that God is going to handle every person's eternal destiny exactly correctly.

How Your Understanding of God Affects Your Worldview

When you look at the world around you, do you see the result of a Big Bang, or do you see the work of the God who has a plan and who works by His eternal power and divine nature?

Do you see the physical world as primarily the result of natural forces or as the handiwork of God? Certainly natural forces are at work, but the Bible says there is a greater Power behind them.

Does the world bring you to an attitude of humility and worship before the one true God, or do your study and understanding lead you to leave no room for God in your thinking and to engage in idle speculation?

Súľov Rocks, Slovakia

Romans 1:20 tells us that our understanding of and response to geography have moral and spiritual consequences. The verse says that the created world—including its geography—reveals God and should result in our worship of God and our submission to Him.

The Creation speaks loudly and clearly to the mind and heart of faith:

The heavens are telling of the glory of God;
And their expanse is declaring the work of His hands.
Day to day pours forth speech,
And night to night reveals knowledge.
Psalm 19:1-2

Assignments for Lesson 65

Worldview — Recite or write the memory verse for this unit.

Project — Finish your project for this unit.

Literature — Finish reading *Kidnapped*. Answer the questions in the *Student Review Book*.

Student Review — Answer the questions for Lesson 65. Take the quiz for Unit 13.

Vologda, Russia

14

Russia

Russia is the largest country in the world. It has a rich and complex heritage. Russia has been a threat to other countries and has felt threatened by other countries. The first lesson in this unit surveys Russian history, geography, and government; its economy, people, and culture; and its relations with other countries. The next three lessons focus on specific aspects of Russian geography: the Ural Mountains, Lake Baikal, and the geographic challenge of building the Trans-Siberian Railroad. The worldview lesson delves into the basic question of whether creation or evolution is the best explanation for how our world began and how it operates.

Lesson 66 - The Bear: Russia
Lesson 67 - Snapshots from the Urals
Lesson 68 - The Deepest Lake in the World
Lesson 69 - The Toughest Project
Lesson 70 - Creation vs. Evolution

Memory Verse   Memorize Genesis 1:1-5 by the end of this unit.

Books Used   The Bible
Exploring World Geography Gazetteer
Lost in the Barrens

Project (Choose One)

1) Write a 250-300 word essay on one of the following topics:
   - What attracts you and what concerns you about Russia? See Lesson 66.
   - Write a persuasive essay that attempts to convince someone to abandon evolution and believe in the Biblical story of Creation. Be confident but respectful in your writing.
2) Write the story of a construction worker on the Trans-Siberian Railroad. Discuss his hardships, the realities of the work, any joys he might have felt, and how it would feel to be away from his family for months. See Lesson 69.
3) Draw a scene in the Urals or on Lake Baikal. See Lessons 67 and 68.

Literature

In the far north of Canada, two teenage boys find themselves stranded alone with winter fast approaching. Jamie grew up in Toronto and only recently came to live with his uncle in the far north. Awasin is the son of the chief of the Woodland Cree. These friends must rely on their own skills and creativity, hard work, and cooperation to tackle a life or death adventure of survival.

Farley Mowat was a prolific writer of fiction and non-fiction who used his platform to speak out for environmental preservation and for fair treatment of indigenous people. He was born in Ontario, Canada, in 1921. He served in Italy during World War II. After returning home, he joined a relative on a trip into the Arctic as part of his recovery from difficult wartime experiences. From that point on, he became an advocate for the people, land, and animals of Canada's Arctic region. He died in 2014.

Plan to finish *Lost in the Barrens* by the end of Unit 15.

Kamchattka Brown Bears, Russia

66

The Bear: Russia

Think about a bear.
    A bear is large.
    A bear is strong.
A bear fiercely defends its own.
A bear is dangerous when it feels threatened.

The symbol of the United States is strong but friendly Uncle Sam. The symbol for Great Britain is rotund and jolly John Bull. The symbol for Russia is the bear. Is that significant?

To think about Russia, think of the word big and its synonyms: vast, huge, immense, and so forth. This will help you begin to grasp the significance of the Russian bear.

Big

Russia (its official name is the Russian Federation) is the largest country in the world in terms of land area. It is almost twice the size of either the United States or China. Russia stretches almost halfway around the world east to west. It covers eleven time zones and is in both Europe and Asia. When people in eastern Siberia are finishing their workday, people in Moscow are getting ready to go to work, all in the same country.

Russia's land mass includes a vast diversity of geography: the treeless, marshy Arctic tundra (about one-tenth of its land area), taiga forests (its largest region), vast steppes, multiple mountain ranges, over two million freshwater and saltwater lakes, volcanoes along its Pacific rim, and intimidating deserts, as well as fertile farmland and numerous large cities. The West Siberian Plain covers over one million square miles and accounts for one-seventh of the Russian land area. Inside Russia are the Volga River, the longest river in Europe; Lake Ladoga, the largest lake in Europe; and Lake Baikal, the deepest lake in the world.

Russia experiences extreme winter weather. The town of Oymyakon in Siberia holds the record for the coldest temperature on earth not in the Arctic or the Antarctic. The temperature there has fallen to -90°F. No records exist of the temperature in Oymyakon rising above zero between December 1 and March 1. The Russian winter helped defeat Napoleon's attempted invasion in 1812 and Hitler's attempted invasion during World War II. However, summer temperatures in parts of Russia can reach over 100°F.

Russia has vast natural resources. For instance, it accounts for about one-fifth of both the world's oil production and the world's natural gas production. It contains the world's largest reserve of coniferous

381

trees. Russia is also a leading producer of steel and aluminum.

However, Russia also has significant geographic limitations. Its northern coast and its Siberian coast are frozen for much of the year. In an attempt to overcome this limitation, after World War II the Soviet Union annexed the city of Kaliningrad and its surrounding district on the Baltic coast from Germany. Kaliningrad, which is geographically separated from the rest of Russia, is the only Russian port on the Baltic that does not freeze in the winter.

Government and Economy

Russia has grand ideas about its government. After Constantinople fell to Muslims in 1453, some Russian Orthodox leaders began referring to Moscow with the majestic title of the Third Rome, after Rome and Constantinople. Ivan IV, grand prince of Moscow, was in 1547 the first person to be proclaimed Czar (also spelled Tsar) of All Russia. The title is the Russian form of caesar. Ivan is also called Ivan the Terrible for his bloody atrocities committed on his family and countrymen.

Some Russian rulers acquired the term "Great" for their reigns: Ivan the Great (1440-1505), Peter the Great (1672-1725), and Catherine the Great (1729-1796). During the Communist era, leaders Vladimir Lenin and Josef Stalin desired to rule the world and took steps to extend Communism to other countries. Stalin and his successors did rule the countries of Eastern Europe through puppet governments that were loyal to Moscow.

However, Russian government has weaknesses. During the time of the tsars, some rulers were strong and effective while others were weak and ineffective; but all were absolute monarchs. During the Communist era, the rulers were totalitarian dictators. Lenin was cold and calculating. Stalin was absolutely brutal and caused the death of millions of his countrymen by political executions and by engineering famines to bring Russians and others to their knees in dependence on their Communist rulers. After the fall of Communism, the eight years of Boris Yeltsin's presidency (1991-1999) were a time of turmoil and incomplete adjustment to capitalism and freedom. The tenure of Vladimir Putin has seen a restriction of personal and political freedom, a trend of government authority flowing into the hands of the leader, and a movement toward an authoritarian regime. In other words, Western-style democracy has never taken firm hold in Russia.

The Russian economy lagged behind the rest of Europe for centuries in terms of industrial output and other measures. Peter the Great worked hard to help Russia catch up during his day. However, in its agriculture Russia relied on the medieval practice of serfdom, in which farm laborers were tied to the land on which they lived and worked as little better than slaves. Russia did not abolish serfdom until 1861, long after the practice ended in the rest of Europe. For centuries, Russia followed the pattern of having a tiny minority who was wealthy with the vast majority of people living in poverty. Only in post-Communist days has a middle class developed, as the country has haltingly moved toward a capitalist economy (although state control of parts of the economy is still in place).

Russia began catching up economically in the late 1800s and early 1900s, but the Communist revolution brought a setback in the name of socialist

Winter Transportation in 16th-century Russia

Russian Farm Workers (1909)

progress. The Communist-led command economy of the 1900s did grow, especially in heavy industry although not so much in consumer goods. The Soviet standard of living was always below that of Western Europe and the United States. Russia did not develop and use its natural resources and abundant farmland well during the inefficient Communist days. Russia still lags behind the West in telecommunications and its road system. Moreover, organized crime has a significant presence in contemporary Russia.

People and Culture

The Russian people are strong. They have accomplished amazing things. Their sacrifice during World War II was immense. They have endured difficult governments, difficult economic times, and difficult weather. Their scientists were the first to launch an artificial satellite into space and the first to put a man into orbit around the earth.

Their most famous writers, Dostoevski, Tolstoy, and Solzhenitsin, wrote huge books about complex issues. Russian composers, such as Tchaikovsky, Shostakovich, and Prokofiev, have created works that are admired and still performed around the world. Russians have set the standard for ballet performance and chess competition. Their athletes consistently win numerous medals in Olympic competition.

However, Russia has significant weaknesses. Its population is less than half that of the United States. Although it is the largest country in terms of area, it has only the ninth largest population of countries in the world.

At the time of this writing, life expectancy in the United States is 77.8 years for men and 82.3 years for women, 45th highest overall in the world. In Russia, life expectancy is 65.6 years for men and 77.3 years for women, 155th highest in the world. Alcoholism is a significant problem in Russia, environmental damage and its consequences have been extensive, and Russian health care is not at the level of many other countries. Russia has had a negative population growth rate (meaning its population is shrinking) for several years, which means it is moving toward a generally older population.

An interesting contemporary trend is the significant growth of homeschooling in Russia. The Russian Orthodox Church supports this movement. The 2018 Global Home Education Conference took place in St. Petersburg and Moscow, Russia.

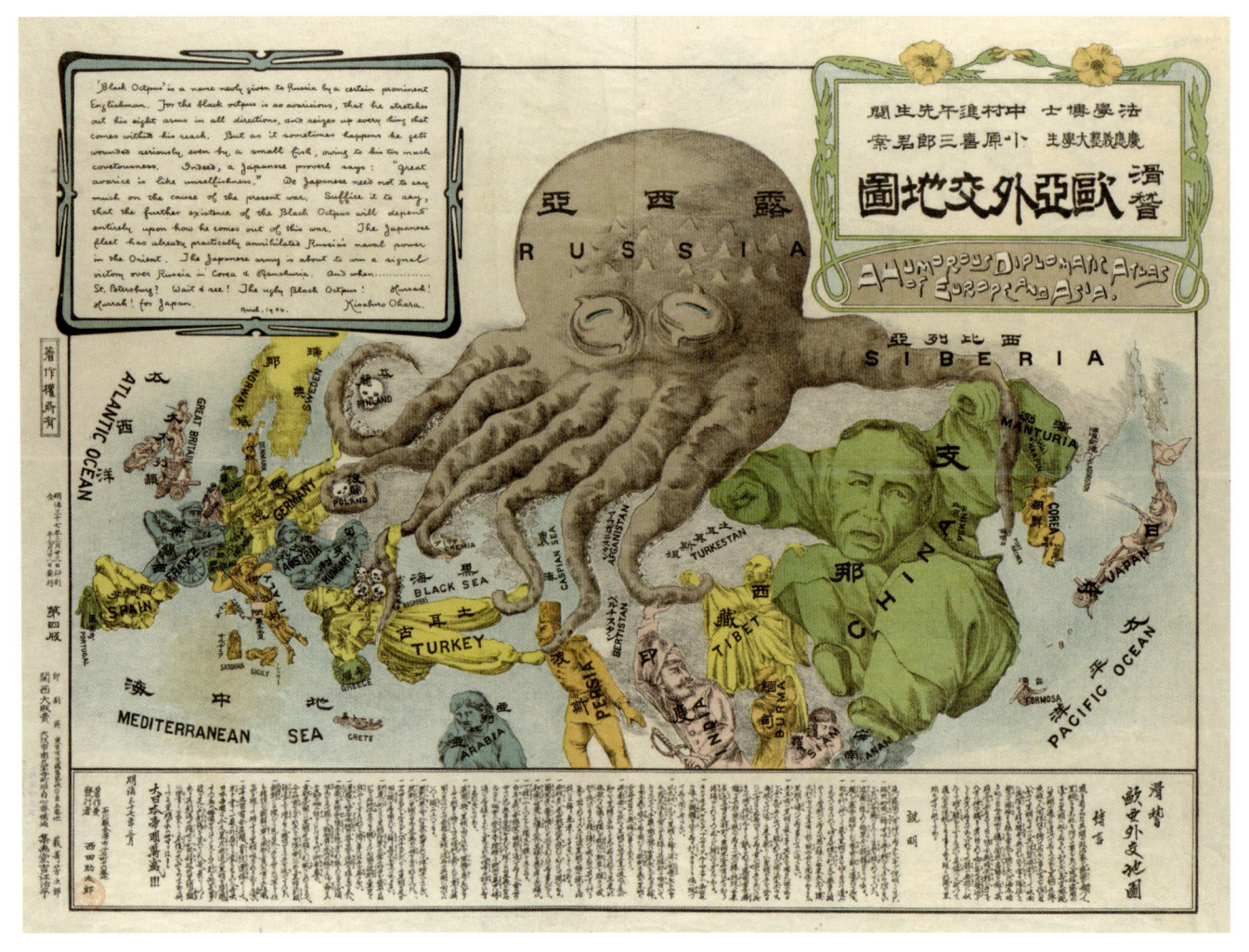

Kisaburo Ohara, a Japanese university student, created this "Humorous Diplomatic Atlas of Europe and Asia" in 1904, during the Russo-Japanese War. It depicts Russia as the "Black Octopus" with its tentacles around other nations.

Russia and the World

Put yourself in the Russian government's historic and geographic shoes. First, Russia is an immense country. This and its rich cultural heritage should (in their minds) lead it to have the status of a world power.

Second, Russia has known glory days in its past. The Russian Empire existed for almost two centuries, until the Communist Revolution of 1917. After that, Communist Russia expanded its power over fourteen other countries to create the Union of Soviet Socialist Republics. The U.S.S.R. also dominated eight countries in Eastern Europe for decades. This Communist empire fell in a matter of two years.

Third, Russia has often been at war, from many smaller conflicts in the days of the tsars to major conflagrations such as the Crimean War of 1853-1856, the Russo-Turkish War of 1877-1878, the Russo-Japanese War of 1904-1905, and the costly involvement in the two world wars of the twentieth century. The Soviet Union supplied weapons to North Vietnam during the Vietnam War and was later engaged in its own frustrating war to prop up a pro-Soviet government in Afghanistan from 1979 to 1989. Post-Soviet Russia seized the Crimean Peninsula from Ukraine, has been involved in the civil war in Syria, and has been involved in several other military conflicts such as the attempted breakaway of the Republic of Chechnya, officially a subject of Russia.

Lesson 66 - The Bear: Russia

Fourth, other countries have often invaded Russia: Poland in 1605, Sweden in 1708, the French in 1812, and the Germans in World War II. Following World War II, the NATO alliance was formed specifically to counter the threat of Soviet aggression. In response, the Soviet Union formed the Warsaw Pact of its allied countries to oppose NATO. Now all of the former Warsaw Pact nations—except Russia—are members of NATO or the European Union. The Russian government has formally stated that it sees the United States as a threat.

Russia, as every nation in the world does, conducts its foreign policy in terms of its own self-interest. Russia wants to be secure economically and militarily. It has an interest in the Arctic region. It has an interest in the countries in Eastern Europe that have a large ethnic Russian population. Russia supplies on average about 25 percent of the oil and natural gas that the countries of Europe use, and Russia takes advantage of this economic reality to influence the positions of those countries on issues pertaining to Russia.

Russian president Putin has described the breakup of the Soviet Union not as a giant step for freedom and justice but as a "geopolitical disaster." He wants to lead his country in regaining the strength and status in the world that it once had. The United States is not a threat to Russia. The United States government and the Russian government have differing worldviews and different approaches to handling international relations. Those differences are what keep matters interesting in the relationship between Uncle Sam and the Bear.

Vladimir Putin has served as either president or prime minister of Russia since 1999. He was elected to a fourth term as president in 2018.

This lesson has described the big picture of the geographic and political realities of Russia. However, in day to day life most everyday Americans and most everyday Russians are not very different in their thinking in some ways. They want to be able to live in peace and security and to be able to fulfill their dreams. Whether a person's country is large or small, powerful or weak, we all have one eternal destiny. That reality outweighs all of the physical characteristics and political maneuverings that we see in the world.

Paul expressed the reality of this destiny that we all face in this way:

Therefore we also have as our ambition, whether at home or absent, to be pleasing to Him. For we must all appear before the judgment seat of Christ, so that each one may be recompensed for his deeds in the body, according to what he has done, whether good or bad.
2 Corinthians 5:9-10

Assignments for Lesson 66

Gazetteer   Study the entry for Russia (page 134).

Worldview   Copy this question in your notebook and write your answer: What do you find most difficult to believe about the theory of evolution?

Project   Choose your project for this unit and start working on it. Plan to finish it by the end of this unit.

Literature   Begin reading *Lost in the Barrens*. Plan to finish the book by the end of Unit 15.

Student Review   Answer the questions for Lesson 66.

Yugyd-Va National Park, Russia

67 Snapshots from the Urals

I—The Mansi

The man wants a wife. He just can't find one in his home village.

The Mansi people who live in the Ural Mountains of Russia are a disappearing ethnic group. The last census indicated a population of less than two hundred, and that number has been shrinking. They face several challenges. One is that they live in remote settlements, a difficult six-hour drive from the nearest town. The largest of these is Yurta Anyamova. Two, just about everybody in the settlement is related. These factors make courtship difficult, to say the least. And three, many of the eligible Mansi young people have been marrying ethnic Russians and moving away.

The regional government, which wants to support and maintain the Mansi people, arranged a matchmaking project with a related Mansi group in a neighboring district. How successful the plan will be remains to be seen. They still have to find some way to get these potential couples together despite the geographic barriers they face.

The Ural Mountains run roughly 1,600 miles from the Arctic tundra south to Kazakhstan. Geographers have traditionally seen the Urals as the dividing line between Europe and Asia. The mountains are covered with thick forests. The Ural region harbors an abundance of mineral wealth, including oil, gold, precious gems, platinum, copper, iron, and mica.

Russia also contains a rich diversity of ethnic groups. About 80% of the people in Russia are ethnically Russian, and almost everybody in the country speaks Russian. However, people from almost 200 other ethnic or national groups live in Russia, but their percentages are relatively tiny compared to the Russian figure. Most of these groups speak their own languages as well as Russian. The great majority of Russians express loyalty to the

Mansi Necklace

387

Russian Orthodox Church, though only about 10-15% are active worshipers. The Russian government officially acknowledges Orthodoxy, Islam, Judaism, and Buddhism as the traditional religions of the country.

Many small ethnic groups struggle to maintain their identity. In their tiny settlements in the Urals, the Mansi struggle to maintain their existence.

II—Russia's Secret Nuclear City

It's code name was City 40. Deep in the Ural Mountains forest, Ozersk was the starting place for the Soviet Union's secret nuclear weapons program in the late 1940s. To keep the place and the project secret, the city of 100,000 people was not shown on any maps. The identities of the people who lived and worked there were removed from the national census.

Today the guarded gates and the barbed wire fences enclosing the city remain. So does the radioactivity. The city enjoys beautiful lakes, flowers, and tree-lined streets, a setting much nicer than the typical Soviet-era community. The water is contaminated, and the food is poisoned; but the residents of Ozersk love their community and live there despite the cost to themselves and their children.

The Soviets began building Ozersk in 1946 around a large nuclear facility. Experts moved there from around the Soviet Union. For eight years, the people who lived there could not leave and could have no contact with the outside world, not even with their own extended families.

The government built attractive private apartments and provided ample food for the stores. While the rest of the Soviet Union suffered shortages because of the failings of the socialist economic system, the people of City 40 had plenty. While most Russians were just getting by, the residents of Ozersk were financially stable.

The tradeoff was that the scientists who lived there developed nuclear weapons for the U.S.S.R. The government told the people of City 40 that they were "the nuclear shield and saviors of the world."

Ozersk (2008)

Lesson 67 - Snapshots from the Urals

What the government didn't tell them were the risks of exposure to atomic radiation as those risks became known. The facility dumped nuclear waste into nearby lakes and streams, which flow into the Arctic Ocean. Many residents died young. The nuclear facility there still houses radioactive materials. Some observers estimate that the residents of Ozersk have been exposed to radiation levels several times higher than that which resulted from the 1986 Chernobyl nuclear reactor accident in the Ukraine, one of the worst nuclear disasters in history. Some people call Ozersk "the graveyard of the earth."

Ozersk is still a closed city. Free access to the city without official permission is still prohibited, and few people leave Ozersk for good. They consider living there to be a privilege. They enjoy their lives in what they consider to be a paradise. It's just a paradise that is slowly killing them.

III—Russian Jewels

The city of Ekaterinburg offers some of the best and definitely the worst memories of the Romanov dynasty of Russia. Its importance involves its geography.

Peter the Great, who ruled Russia from 1682 to 1725, strove to bring Russia into the modern age. Because of the iron mines in the Urals, Peter ordered an iron works built on the eastern slope of the Urals in 1721. Workers built a fortress there the next year. In 1723 the location was named Ekaterinburg for Peter's wife, who later ruled Russia as Catherine I. It is now the fourth largest city in Russia and has become a major center for heavy industry.

Another, also important but quite different industry in the city is also related to the Urals. Because of the precious stones mined from the mountains, Ekaterinburg became a center for expert stone cutting for jewelry. The master Russian jeweler Carl Faberge employed stonecutters in the city for some of his beautiful works in the late 1800s and early 1900s. Faberge obtained most of the precious stones he used from the Urals.

Ekaterinburg (1789)

The city became a major transportation hub when Russia built the Great Siberian Highway through it in 1783. Its importance increased when the Trans-Siberian Railroad came through in 1878. The city is also a major cultural center, with many universities and other professional schools—including, of course, a mining institute.

In 1918 the Communists who had seized control of Russia the previous year held the last Romanov tsar, Nicholas II, and his family captive in a private home in Ekaterinburg. The captors murdered the royal family in the basement of that home. The house was demolished in 1977 but the basement remained. After the fall of Communism, a Russian Orthodox Church building was constructed on the site. It was completed in 2003 and named The Church on Blood in Honor of All Saints Resplendent in the Russian Land. The basement of the original house was included as part of the structure.

To the west of the Ural Mountains live about three-fourths of the Russian people. To the east lies about three-fourths of the Russian land area. The Urals have played and continue to play a crucial role in the history, geography, culture, and economic life of the country.

Mountains portray strength, endurance, and majesty. The prophet Isaiah described the reign of the Lord as the establishment of His mountain as chief of the mountains.

*Now it will come about that
In the last days
The mountain of the house of the Lord
Will be established as the chief of the mountains,
And will be raised above the hills;
And all the nations will stream to it.
And many peoples will come and say,
"Come, let us go up to the mountain of the Lord,
To the house of the God of Jacob;
That He may teach us concerning His ways
And that we may walk in His paths."
For the law will go forth from Zion
And the word of the Lord from Jerusalem.
Isaiah 2:2-3*

Assignments for Lesson 67

Gazetteer Read "How Much Land Does a Man Need?" (pages 281-291). Answer the questions about the story in the *Student Review Book*. The questions come after the review questions for this lesson.

Project Continue working on your project for this unit.

Literature Continue reading *Lost in the Barrens*.

Student Review Answer the questions for Lesson 67.

Nerpa Seal, Lake Baikal

68 The Deepest Lake in the World

Runners from all over the world gather to compete in this marathon. It occurs every March; but it doesn't take place on a track, or city streets, or across open fields. Instead, competitors run 26.2 miles from one bank to the other across the deepest lake in the world while it is frozen in winter.

Lake Baikal

The deepest of Russia's two million freshwater and saltwater lakes is Lake Baikal in southeastern Siberia, just north of Mongolia. It is 2,700 miles east of Moscow. The lake has a surface area of 12,200 square miles, which is less than half the surface area of Lake Superior in North America. Baikal is dotted with 45 islands and islets, the largest of which covers 275 square miles. Over 330 rivers and streams feed into Lake Baikal; but only one river, the Angara, flows out of it. Waves on the lake can reach fifteen feet in height.

Lake Baikal is the deepest lake in the world. It holds about one-fifth of the world's supply of fresh water. Its deepest point is 5,315 feet below the surface. Mountains stand next to it that rise over 6,600 feet above it. This means that within a short distance the difference in elevation is over two miles.

In 2009 Russian President Vladimir Putin entered a minisubmarine and descended to the lake's bottom. He described the water as pure but "a kind of plankton soup."

Lake Baikal has a remarkable environment. The region is home to over 1,500 animal species. The nerpa seal, one of only three species of freshwater seals in the world, is only found here. During the winter, the seals dig breathing holes in the ice with their sharp claws and teeth. Pregnant female nerpa spend much of the winter on top of the ice.

Some 80% of the animals that live around the lake are found nowhere else on the earth. The lake setting is also home to hundreds of plant species, about 370 bird species, and 50 species of fish. Because of its geography and unique environment, UNESCO named it a World Heritage Site in 1996.

The Baikal Ice Marathon

Baikal freezes in January and thaws in May or June. The tourist town of Listvyanka on the western shore hosts an annual Winter Games Festival, which includes competition in ice fishing, ice golf, and ice skating. In 2005 festival organizers approached Aleksey Nikiforov about organizing a marathon along the lake. He thought, "Why not across it?"

Listvyanka, Russia, on Lake Baikal

The Baikalsky Nature Reserve sits on the eastern shore, 23.2 miles away from Listvyanka. Nikiforov's challenge every year is to find a winding path over the lake that adds the additional three miles required for a marathon but avoids dangerous cracks and uneven mounds of ice. Flags mark off the route for the runners. Aid stations with tea and snacks are positioned every four miles. Restrooms (the women's is heated) stand at the halfway mark.

Nikiforov also has to measure the thickness of the ice to make sure it is safe. A thickness of 17 inches will support a military tank. In 2019 the ice was 27 inches thick. Still, cracks do appear and sometimes existing cracks close up. These cause the ice on the lake to shudder, which can be disconcerting to runners. Most of the frozen lake is covered with snow, but occasional patches of bare ice show up where the wind has blown away the snow. It is on this surface that runners compete for 26.2 miles.

Many runners want to take on unusual marathon courses as a challenge to themselves and their athletic skills and endurance. In 2019 a total of 97 men and 30 women representing 23 countries gathered on Lake Baikal to run. A fourth were from Russia. Twenty-four of these ran a half-marathon. The number of runners is small compared to the 27,360 who started the Boston Marathon in 2019, but Nikiforov keeps the number small. For one thing, this makes evacuating the runners easier if bad weather arises suddenly. In 2018 a strong wind blew in during the race. It stirred up the snow and reduced visibility to a few yards, forcing Nikiforov to cancel the race and evacuate the participants. This had never happened before.

Recommended (and enforced) racing attire includes a balaclava (ski mask), athletic tape covering exposed facial skin, goggles or glasses, a lightweight windproof jacket and pants, two layers of thermal underwear, and thick gloves. Spiked running shoes are a must.

Organizers carry the runners to the starting point by means of an hour ride in hovercraft. Since the course is basically straight and flat, the finish line is visible even from the starting point, but as

Lesson 68 - The Deepest Lake in the World

the runners approach it they have no perspective on how far away it really is. At times the runners can hear the ice crack, and they have to wonder how those changes might affect the race course. In 2019 the temperature during the race reached 26 degrees, relatively tropical for Siberia in March.

Over three hours after the 2019 race started, the winners crossed the finish line. The men's winning time was 3:05:05; the women's was 3:49:30. These times compare to Olympic marathon records of 2:06:32 and 2:23:07, respectively.

Environmental Issues

As with so many places around the world, environmental issues beset Lake Baikal (the ice marathon is officially titled the Clean Water Preservation Run). The continuing conflict pits environmentalists against those who want to see more industry and more tourism come to the lake.

Issues with industrial waste dumped into the lake caused a paper and pulp mill to close in 2013. In 2017 the World Bank froze its funding process for construction of a hydroelectric dam on the

The Volga

The other great body of water in Russia is the Volga River. Flowing 2,300 miles, it is the longest river in Europe. Starting northwest of Moscow, it runs south to the Caspian Sea. The Volga waters one-fourth of Russia's agriculture, and forty percent of the country's population lives within its basin.

In winter the Volga freezes, but the rest of the year it is a vital economic highway for the country. Communist leader Joseph Stalin ordered a canal to be built to connect the Volga to Moscow. The canal helped the growing city have the water it needed, but thousands of workers died building it. Other canals have created a transportation network that utilizes the Volga to connect Russians to the White Sea to the north, the Baltic and Black Seas, and the Sea of Azov, as well as the Caspian.

However, forests along the Volga are being cut down, which affects runoff. Smaller rivers that flow into the Volga are drying up. Pollution is harming the Volga habitat. This photo shows the Volga near Samarskaya Luke National Park.

Selenga River in Mongolia because of concerns that the changed water flow on the Selenga might affect the habitat of Baikal, into which the Selenga flows.

In 2019 the Russian government proposed new standards that would allow higher levels of industrial pollutants to be dumped into the lake. Environmentalists opposed this change. In the same year, a Russian court ruled against construction of a Chinese water bottling plant on the lake because of questions about the environmental impact statement that the Chinese company had submitted.

The omul fish, a species of salmon found only in Lake Baikal, is disappearing, perhaps from overfishing. The Russian government has banned commercial fishing of the omul.

God provided for us the water, air, and land that we need in order to live. We need to treat these well in order to live well on the earth. Jesus once described the new life He offered to people as the Holy Spirit, who would be as living water flowing from a person's innermost being.

Now on the last day, the great day of the feast, Jesus stood and cried out, saying, "If anyone is thirsty, let him come to Me and drink. He who believes in Me, as the Scripture said, 'From his innermost being will flow rivers of living water.'" But this He spoke of the Spirit, whom those who believed in Him were to receive; for the Spirit was not yet given, because Jesus was not yet glorified.
John 7:37-39

Assignments for Lesson 68

Worldview Copy this question in your notebook and write your answer: How have you seen society at large take the theory of evolution taken for granted?

Project Continue working on your project for this unit.

Literature Continue reading *Lost in the Barrens*.

Student Review Answer the questions for Lesson 68.

Railway Tunnel Along Lake Baikal

69 The Toughest Project

How would you build a railroad? How would you build a railroad over five thousand miles long?

How would you build a railroad over five thousand miles long across some of the most difficult geography in the world?

How would you build a railroad over five thousand miles long across some of the most difficult geography in the world without the machinery that is available today?

The Russians did it.

The Challenge

The straight line distance between Moscow in the west to Vladivostok on the Pacific Coast is 3,986 miles. Several mountain ranges and many wide rivers criss-cross all of Russia.

The European kingdom of Russia based in Moscow began sending explorers, armies, and pioneers east of the Urals into the region known as Siberia in the first half of the 1600s. By 1639 Russians had reached the Pacific coast. At the time, a Muslim Khan ruled Siberia. Russian forces overpowered the khan and his armies and assumed control of Siberia in 1640.

Northern Siberia is mostly tundra. Tundra is swampy, grassy area when it is not frozen in winter. Further south is the taiga or forested land. The southwestern part of the region contains grassy steppes. Siberia is rich in lumber and minerals, and farming takes place in the southern part. Winters in Siberia are notoriously brutal.

When the Russians came, Siberia was sparsely populated with small settlements of people from many ethnic groups. The Russian government faced the challenge of exerting control over Siberia and its people, incorporating it fully into the Russian state, and extracting its natural resources for domestic use and international trade. Roads were few. Most rivers in Siberia flow south to north (when they aren't frozen solid) and empty into the Arctic Ocean. In the 1800s, the best chance for Russia extending control over Siberia and developing its resources lay with the building of railroads.

The Concept

People first developed usable rail transportation systems in the early 1800s. By the 1830s rail lines were in use in several countries. In 1851 the governor of Eastern Siberia proposed that Russia build a transcontinental railway. There the idea sat

395

Soldier Riding a Reindeer at a Railway Survey Camp (1895)

in Russia for over twenty years. In the United States, by contrast, the idea emerged in the 1850s. President Abraham Lincoln encouraged it, Congress enacted legislation enabling it, and construction began in 1863. The American transcontinental railroad was completed in 1869.

The process moved much more slowly in imperial Russia. Tsar Alexander II sent surveyors out to consider three possible routes between 1872 and 1874. Then more delays ensued, but meanwhile people did build shorter lines connecting cities. A few of these even crossed the Ural Mountains. Some of these shorter lines became part of the Trans-Siberian line. In 1880 workers completed a railroad bridge across the Volga River. It was not until 1892 that work actually started on what people envisioned as a railroad line to cross the country of Russia from Moscow to Vladivostok.

In fairness, we need to remember that, for all the challenges that American builders faced, Russian civil engineers and builders faced even more difficult challenges. The distance to be covered was greater. The geographic challenges of the land to be traversed were greater. The Americans did not have to cut through the dense forests that the Russians faced. American companies had a few small rivers to cross; while the Russians had many large rivers to cross, and they were usually frozen for much of the year. As nasty as the winter can be on the northern U.S. prairie, winter in Siberia can be almost unbearable. The American route had at least some small towns along it, but Siberia had few towns and almost no roads.

The U.S. project had the cooperation—some would say collusion—of the federal government and the private companies that built it. The Russian government supported the project but only minimally, as we shall see. The entire project was a government operation, which can sometimes lead to inefficiencies and poor work.

The Construction

Officials of the Russian government instructed the civil engineers in charge of construction to build the line as cheaply as possible. This meant that the road bed was thinner than the standard beds used even in Russia. They built the main line using lighter weight rails, which were the standard for a light branch line. This was a poor decision, as usage demonstrated in later years.

Construction took place in three different sections (western, middle, and eastern) at the same time, sometimes creating completely new lines and sometimes linking up existing shorter lines. These linking projects added to the cost and time needed for construction as compared to a new, straighter line. In flat or low-lying areas, workers had to build an embankment for the track several feet high to avoid problems with flooding. As they built through the steppe grasslands, workers had to import stone and timber for crossties, since stone was absent and native trees were too small. Even potable (drinkable) water was scarce, so workers had to bring in heavy pumping machinery to dig the needed wells.

Builders had to construct numerous bridges to span the many rivers, some of which were a quarter-mile or a half-mile wide. Many smaller bridges were

Lesson 69 - The Toughest Project

made of wooden trestles, since steel and stone were scarce. On many bridge projects, the piers holding up the track were built with triangular buttresses that pointed upstream (see photo on the next page). These buttresses were intended to break up the ice that floated downstream in winter and threatened to destroy the piers.

Engineers constantly had to make allowance for the weather. For instance, they had to make allowance for expansion and contraction of the rails and sleepers (ties) as a result of extreme weather conditions. When they built masonry piers for bridges in winter (and winters were too long for the workers simply to sit idle), they had to protect the concrete and mortar from the intense cold so the materials would set properly. They did this by building wooden sheaths around the piers to keep out the cold, sometimes using heat sources inside the sheaths to hasten the process. In the eastern section, subsoil permafrost in some areas meant continuous blasting to create the rail bed required for the line.

Through the dense taiga forests, across immense grassland steppes, over wide rivers, and through challenging mountain ranges the men continued their exhausting work. In one section of the Mid-Siberian line, workers needed eighteen months to lay sixty-two miles of track. At Lake Baikal in eastern Siberia, the project required an ice-breaker train ferry. This ferry was constructed in England, sailed to St. Petersburg on the Baltic Sea, disassembled into smaller pieces, transported by train, wagons, and sleighs across Russia and Siberia, loaded onto a river steamer, and carried to the lake where it was reassembled.

But that ice-breaker and a second one that was brought in were not enough to get the needed supplies to both sides of the lake. As a result, workers laid a temporary rail line across the ice on the lake (that was really thick ice!). But then the engineers made the decision to lay the line around the southern tip of Lake Baikal, which involved blasting away huge rock outcroppings in the mountains and digging tunnels through cliffs.

Working on the Trans-Siberian Railway (1895)

East of Lake Baikal, the builders constructed a branch that connected with the Chinese Eastern Railway which ran through Manchuria. Russian builders later completed a route that remained completely in Russian Siberia. Eventually workers connected all of the sections of the Trans-Siberian Railroad, and the line opened in 1904. The route that avoided Manchuria opened in 1916. Thus a traveler could go from England to Japan in three weeks, and around the world in thirty-three days.

The American companies that built the Transcontinental Railroad employed about ten thousand Chinese workers and thousands of Irish workers. The Trans-Siberian Railroad used about two hundred thousand Chinese workers. The construction teams also included thousands of Russian convicts. Many men died on the American project; many more died on the Russian project.

The Aftermath

The Trans-Siberian Railroad was immediately inadequate for the demand placed on it. Lines of wagons carrying cargo destined for a train sometimes waited for months. Additional tracks and siding lines built over the years have increased its efficiency.

It is still the longest rail line in the world. Newer branches and connections give the traveler several options for making the journey. As a result, mileage estimates vary, but the line is at least 5,338 miles long and takes around six days to complete. The route passes through eight time zones.

The building of the Trans-Siberian Railroad involved many difficulties and uncertainties. The book of James teaches us to be humble in planning business activities, since we do not know what will happen tomorrow.

Railway Bridge Over the Kama River (c. 1910)

These women are selling food at a railway station in Siberia in 1919. Travelers on the Trans-Siberian Railway today are still able to purchase a variety of foods from vendors at stops along the way.

Come now, you who say, "Today or tomorrow we will go to such and such a city, and spend a year there and engage in business and make a profit." Yet you do not know what your life will be like tomorrow. You are just a vapor that appears for a little while and then vanishes away. Instead, you ought to say, "If the Lord wills, we will live and also do this or that."
James 4:13-15

Assignments for Lesson 69

Geography — Complete the map skills assignment for Unit 14 in the *Student Review Book*.

Project — Continue working on your project for this unit.

Literature — Continue reading *Lost in the Barrens*.

Student Review — Answer the questions for Lesson 69.

Gulf of Alaska

70 Creation vs. Evolution

Those who believe that the world came into existence as described in Genesis, and those who believe that our world came into being without God and developed through a long process of evolution all have worldviews. Many in both groups want to influence you to see the world as they do. People in each group believe that theirs is the only correct and logical explanation for the world.

The debate between creation and evolution follows lines similar to the differences described in Lesson 60 between those who believe in God and those who do not. The debate over this issue is a specific application of those differing worldviews.

What Evolutionists Want Us to Believe

I speak as an outsider to the evolution group, so I might not express their ideas the same way they do; however, from my perspective, this is what evolutionists want us to believe.

Evolutionists, those who believe in macro-evolution or evolution across species, want us to believe that nothing created our immense, complicated universe—no outside intelligence, no guiding hand; it all just happened.

Some evolutionists believe that matter is in some way eternal or that it is somehow self-creating; in other words, if matter did not come from nothing, it has always existed.

Whether they believe that the universe came into existence or that it has always existed, evolutionists want us to believe that our universe, which reflects order and which reason can understand, has evolved as it has through a mindless, purposeless series of chance events. We can't really call it a process because by definition evolution is non-directed.

Evolutionists want us to believe that all life is descended from a common ancestor. They believe that simple life forms which multiply by asexual reproduction evolved into complex life forms that reproduce with male and female forms, and that this stage in evolution occurred in just the right way at just the right time.

Evolutionists want us to believe the descendants of one kind of creature can be completely different kinds of creatures because of a series of genetic mutations.

Evolutionists apparently want us to believe that human life can have purpose in a purposeless universe. I say apparently because some evolutionists might not be concerned about purpose, even though they apparently see themselves as having a purpose.

400

Lesson 70 - Creation vs. Evolution

Problems with the Evolutionary Theory of Origins

No aspect of our created world is the result of something arising from nothing with no cause.

No matter that we know is eternal.

Nothing in our universe that is the result of chance, irrational events reflects order and can be studied and explained rationally. As a result, it is hard to believe that the universe itself came about in this way.

As F. LaGard Smith demonstrates in his 2018 book *Darwin's Secret Sex Problem*, the evolution from asexual reproduction to sexual reproduction as required for the theory of evolution to work is simply impossible. If species reproduction is impossible, evolution cannot happen.

Mutations in the genetic code of organisms happen, but the evidence does not explain how this gives rise to completely new types of organisms.

In other words, the advocates of macro-evolution want us to believe that what cannot happen and what we have no evidence of happening, did in fact happen and continues to happen.

That takes a lot of faith!

These problems, taken individually or together, prove fatal to the theory of evolution. The theory of evolution cannot, logically or on the basis of scientific evidence, be considered as true.

Charles Darwin observed variations that had occurred within types of animals and plants. He extrapolated (guessed) that these variations led to the diversity of life we observe. He also guessed that this evolution progressed from simple forms to more complex forms and that evolution took place by means of natural selection or the survival of the fittest. It was a guess. Darwin was wrong.

In addition, the theory of evolution has changed so often (dare we say that it has evolved?) and now has so many different variations that it is difficult to know which one a speaker or writer means. Did evolution take place as a long series of tiny changes (gradual evolution) or as occasional large changes (punctuated equilibrium)? Did the universe begin with a Big Bang or in what is called a steady state? Proponents of the latest version of evolution always seem to submit their idea as the final, assured truth—that is, until the next variation comes along.

All dogs share essentially the same genetic code. The amazing differences among dog breeds show how much variation is possible within the genes of one kind of animal.

Geography influenced Charles Darwin's theories as he tried to explain the distribution of varieties of plants and animals in different places. He considered the marine iguanas of the Galápagos Islands to be ugly but was intrigued by their ability to swim.

Finally, evolution devalues mankind. According to evolution, humans are just another physical object in the universe, without a purpose and without a soul. Any personal conscience, any societal standard of morality, is just the result of social or cultural norms, which might change or, in some times and places (because the judgment of man is an unreliable standard), might actually be harmful.

What Creationists Believe

Creationists believe that God created the world and continues to guide it. This is possible. We know power, and God is the ultimate power. We know intelligence, and God is the ultimate intelligence. The world is not evidence of something arising from nothing without cause or by irrational chance; it is evidence of design and order.

The most commonly cited reservation about the belief that God created the universe involves the issue of why, if God is the perfect Creator, so much pain and suffering exist in the world. If God is all-powerful, the question usually goes, why does the world He created seem so imperfect—especially humans, who are supposedly the pinnacle of His creation?

The question is understandable as humans try to figure out and live in the world that exists. The standard answer that people who believe in the truth of the Bible offer is that mankind's sin has corrupted the world. This world longs for redemption, and until that happens the world will experience pain, suffering, and disappointment.

For believers, as difficult and disappointing as this life can be, the alternative—a world in which God does not exist—is even more difficult and disappointing to contemplate. For believers, complete understanding of this world rests with God, not with people. People who believe in God trust that He knows and is in charge, even when it looks to us as though He isn't.

Maybe It's True Anyway?

Die-hard evolutionists, faced with the numerous impossibilities and logical failings of their position, sometimes reply, "Perhaps evolution is true anyway, even if the evidence we know about doesn't support

Geography also influences the understanding of scientists who believe in God as Creator. Folded rocks, such as these in Provo Canyon, Utah, suggest the work of catastrophic forces that could have been caused by the flood, for example.

it now." In other words, they have faith (trust in the unseen) that their position is true and that one day what we see today as impossible will have an explanation.

This is little more than a religious faith, a trust in doctrinal statements as true despite the lack of evidence to support them. If the creation-evolution debate is really a religious debate, we should recognize it as such.

Theistic Evolution

Some people who believe in God also believe in evolution. We call the process they believe in theistic evolution. They believe that evolution is the way that God has developed the world in which we live.

The major problem with this line of thinking is that it says the all-wise God has used an irrational approach to bring about an orderly world. They might fall back on the attempted defense that the process God used is logical but it is simply one we don't understand yet. The argument tries to combine faith in God with belief in evolution, but it doesn't hold together.

The Impact on Worldview

The practical significance of the creation-evolution debate is similar to the difference the existence of God makes that we discussed in Lesson 60. Did the God of the Bible create you in His image with the worth and purpose that He gives you, or are you the accidental result of a mindless series of accidents with no innate value and no more purpose than the value and purpose of a rock or a lizard? How you see yourself influences how you see the world in which you live and how you live in this world.

If life is merely the survival of the fittest, then a group that sees itself as the fittest can justify destroying all others. This is exactly what the Nazis did in the 1930s and 1940s. Many people understand that Naziism was a logical result of the theory of evolution.

If humans are the pinnacle of evolution, then whoever is the strongest person or group can decide what to do about the environment, the unborn and the elderly, and anything and anyone that the strongest sees as an issue to be dealt with.

If there is no Creator, then the most powerful group sets the contemporary standards (they would not say the absolute standards of right and wrong) that people must follow. If and when another group gains supremacy, their standards will prevail, imposed by force if that new, more fit group considers such force necessary. Laws, morality, school grading standards, and what constitutes acceptable thinking would all be what the most powerful group says they will be.

In short, if we are the result of chaos, the result of life will be chaos.

Which Takes More Faith?

The eye of faith in God looks at the Creation and sees evidence for the existence of God, His eternal power, and His divine nature.

The eye of unbelief looks at the same evidence and sees only a big bang and its aftermath: a mindless, purposeless accident. In this view, the entire basis for life is a chance process.

These different responses to the evidence arise because, even though evidence is important, evidence will only carry you so far. The skeptic who refuses to put his faith in Jesus as Lord of his life will have a hard time accepting evidence. On the other hand, a minimal amount of evidence will convince the person who is ready to put his faith in Jesus.

A Biblical, God-centered, Christ-oriented worldview involves an intellectual acceptance of truth; and it also includes the conviction that Christ is the Son of God and the commitment to follow Him as Lord of one's life in his or her everyday thoughts, desires, words, and actions.

Which idea of how the world came into existence and continues to work today is more plausible? Which requires more of a logical stretch, more "faith" in what is unseen?

The Bible says:

In the beginning God created the heavens and the earth.
Genesis 1:1

Assignments for Lesson 70

Worldview   Recite or write the memory verse for this unit.

Project   Finish your project for this unit.

Literature   Continue reading *Lost in the Barrens*.

Student Review   Answer the questions for Lesson 70. Take the quiz for Unit 14.

Aurora Borealis (Northern Lights) in Greenland

15   The Arctic and The Antarctic

The Arctic is largely a frozen ocean surrounded by continents. The Antarctic is largely a frozen continent surrounded by an ocean. God created them as part of the "just right" nature of the earth. As remote and challenging as these regions are—or perhaps because they are remote and challenging—people have sought to explore them. Lessons 71 and 74 tell the stories of two different attempts to reach the poles. Lesson 72 surveys the current conditions in the Arctic and examines why it has become a source of potential conflict among some nations. Lesson 73 describes Nunavut, the newest Canadian territory that is a homeland for the Inuit. The worldview lesson summarizes the story of Job as a testimony to the wonders of Creation.

Lesson 71 - To the North Pole and UNDER
Lesson 72 - Sitting on Top of the World
Lesson 73 - A Homeland for the Inuit
Lesson 74 - *Endurance*
Lesson 75 - "Have You Entered the Storehouses of the Snow?"

Memory Verse — Memorize Job 38:22-24 by the end of this unit.

Books Used — The Bible
Exploring World Geography Gazetteer
Lost in the Barrens

Project (Choose One)

1) Write a 250-300 word essay on one of the following topics:
   - What is the appeal of exploration—on land, on the sea, under the sea, or in space? Discuss the obstacles, the difficulties, and the potential benefits of venturing into uncharted territory.
   - Summarize the attempts people have made to reach the North and South Poles.
2) Prepare a report, a display, or a video on one of these subjects:
   - What is it like to serve on a modern American submarine.
   - Tell about the history, cultural practices, and current status of the Inuit.

USS Nautilus (1955)

71   To the North Pole and UNDER

When William Anderson began his studies at the U.S. Naval Academy in Annapolis, Maryland, in 1939, he knew that he wanted to serve on submarines. He had no idea where his service would take him.

A Submarine with Nuclear Power

Anderson was a member of submarine crews in the Pacific during World War II and the Korean War. Then an exciting development in submarine technology took place in the early 1950s with the construction of the world's first nuclear-powered submarine. The USS *Nautilus* took two and a half years to build, and the Navy launched it in 1955.

The *Nautilus* was built and based in Connecticut. At the time of its commissioning, the state's general assembly unanimously passed a resolution praising the sub as "a mighty force for the preservation of peace" and stating that it was the assembly's "fervent prayer that God will pilot her and the country that built her." Two years later, the U.S. Navy assigned Anderson to be the second captain of the *Nautilus*.

Conventional, diesel-powered submarines had to come to the surface daily, but a nuclear-powered sub could remain underwater indefinitely, for months at a time. Anderson immediately began thinking about the possibility of an undersea mission to the North Pole.

Troubled Times and a Dangerous Mission

The United States faced many challenges in 1957. The country was locked in a Cold War with the Communist Soviet Union. In October of that year, the Soviets launched Sputnik, the first artificial satellite to orbit the earth. Americans were embarrassed that the Russians were the first to go into outer space. It appeared that the Soviets were ahead of the Americans in science and technology. Even worse, many Americans feared what the Soviets might do with that advantage. Would they spy on us? Would they attack us?

Domestically, the U.S. was struggling through the civil rights movement, epitomized by the Little Rock school integration crisis that year. Just over a decade after the end of World War II, and with the military deadlock in Korea even more recent, the United States appeared to have little to cheer about.

Anderson wondered what significant mission the *Nautilus* might achieve for the U.S. He had the opportunity to discuss the possibility of a subpolar mission with Naval officials in Washington. The idea was fraught with serious questions and possible dangers. No one knew how thick the polar ice cap was, how deep the Arctic Ocean was, or the nature of the Arctic Ocean floor. No reliable maps or charts of the Arctic Ocean or the ocean floor existed. What if the *Nautilus* went under the ice cap and got stuck or lost or had onboard difficulties and could not get out? What would it mean to have a nuclear reactor on a submarine with over one hundred sailors? When the sub was underneath the ice cap, it could not communicate with the outside world.

In addition, navigation in the Arctic using a magnetic compass was uncertain. The magnetic North Pole, which lies hundreds of miles south of the geographic North Pole, renders magnetic compass readings inaccurate. The *Nautilus* would also have gyrocompasses that depended on the rotation of the earth for accurate readings. However, because the surface of the earth does not spin as much near the poles, the accuracy of the gyrocompass was questionable. The sub could be under the ice cap and heading for a land mass or going in circles, and the crew might not realize it. The crew could not navigate with a sextant because they could not see the stars while they were under the ice.

In terms of foreign relations, the Russians considered the Arctic Ocean to be their backyard. Over one-fourth of Russian territory lies above the Arctic Circle, and the Russians have long wanted to control the Arctic. The Soviets could have looked upon an American Navy vessel sailing through the Arctic as a military provocation. Any military venture into the Arctic could have ramifications that would be unknown at the time. We did know that any missile attack that the United States or the Soviet Union launched against the other would pass over the Arctic region; a U.S. military presence in the Arctic seemed like good policy. Besides, America's allies wanted to know that they could depend on American military capabilities to defend them if necessary.

The United States government wanted any Arctic undersea mission to be top secret until its

Bunks on the Nautilus

successful conclusion. Some officials flatly opposed it as too great a risk to take with the nation's only nuclear-powered sub. In addition, American efforts to launch an earth satellite were repeatedly failing. No one wanted an unsuccessful polar mission to make the U.S. appear to be even more scientifically and technically lacking.

First Attempts, 1957

Anderson received orders to explore under the polar ice cap from the Atlantic, with no definite limit as to how far he could go. Anderson hoped that the *Nautilus* could reach the Pole and return to the Atlantic. During the first attempt, the *Nautilus* went 150 miles under the polar ice cap. Cruising depth was 350 feet under ideal conditions, but the captain had to order several changes of depth as undersea conditions changed. The thickness of the ice varied considerably, and the ocean floor had great variations also. They were literally sailing into uncharted waters.

Anderson depended greatly on what he could see through the sub's periscope. At one point, he ordered the craft to surface in what he thought was an opening in the ice. However, there was no opening. The ice turned out to be solid, and it damaged the sub's two periscopes beyond usefulness. The *Nautilus* made it back into the open sea, where crewmen fought bitterly cold weather to repair one periscope. On the

Lesson 71 - To the North Pole and UNDER

second attempt, the sub penetrated 240 miles further in, but the navigational equipment failed. Around the 87th parallel of latitude, Anderson ordered the *Nautilus* to turn around. During the trip out, at one point the *Nautilus* was simply lost; but the sub did manage to emerge from the polar ice pack. During a third attempt, Anderson reported peering through the periscope and feeling "an overpowering sense of awe at the majesty of it all."

The *Nautilus* did not reach the North Pole, but it went further under the ice cap than any craft had ever gone. The sub got within 180 miles of the Pole, and the crew was able to gather important scientific and navigational data.

Following these attempts, Anderson reported to his superiors in Washington. While there, he ran into an old friend who was Eisenhower's naval aide. Anderson credited "the hand of the Almighty" with this meeting, because it gave him the opportunity to suggest a transpolar voyage that would start from the Pacific, cross under the North Pole, and end in the Atlantic. Anderson believed that demonstrating the ability to make a transpolar voyage would be even more important than simply getting to the North Pole. Most scientists thought that the approach through the Bering Strait was more shallow and thus more difficult, so Anderson thought that starting from the Pacific side would get the more difficult part over first. Could the *Nautilus* complete such a mission? There was only one way to find out.

Eisenhower gave final approval, but he made one thing clear. Because a Pentagon official had leaked word of the previous mission before the sub had completed it, Eisenhower insisted that he alone be the one to decide when and where the government would announce the voyage.

Commander Anderson (right) and Crew on the Bridge of the Nautilus *in the Arctic Sea*

Another Year, Another Attempt

The United States finally achieved success in its space program when it launched its first satellite into earth orbit in January 1958. A few weeks later, however, another launch attempt failed. In June of 1958, the *Nautilus* left Seattle for a Pacific-side attempt to cross the Arctic Ocean underwater. The crew had painted over the sub's number and name on its hull. Since it was a secret mission, American officials wanted to keep it from the Russians AND the Americans. Painting over the huge signage on the hull would help the sub attract less attention.

The crew discovered that the Bering Sea and Bering Strait approach was indeed more shallow than the Atlantic side, and the ice was thicker. Some protrusions of ice extended much further down into the water, and the ship had to maneuver quickly to avoid these. Soon Anderson decided to turn back. The difficulties and uncertainties were simply too great. He compared the mission to "driving your car down a building's hallway with your eyes closed, relying on someone else to tell you which way to steer" at 30-35 miles per hour. Anderson hoped that the ice would melt more later in the summer, which would make another attempt possible. The captain "thanked God for pulling us through and out of danger." He also thanked God for the crew and for his ability to assess the situation that the sub encountered.

As the *Nautilus* headed for dock at Pearl Harbor, Hawaii, the crew painted the sub's name and number back onto the hull so that it would once more have its normal appearance. After arriving at Pearl Harbor, Anderson flew to Washington to consult with his superiors about making another attempt. While there, he learned that the skipper of a new nuclear sub, the *Skate*, wanted to beat the *Nautilus* to the Pole. This provided an additional motivation for the captain of the *Nautilus* to try again.

On July 22, the *Nautilus* departed from Pearl Harbor. The official story was that the sub was going to make a submerged endurance run to Panama, but it actually went in the opposite direction. Once

Checking the Position of the Nautilus

again, the crew painted over the name and number. The sub had new navigational equipment and a new closed-circuit television system that provided the captain with a better view of what was outside the sub. The crew found much less ice, but they had to zigzag around the ice they did find to avoid a collision. Icebergs are rare in Arctic waters because icebergs are calved from glaciers that are on land. More common in the Arctic are ice floes that wash up on each other because of the waves of the sea. This action pushes large pieces of ice down several feet into the water. The crew figured that at one point the top of the sub was about eight feet below the underside of the ice.

Nautilus 90 North

As the *Nautilus* got closer to the Pole, the sub finally entered deep water. Anderson commented, "I quietly thanked the Almighty" after they had sailed through such difficult waters for so long. They encountered some rugged terrain on the ocean floor, including peaks that they estimated were 9,000 feet high. The ice on the surface was almost completely solid. An earlier Arctic explorer had compared the hardness of the ice cap to granite.

The scientific crew determined that the geographic Pole wanders as the earth rotates, but only slightly, drawing an irregular circle perhaps twenty-five feet in diameter. As the *Nautilus* approached the

Lesson 71 - To the North Pole and UNDER

Pole, the crew was tense but excited. They constantly and carefully checked the equipment readings to make sure they were on course.

On the evening (Seattle time) of Sunday, August 3, 1958, just before the *Nautilus* reached the North Pole, Captain Anderson spoke on the sub's public address system: "As we approach the Pole, I suggest we observe a moment of silence dedicated, first, to Him who has guided us so truly." A few moments later, the *Nautilus* "pierced the Pole," as one crew member put it. Then the party started. The crew had staged a contest to create a postage cancellation stamp for envelopes to be "mailed" from the North Pole. A crewman crafted the winning entry from gasket material. The postmark read, "North Pole Post Office." One of the envelopes is now in the National Postal Museum, which is part of the Smithsonian. Another crewman appeared in a Santa Claus suit made from medical cotton and red flag bunting. "Santa" welcomed everyone to his "neighborhood." Sonar readings indicated that the ocean depth at the Pole was 13,410 feet.

The *Nautilus* did not stop at the North Pole but continued on its course. It surfaced in the Greenland Sea and continued to a point near Reykjavik, Iceland. There a helicopter picked up Anderson and flew him to a plane in Reykjavik, which transported him to Washington for the formal announcement with President Eisenhower. Anderson then returned to the sub, which had continued to Portland, England.

The Significance of the Voyage

The feat of the *Nautilus* was huge news around the world. The sub and its crew received an enthusiastic welcome in England and then made a triumphal entry into New York harbor. A ticker tape parade through New York City, witnessed by perhaps one-third of a million people, followed. A spokesman for the company that built the *Nautilus* commented that the sub had made the incredible journey without a scratch (although it did have a bit more paint on it than when it was launched originally).

The Nautilus *in New York Harbor*

The amazing voyage of the USS *Nautilus* achieved its goals. The submarine had gone where no human and no naval craft had gone before. It accomplished the transoceanic Northwest Passage that people had sought for centuries. It demonstrated the United States' technical, scientific, and military expertise and restored national pride regarding what Americans could accomplish. The trip served notice to the Soviet Union that the United States was capable of defending its homeland and its allies and of responding militarily to any attack the Russians might launch.

The actual center of the polar ice pack, which lay a few miles to one side of the Pole, had been called the pole of inaccessibility. Anderson realized that the spot was inaccessible no longer because the *Nautilus* had passed it. What other feats might Americans accomplish? One indication came on August 12, when the *Skate* became the second American nuclear submarine to reach the North Pole.

The story of the USS *Nautilus* and its voyage under the North Pole is a prime example of human geography: the interaction between humans and the geography of the earth. The next lesson will discuss how the voyage of the *Nautilus* was not just a stunt. It had profound meaning for how people are interacting with each other in the Arctic and with the Arctic region today.

Captain Anderson returned to his home state of Tennessee and received additional honors. Later he remembered thinking on one particular day, "God had blessed me in many ways, and I was feeling particularly grateful that evening for each and every one." Anderson later served four terms as a U.S. congressman from Tennessee, then went into private business.

Anderson died in 2007 and was buried in Arlington National Cemetery. Shepherd Jenks, the navigator on the polar expedition, delivered the eulogy at the memorial service for Anderson. Another crewman wrote in the commemorative booklet that was given out at the memorial service, "I would have followed him anywhere."

Give thanks to the Lord, for He is good,
For His lovingkindness is everlasting.
Psalm 136:1

Assignments for Lesson 71

Geography — Study the map of the Arctic (page 137).

Worldview — Copy this question in your notebook and write your answer: What is something in the created world that fills you with wonder, and why?

Project — Choose your project for this unit and start working on it. Plan to finish it by the end of this unit.

Literature — Continue reading *Lost in the Barrens*. Plan to finish it by the end of this unit.

Student Review — Answer the questions for Lesson 71.

Franz Josef Land (Russia)

72 Sitting on Top of the World

On August 2, 2007, two Russian mini-submarines descended to the seabed at the North Pole. The skipper of one of the mini-subs used a mechanical arm to plant a titanium Russian flag on Lomonosov Ridge at the site. Then the crafts returned to the icebreaker and the research vessel from which they had emerged, carefully finding the opening in the ice through which they had descended.

All of Russia cheered the accomplishment, including the cosmonauts who were aboard the International Space Station at the time. This was the first descent by manned underwater vessels to the ocean floor at the North Pole, but that was not the main cause for the Russian celebration. Russia believes that the Lomonosov Ridge is a continuation of the Siberian continental shelf. Russia says that this fact gives it rightful claim to that seabed area, and to any minerals that lie beneath it. In fact, a Russian think tank has proposed renaming the Arctic Ocean the Russian Ocean.

Other countries were skeptical of the Russians' flag-planting and considered it an attempted land grab. The U.S. State Department said the action had no legal standing. The Canadian foreign minister responded: "This isn't the fifteenth century. You can't go around the world and just plant flags and say 'We're claiming this territory.'"

So go the tensions and conflicts in what some call the scramble for the Arctic. The Arctic was once a frozen, largely inhospitable, mostly ignored place that cost the lives of several explorers. Now changing conditions and changing interests have increased the significance of the Arctic.

The Battleground

The word Arctic comes from the Greek word that means "near the bear," referring to the Ursa Major

---

The Russian MIR 1 and MIR 2 submersibles, built in 1987, have completed many deep-sea missions. Here is an image of MIR 1 taken by a crew member in MIR 2 on a 2003 mission to the wreck of RMS Titanic.

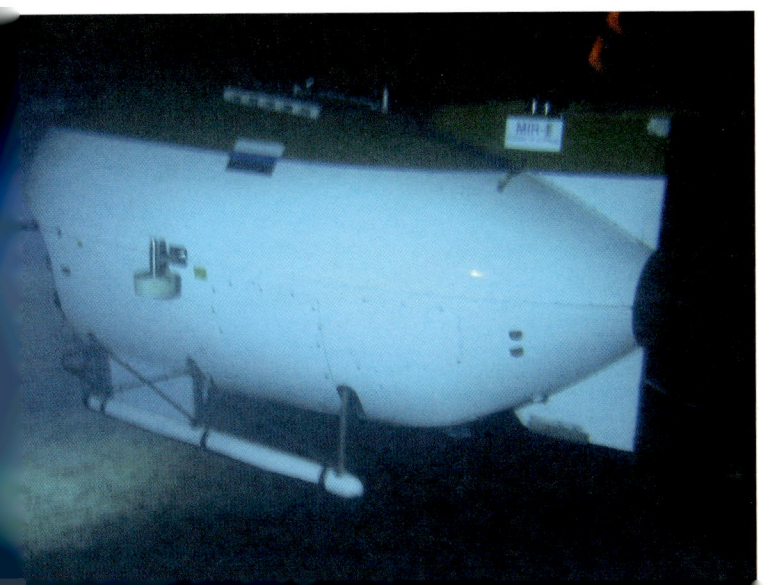

Arctic Iceberg North of Russia

and Ursa Minor constellations. The Arctic Ocean is one and a half times the size of the United States and five times as large as the Mediterranean Sea. In winter the temperature can dip to forty-five degrees below zero. Unlike the continent of Antarctica, the Arctic has no landmass; an ice cap simply floats on top of the water. The ice can range from between six to ten feet thick, and the water can have waves up to forty feet high.

Only a few countries have land area within the Arctic Circle: Canada, Finland, Denmark through its ownership of Greenland, Iceland, Sweden, Norway, Russia, and the United States (through Alaska). These countries make up the Arctic Council, which attempts to arbitrate disputes between member nations. Another twelve countries have observer status with the council because of their interests in the Arctic related to navigation, security, and access to natural resources.

According to the United Nations Convention on the Law of the Sea (UNCLOS), a nation can claim the land under the sea (and the minerals within it) to a distance of 200 nautical miles from its shore as an exclusive economic zone (EEZ), provided that it does not conflict with the claims of other countries. A nation can claim up to 350 nautical miles from shore as an EEZ if it can prove a legitimate connection to that additional territory. This is the additional expanse that Russia claims. Although the United States generally abides by the UNCLOS, it has not formally ratified it.

During the Cold War between the former Soviet Union and the United States, the Arctic had strategic

Unit 15: The Arctic and The Antarctic

importance because of the belief that a missile attack would come by way of the shorter polar route instead of around the side of the globe. The Arctic's primary economic value to the world involved its being the route that airliners took on many transcontinental flights

Less Ice

The Arctic still has importance for these reasons, although the threat of military attack is significantly less; but other reasons for nations taking an interest in this region have emerged. One key factor is the shrinking of the polar ice cap. We might debate whether this is the result of man-made climate change or simply reflects a long-range cyclical weather pattern, but the fact is that the Arctic Ocean has less ice coverage than it did a few years ago. This means that more human activity is possible in the Arctic.

For one thing, the Northwest Passage has become a reality for part of the year. For centuries European explorers tried to find a water passage through the American landmass but were unable to do so. Now with less ice coverage, a nine-hundred-mile route has emerged through Canada's Arctic Archipelago to the north of the continent. It runs five hundred miles north of the Arctic Circle and is twelve hundred miles south of the North Pole.

The Franklin Strait passes between Prince of Wales Island and Somerset Island in Canada.

Polar Bear Near Svalbard (Norway)

This creates a route for some ships that is forty percent shorter and much deeper than the route through the Panama Canal. This means that ships can carry more cargo, use less fuel, give off fewer emissions, and arrive sooner than when taking traditional routes. Ships that take this route must still negotiate many icebergs that float at either end of the route. Of course, the Northwest Passage has controversy. The United States claims that it is an international transport route and thus open to all nations. Canada says that it is an internal waterway which it should control.

The shrinking polar ice coverage means that the Northern Sea Route to the north of Russia is also open for a longer period of the year. The receding ice also means that the Arctic tundra is ice-free more of the year, which affects the habitat of flora and fauna and also makes cultivation of crops more possible within a brief window of time.

Access to Mineral Deposits

Less ice is also a factor in enabling seabed mineral exploration. The United States Geological Survey has estimated that significant oil, natural gas, and other mineral deposits could lie beneath the floor of the Arctic Ocean. This is the main reason for the increased interest that many countries and exploration companies have in gaining access to the Arctic, and the real reason why Russia staked its claim by planting a flag.

Russia has taken the most active interest in exploring the Arctic. It has the largest fleet of icebreakers in the world and is building more. In 2018 Russia launched a floating nuclear power plant than can provide power for offshore drilling activities. The environmentalist group Greenpeace called it a "floating Chernobyl," referring to the site of the 1986 Soviet nuclear reactor accident. China has a small fleet of icebreakers, including a nuclear-powered one. China does not have territory within the Arctic Circle, but China wants in on the action that is happening there.

The (Magnetic) North Pole Is Moving!

A change that actually has less significance than the ones mentioned above is the increased movement of the magnetic North Pole. People have long known that the geographic North Pole, at the top of the world, is not the same as the magnetic North Pole to which compasses align. Scientists in the 1800s discovered that the magnetic Pole tends to drift, on average about nine miles per year. In 2018 the magnetic pole drifted thirty-four miles and crossed the International Date Line into the Eastern Hemisphere.

Scientists believe that this increased movement is the result of greater activity in the earth's liquid-iron outer core as well as the magnetic minerals in the crust and upper mantle of the planet. The change affects the navigation accuracy of ships and planes and people who use Google maps, but since we know about the change, navigational experts can adjust measurements to compensate. Scientists recalculate the map of the earth's magnetic field every five years but check it every year. They found that the 2015 map was out of date by 2018. There is no indication of any human influence on the changes in the position of the magnetic North Pole. It is simply another example of how geography doesn't stand still.

Paying Attention

The emergence of the Northwest Passage, the possibility of mineral exploration, the movement of the magnetic North Pole, and the conflicts and controversies surrounding the Arctic Ocean all remind us that geography is not a static subject and that human interaction with the earth is a dynamic that affects people around the world.

Job described how God created the earth, including the northern reaches:

He stretches out the north over empty space
And hangs the earth on nothing.
Job 26:7

Assignments for Lesson 72

Worldview — Copy this question in your notebook and write your answer: What is one thing the book of Job teaches you about God?

Project — Continue working on your project.

Literature — Continue reading *Lost in the Barrens*.

Student Review — Answer the questions for Lesson 72.

Iqaluit, Nunavut

73 A Homeland for the Inuit

Paul Okalik was a welder who became a lawyer, the first lawyer of his ethnic group in his home territory.

Eva Aariak is an expert in languages. She developed the word for Internet in her people's language: Ikiaqqivik, which translates "traveling through layers."

Peter Taptuna worked in the oil fields until his back gave out. Then he took some college courses and embarked on a new career.

Joe Savikataaq spent thirty years as a conservation officer.

These people with diverse backgrounds have at least two things in common. They are all part of the Inuit people of Canada, and they have all served as premier of Nunavut Territory in Canada.

The Circles

The earth is tilted a little more than 23 degrees from perpendicular relative to the plane of the earth's orbit. This enables the earth's temperatures, daylight and darkness, and seasons to be hospitable for humans and other living things. The Arctic Circle is a line of latitude a little more than 66 degrees latitude north of the equator. Because of the tilt of the earth, the Arctic Circle is the line above which the sun does not set for at least one day per year and above which the sun does not rise for at least one day per year. As one goes north, the number of such days increases until, at the North Pole, there are six months of daylight and six months of darkness.

The Antarctic Circle is a line of latitude approximately 66 degrees latitude south of the equator. The same periods of daylight and darkness occur within that circle, just on the opposite days of the year.

Hasn't God created a marvelous system of daylight and darkness with the tilt and rotation of the earth?

The Inuit

Antarctica has no indigenous people. By contrast, indigenous people do live within the Arctic Circle, in the northern extremities of the countries that extend into the Arctic Circle. These people include the Sami of northern Europe and the Inuit and other tribes of Greenland, Canada, the United States, and Russian Siberia.

The Inuit of Canada, like other minority ethnic groups, have often been the victims of prejudice and of being treated like children. The traditional lifestyle of harpoons and dogsleds has given way to modern practices such as rifles and snowmobiles. The Inuit live in towns and cities and work in mines, oil fields, stores, and offices.

Nunavut

A major part of the identity of a people is having a land which they can call home. After European colonization of the area we know as Canada, the Inuit did not have a place they could call home. In 1992 the residents of the Northwest Territories of Canada voted to create from their land a new territory that could honor Inuit identity. The Canadian national government approved the idea, and in 1999 the Nunavut Territory came into being in the north central region of Canada.

Nunavut contains a significant portion of the Canadian Shield, which refers to the mainland and the islands around Hudson Bay, and the Arctic Archipelago to the north. Much of the Archipelago is permanently covered with ice and snow. A significant part of Nunavut Territory extends north of the Arctic Circle

Nunavut has several significant geographic features. Covering just over one-fifth of Canada's land area, it is the largest administrative unit in the country. Its 680,000 square miles are larger than Mexico and about one and a half times the size of Alaska. Much of it lies north of the tree growth line, so it is largely covered with tundra and the small plant growth that characterizes it. Beneath the surface lies significant mineral wealth, in the form of iron, precious metals, petroleum, and natural gas; but extracting it is difficult and expensive because of the frigid weather and distance from population centers.

One Nunavut community, the Canadian Forces Station named Alert, is the northernmost community in North America. At 82 degrees north latitude, the community of about two hundred persons is some five hundred miles north of the next nearest community and about the same distance south of the North Pole. Everyone there is a member of the Canadian military, a worker for a civilian contractor, or serving in the government's environmental agency. Winter temperatures drop to about sixty degrees below zero. A main safety concern is staying on the lookout for polar bears that occasionally wander into town.

The DEW Line

The territory that became Nunavut was involved in the Distant Early Warning (DEW) Line that the United States and Canada maintained for decades. Because of the threat of a missile or bomber attack by the Soviet Union, the U.S. and Canada worked together to build a line of over sixty manned radar installations that stretched from Alaska east to Baffin Island, which lies between Greenland and Canada and today is part of Nunavut. The first stations became operational in 1957.

These radar stations were equipped to detect a Soviet attack and alert U.S. and Canadian defense forces that could have destroyed incoming missiles or bombers and retaliated with a massive attack on the Soviet Union. Over time more than 25,000 workers, many of them employees of contractors

Stop Sign in Inuktitut, English, and French

Pond Inlet, Baffin Island, Nunavut

hired by the governments involved, built roads, residences, airplane hangars, and communication facilities.

Starting in the 1980s, newer and improved communication and detection technology and a diminished threat from the Soviet Union led to the DEW Line facilities being either abandoned or removed.

The People of Nunavut

In 2019 the population of the huge Nunavut Territory was only about 40,000 people, about 80-85 percent of whom were Inuit. The territory is one of the most sparsely populated habitable regions on earth, which creates challenges in governing and providing services for the people. However, Nunavut has the fastest rate of population growth of any province or territory in Canada. About one-third of the territory's population is under fifteen.

In 2019 Nunavut observed twenty years as a Canadian territory. The events of its first two decades were a mixed bag. The people have experienced a renewed appreciation for Inuit culture, and special events have highlighted this background. Educational opportunities have increased.

However, Nunavut has experienced some of the same problems that reservations in the United States have experienced, as the territory has been heavily dependent on funding from the national government. Substance abuse, crime, and unemployment have been serious issues. Thus all those young people in the territory face many uncertainties regarding their future.

In addition, the territorial government has not functioned well, with numerous conflicts and changes in leadership taking place. Use of the main Inuit language, Inuktitut, has declined somewhat as English has increased, which concerns many government and cultural leaders. Instruction in public schools takes place in English, and most public school teachers do not speak Inuktitut.

Still, the Inuit of Nunavut have something of which they can be proud. As one official put it, "It was Inuit who changed the map of Canada, all in a peaceful manner." After all, Nunavut in Inuktitut means "Our Land."

God's gift of the land of Canaan to the people of Israel was a significant part of their identity. The Lord showed the land to Moses:

Now Moses went up from the plains of Moab to Mount Nebo, to the top of Pisgah, which is opposite Jericho. And the Lord showed him all the land. . . . Then the Lord said to him, "This is the land which I swore to Abraham, Isaac, and Jacob, saying, 'I will give it to your descendants'. . . ."
Deuteronomy 34:1a, 4a

Assignments for Lesson 73

Geography — Complete the map skills assignment for Unit 15 in the *Student Review Book*.

Worldview — Copy this question in your notebook and write your answer: How does a person's view of suffering influence his or her worldview?

Project — Continue working on your project.

Literature — Continue reading *Lost in the Barrens*.

Student Review — Answer the questions for Lesson 73.

Flags at the South Pole

74 Endurance

Ernest Henry Shackleton was born in Ireland in 1874. A few years later his family moved to England. Ernest's father was a physician and wanted his son to go to medical school. However, Ernest loved the sea, so at age 16 he joined the British merchant marine. He later became a junior officer in the Royal Naval Reserve. These decisions set his course for becoming an explorer and adventurer.

Antarctica

The Arctic is an ocean covered with ice. Antarctica (which means "opposite to the Arctic") is a landmass that is covered by the largest ice sheet in the world. Antarctica is the fifth largest continent. The Antarctic region, which includes the ocean and islands around the continent itself, covers about 20% of the Southern Hemisphere. Some of the mountain peaks on the continent reach over 14,000 feet in height. Antarctic ice shelves are ice sheets floating on the sea next to the continent.

Antarctica is cold. Coastal temperatures in summer (when it is winter in the Northern Hemisphere) are usually around freezing but have been recorded as high as 65° Fahrenheit (F). In winter, the temperature along the coast varies from +14° to -22° F. In the interior of the continent, temperatures usually vary from -4° in summer to -76° F in winter. The coldest temperature ever recorded at a weather station on earth was -128° F at the Soviet Vostok station in Antarctica in 1983. Satellite measurements have since recorded temperatures as low as -144° F.

The land under all that ice is mostly desert. The interior of the continent only receives about 2 to 4 inches of precipitation per year, all in the form of snow. It is one of the driest deserts in the world.

Despite the extreme weather conditions in Antarctica, we shouldn't be surprised that God has a plan for this frigid region. Antarctica plays a vital part in the earth's heat balance, which is the balance between solar heat absorbed by the atmosphere and solar heat reflected back into space. All that ice reflects a great deal of heat away from the earth's surface and keeps the planet's heat in balance. Also, the oceans around Antarctica have a vital role in the circulation of the ocean's waters around the globe.

The Antarctic Treaty of 1961, signed by dozens of countries, recognized claims of territory on the continent that seven countries had already made. The treaty says that all of the continent is open for scientific investigation to countries that have signed the treaty. The treaty states that no new territorial claims will be recognized. No country can use

Antarctica for military purposes or to dispose of radioactive waste. The continent has no indigenous human population, but it has numerous scientific stations where people live and work for periods of time.

In 1773 Captain James Cook of England crossed the Antarctic Circle for the first time, but he did not see the continent of Antarctica. The Russian explorer Thaddeus Bellingshausen is credited with being the first to see Antarctica, doing so in 1820. A group of Norwegian whaling men had the first confirmed landing on the continent in 1895.

The Heroic Age

The late 1800s and early 1900s were called the Heroic Age of Polar Exploration. Men from several countries set their sights on reaching the North and South Poles. In September 1909, Dr. Frederick Cook, a physician from New York, claimed to have reached the North Pole. A week later, Robert Peary, a civil engineer and U.S. Navy commander, announced that he had reached the North Pole on his third attempt, accompanied by his black assistant Matthew Henson, and that Dr. Cook was a fraud.

At the same time, men were working to reach the South Pole. Captain Robert Scott chose Ernest Shackleton to take part in his attempt to reach the South Pole in 1901. However, Shackleton became ill and was sent home early. Scott came within 463 miles of the Pole in 1904 when he had to turn back.

Shackleton led another attempt to reach the South Pole in 1907. The effort lasted until 1909. Shackleton was able to get within 97 miles of the Pole but again had to turn back. For this effort Shackleton was made a knight.

Roald Amundsen of Norway won the fierce competition to get to the South Pole first on December 14, 1911. He arrived five weeks before Scott and his crew, who were making another attempt, got there. Sadly, Scott and all of his men died on their return trip from the Pole.

The Voyage of the *Endurance*

The goal of being first to the South Pole had been accomplished, but Shackleton set a new goal for himself: to cross Antarctica overland by way of the South Pole. He created the British Imperial Trans-Antarctic Expedition. Shackleton purchased a ship made in Norway powered by both sails and steam that he christened *Endurance*. It weighed 300 tons, was 144 feet long, and its hull was a thick construction of oak, fir, and greenheart wood. It had never sailed before.

Shackleton assembled a crew of 26 men, several of whom had been to Antarctica before. These men included sailors, land explorers, a meteorologist, a geologist, a biologist, two surgeons, and a carpenter.

Frank Hurley documented the Endurance *expedition with still photographs, including many in color using the Paget Colour Plate system. Here is a photo of Hurley with his cinematograph, an early motion-picture camera.*

Crew Members Raising a Signal Flag (left) and Mending a Net

Shackleton also knew the value of photographs of the adventure. He sold the publication rights for the pictures of the expedition as part of his fundraising; and he brought along a skilled photographer, who brought along his forty pounds of photography equipment to record the journey. Someone once said that Antarctica was the first continent to be discovered by camera. After they departed, the crew discovered a stowaway, who was added to the crew, making a total of 28 on the ship.

After years of training, preparation, and fundraising, Shackleton sailed from London on August 1, 1914. However, an even bigger event overshadowed their departure. On that day, Germany declared war on Russia. This began the fighting in World War I. Shackleton stopped in Plymouth, England, and offered his ship to his country to serve in the impending conflict; but the king ordered him to continue on his way.

Shackleton left Plymouth on August 8 and arrived on South Georgia Island on November 5. The Norwegian whaling men who were there convinced Shackleton to wait until the weather got warmer, so Shackleton stayed there a month. "Warmer" in Antarctica is a relative term. The *Endurance* and her crew still had over a thousand miles to go in notoriously rough seas to reach Antarctica. Once there, the men might still encounter winds as high as 200 miles per hour and temperatures approaching -100 degrees Fahrenheit.

Trapped in the Ice

The ice in the Southern Ocean was worse that year than it had been in many years. On January 18, 1915 (the height of the Antarctic summer), the ice pack trapped the *Endurance* about 100 miles short of the desired landing site on Antarctica. As the winds and ocean currents buffeted the ice pack, the ship slowly drifted north—away from the expedition's goal but closer to a possible rescue. The *Endurance* was trapped for months. The ice was not just a smooth sheet. Instead, the large, jagged blocks of ice roared and cracked and heaved the ship up and down. The *Endurance* and its crew never made it to Antarctica.

The ice eventually crushed the *Endurance,* and it began to sink on November 21, 1915. The men had enough warning that they were able to take a good portion of their supplies off the ship—the supplies

At left you can see the Endurance *trapped in the ice. The upper left photo shows the frozen deck of the ship. The image above was taken at noon on a winter day in 1915.*

Shackleton decided to leave 22 men there and take five men with him in one of the 22 1/2-foot boats. They traveled 800 miles over the open, dangerous winter sea to South Georgia Island to contact rescuers. The journey took sixteen days and is considered one of the greatest boat journeys in human history. The 22 men on Elephant Island built a hut out of the two remaining boats. This removed any possibility of using the boats to leave the island, but they had to meet the definite need of shelter while they waited. The return of the other members of their party was their only hope.

Rescued!

The six men landed on South Georgia Island, but on an uninhabited part of it. Shackleton decided to take two men and venture 22 miles over the mountainous, glacier-filled interior of the island to the whaling station. This had never been done. There were no maps of the interior of the island. The trip took 36 hours, but they made it on May 20, 1916. The men saw human beings other than their fellow crewmen for the first time in nearly eighteen months.

Shackleton sent one of his crew with some Norwegian whaling men to rescue the other three

they were expecting to use in Antarctica. Their goal became survival and rescue. They set up camp on the ice near the stranded ship. Shackleton decided that they would proceed 200 miles toward Snow Hill or Robertson Island, pulling all of their supplies and equipment including three smaller boats. When this proved impossible, they loaded the three small boats and set sail for Elephant Island on April 9, 1916.

They landed on Elephant Island on April 16, touching actual land for the first time in 497 days. Elephant Island, however, was not inhabited.

Lesson 74 - *Endurance*

on the opposite side of the island while he made plans to rescue the 22 crew members on Elephant Island. Three rescue attempts failed when the ship had to turn back. A fourth attempt finally made it.

The 22 stranded crew members had spent four months on Elephant Island. One of the rescued men recorded in his diary, "I felt jolly near blubbing for a bit & could not speak for several minutes." The expedition of the *Endurance* officially ended on October 8, 1916, when the crew was welcomed in Buenos Aires, Argentina. Not one of the 28 men on the expedition lost his life during the entire ordeal that lasted over two years.

Meanwhile, World War I had been going on. Many of the men from the *Endurance* enlisted in the British military, and two of them lost their lives in the war. Ernest Shackleton embarked on another Antarctic expedition in 1921 but again failed to reach his destination. While making final preparations on South Georgia Island in early 1922, he died of heart failure at the age of 47. The last surviving member of the *Endurance* crew died in 1979, having lived long enough to see men walk on the moon.

In 2019 an Antarctic expedition hoped to locate the wreckage of the *Endurance*, but the attempt was called off when an autonomous underwater vehicle it was using was lost under the ice.

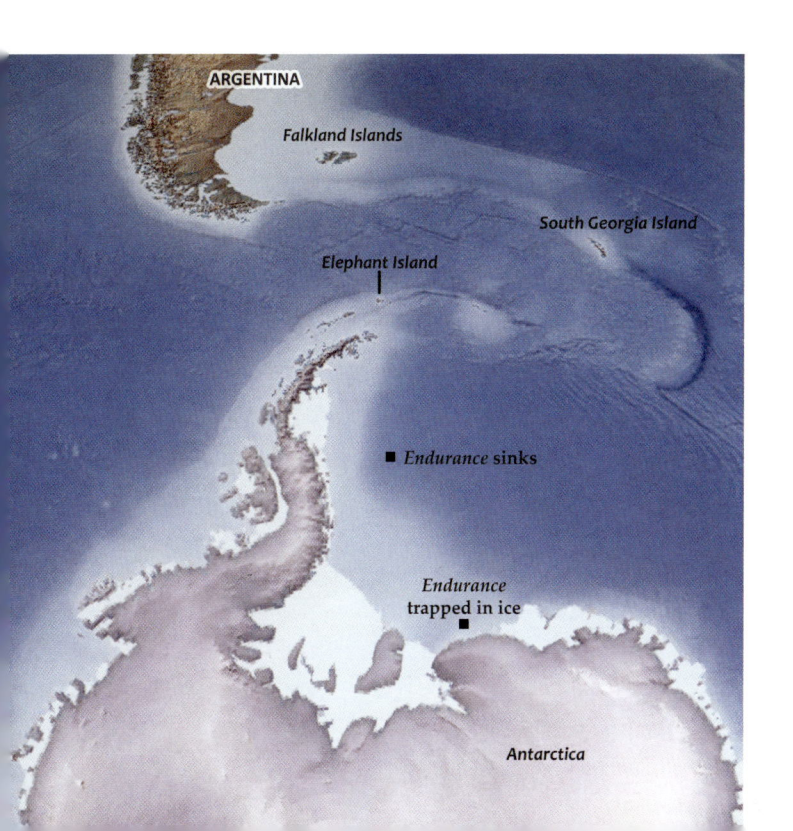

The men who accompanied Shackleton on his expeditions credited him with being an excellent leader and motivator. Surely he showed considerable courage and a willingness to share in his men's difficulties, and he was always willing to take the most difficult jobs for himself.

Part of his leadership ability involved being able to change his definition of success when the situation called for it. He had hoped to be the first to reach the South Pole, then he had hoped to travel across Antarctica by way of the South Pole. When those feats became impossible, he redefined success to mean saving the lives of all his men. Given his circumstances, this was a remarkable accomplishment.

Shackleton also had the eye of faith. Reflecting later on the overland crossing of South Georgia Island to get to the whaling station, Shackleton wrote:

> When I look back at those days I have no doubt that Providence guided us, not only across those snowfields, but across the storm-white sea that separated Elephant Island from our landing-place on South Georgia. I know that during that long and racking march of thirty-six hours over the unnamed mountains and glaciers of South Georgia it seemed to me often that we were four, not three. I said nothing to my companions on the point, but afterwards Worsley [one of the other men] said to me, "Boss, I had a curious feeling on the march that there was another person with us." Crean [the third man] confessed to the same idea.

About the entire expedition of the *Endurance*, Shackleton said:

> We had seen God in His splendor,
> Heard the text that Nature renders.

Modern Photo of Penguins and Mountains on South Georgia Island

"Ah Lord God! Behold, You have made the heavens and the earth by Your great power and by Your outstretched arm! Nothing is too difficult for You. . . ."
Jeremiah 32:17

Assignments for Lesson 74

Gazetteer — Study the map of Antarctica (page 138).
Read about the Amundsen-Scott South Pole Station and the McMurdo Station in Antarctica (pages 292-294)

Project — Continue working on your project.

Literature — Continue reading *Lost in the Barrens*.

Student Review — Answer the questions for Lesson 74.

Pleneau Bay, Antarctica

"Have You Entered the Storehouses of the Snow?"

[left column obscured by notebook paper overlay]

...o had it all: a big
...every day, material
...titude before God.
...worldview.
...conversation. God
... Satan replied by
...her worshiper, only
...im. So God allowed
...b, which Satan did.
...r blame God.
...nother conversation.
...ithfulness, and this
...o suffered physically
...God. So God allowed
...ng, but again Job did
...to curse God and be
...would not do it. Job's
...reak him.
...book of Job reveal to
...connectedness of the
...What happens in this
...Events in the physical
...nces.
...with Job in his misery.
...th Job in silence. Then
...as born. He struggled
...to endure all that had

happened to him. His friends thought they had it all figured out, and they proceeded to give Job their explanation. Their words to Job revealed their worldview.

It was clear to them that his suffering was punishment for sin. If Job would simply repent, they said, God would be satisfied. Job denied that he deserved such punishment. Job thought he had it all figured out, and he expressed a desire to argue his case before God. So Job's friends claimed to understand God's ways, and Job claimed that he could convince God of a better way to treat him.

The friends' arguments and Job's counter-arguments went round and round, as such conversations often do. Each side was convinced that their understanding of the world and God's actions was correct. During this back and forth, Job had an "Aha!" moment, as recorded in chapter 28, in which he shared an insight into man's existence.

"Where Can Wisdom Be Found?"

People seek wealth and success, and they use the geographic resources of the earth to achieve it. Mankind has decided that certain minerals have great value, so people spare no effort to obtain them. Chapter 28 mentions silver, gold, iron, copper,

Sisimut, Greenland

sapphires, onyx, coral, crystal, pearls, and topaz as objects of man's labor. People also use the ground to plant and harvest crops in the hopes of obtaining wealth. Apparently people are willing to do almost anything to acquire what they perceive as valuable.

But what about what is truly valuable, namely wisdom and understanding? Can they be mined from the deep or harvested like crops? No, for "God understands its way, and He knows its place" (verse 23). God controls our physical setting on the earth, including the seas and the weather that sometimes appear out of control. In all our striving and all our seeking to understand this physical world and to acquire its riches, we must never forget that God truly sees, understands, and controls all.

So what is there of real, lasting value that we can obtain? Job apparently had it all; but all that he had acquired, even his health, he lost in a matter of moments. The insight that Job had is that the greatest treasure lies with God. None of the usual efforts that men employ to gain what is valuable—not work, mining, building, planting and harvesting, or anything else—will bring true wealth to us permanently. Instead, "the fear of the Lord, that is wisdom, and to depart from evil is understanding" (verse 28). This is a truly significant worldview. How different the history of the world and life in this world would be if people pursued these true riches instead of doing what so many have done and continue to do to acquire material power and wealth.

God's Reply

After this discourse, Job returned to his posture of defending himself and stating his case. A fourth friend appeared and made his attempt to explain it all. Finally, beginning in chapter 38, God silenced everyone with His reply. He used elements of geography to prove His case.

Lesson 75 - "Have You Entered the Storehouses of the Snow?"

God asked, in effect, "What do you know?" As a mere human, you have only a limited perspective on the world, God said, let alone a grasp of such deep subjects as suffering and the mind of God. "Where were you when I laid the foundation of the earth?" God asked (38:4). Who created the earth and sea, sky and weather, rain, snow and hail, dew and ice, daylight and darkness? Can you summon the dawn? Do you control the constellations? "Have you understood the expanse of the earth? Have you entered the storehouses of the snow, or have you seen the storehouses of the hail?" (38:18, 22).

Can you explain the wasteland and the sprouting of seeds? (38:27). Modern science has given us some insights into the operation of the physical world, but, "Has the rain a father? Or who has begotten the drops of dew?" (38:28). In other words, perhaps you can explain some things about the world, but how did it all begin? We certainly cannot take credit for that. The knowledge and understanding of the One who did create it and who crafted its workings and its purpose is far above the understanding of anyone who might try to explain how it all works.

"Do you know the ordinances of the heavens?" God asked (38:33). Can you command the clouds and lightning (38:34-35)? "Who has put wisdom in the innermost being or given understanding to the mind?" (38:36). Did a process of mindless, random chance produce wisdom? It doesn't seem possible. Did mankind produce the wisdom to create and maintain the universe? Obviously not. It must be the work of God, so who are we to question Him?

Do you control and provide for the wild animals and understand their ways? No; God does that. God's speech convicted Job—and convicts us—that our human understanding of the world, which we think is fairly advanced, is actually pretty limited. This is another important worldview realization. So "Will the faultfinder contend with the Almighty?" (40:1). We are not on the same level of power, wisdom, and knowledge that God is. Job expressed his repentance (40:3-5), but God was not through stating His case. The Lord pointed to two large creatures, Behemoth and Leviathan (40:15-41:34). Bible scholars disagree about the identity of these creatures; but whatever they are, man did not create them and cannot control them. God made them and controls them.

Chapter 42 tells us that Job again expressed his repentance and humility, and then the merciful Lord gave instructions about sacrifices that Job was to make on behalf of his friends, who "have not spoken of Me what is right as My servant Job has" (42:7, 8). The narrative concludes with the Lord restoring and even increasing the fortunes of Job, and we read that Job lived out a richly blessed life.

How Job Teaches Us a Correct Worldview

The immediate cause of the dramatic interchange of ideas in the book of Job is the suffering that Job endured. How you understand and deal with suffering is an important element of your worldview. Do you think you don't deserve it, as Job thought? What do you think about the God who allows suffering and disciplines people through it? The book of Job says that God believes in us, and that is why He allows us to suffer. Do you have this perspective of God in your worldview? Often we think about

Wapusk National Park, Manitoba, Canada

ourselves—our weakness and our sinfulness—when we suffer. The book of Job leads us to think about God in such situations.

When Job felt richly blessed, he was humble and grateful before God. When he suffered, he began to question God and wanted to make the case to God that he didn't deserve what had happened to him. Job's friends gave him their own perspectives about what had happened. When God spoke, He didn't explain everything about the immediate issue; instead, in effect He asked, "You don't have my perspective on life—on your life—and on the world. Do you trust Me to know what I am doing?" Job realized that God does know what He is doing, and so he humbled himself before the Lord.

What do you see—in other words, what is your worldview—when you consider the world, including its geography—both physical and human—and your life in it? As Job found out, some things that happen in this world are hard to accept. Nevertheless, who are you—with your limited understanding and as someone who had nothing to do with creating the world and has nothing to do with guiding everything that happens—to question God and to think you have a better idea?

The worldview that the book of Job encourages us to have is to look at the world and see the work of a loving, powerful God who is in control and who has your best interests at heart. In your education and your life, you will learn much about geography, history, science, mathematics, human society, the plant and animal populations, and the fearfully and wonderfully made person you are. Although the totality of what you learn will be minuscule compared to God's understanding, this knowledge should only deepen your commitment to this worldview.

Do you have an arm like God,
And can you thunder with a voice like His?
Job 40:9

Assignments for Lesson 75

Worldview — Recite or write the memory verse for this unit.

Project — Finish your project for this unit.

Literature — Finish reading *Lost in the Barrens*. Read the literary analysis and answer the review questions in the *Student Review Book*.

Student Review — Answer the questions for Lesson 75.
Take the quiz for Unit 15.
Take the third Geography, English, and Worldview exams.

Exterior of the Library of Alexandria, Egypt

Sources

Lesson 1

www.answersingenesis.org

www.nasa.gov

Shelly, Rubel. *Prepare to Answer: A Defense of the Christian Faith*. Nashville: 20th Century Christian, 1990.

Calderone, Julia. "Something Profound Happens When Astronauts See Earth from Space for the First Time." www.businessinsider.com, August 31, 2015, accessed January 9, 2018.

Lesson 3

Complete Works of Strabo. London: Delphi Classics, 2016.

Lesson 4

www.earthsky.org

www.surtsey.is

www.unmusuem.org

www.geology.sdsu.edu

Gregory, Ted. "The Precarious Case of Kaskaskia." www.seattletimes.com, March 14, 2011, accessed April 4, 2018.

Lesson 5

Lewis, C. S. *The Abolition of Man*. New York: Macmillan, 1947, p. 35.

Perman, Matt. *Unstuck: Breaking Free from Barriers to Your Productivity*. Grand Rapids: Zondervan, 2018.

Lesson 6

Miller, Greg. "Inside the Secret World of Russia's Cold War Mapmakers." www.wired.com, July 2015, accessed June 27, 2019.

Lesson 8

Jacobs, Frank. "The Border That Stole 500 Birthdays." www.nytimes.com, July 31, 2012, accessed July 10, 2019.

Lesson 9

www.randmcnally.com/about/history, accessed January 5, 2018.

Lesson 10

Sire, James W. *The Universe Next Door: A Basic Worldview Catalog*, 5th edition. Downers Grove, Illinois: InterVarsity Press, 2009.

Lesson 11

Geoff Emberling, "The Geography of the Middle East." lib.uchicago.edu. Article updated December 29, 2010. Retrieved August 24, 2017.

Jack P. Lewis, "Biblical Archaeology and Geography," in *The World and Literature of the Old Testament*, John T. Willis, ed. Austin: Sweet Publishing Company, 1979.

Jimmy J. Roberts, "The Geography of Palestine in New Testament Times," in *The World of the New Testament*, Abraham J. Malherbe, ed. Austin: Sweet Publishing Company, 1967.

Lesson 12

Christianson, Scott. "The Origins of the World War I Agreement That Carved Up the Middle East." *Smithsonian Magazine*, www.smithsonianmag.com, November 16, 2015, accessed March 2, 2019.

Miller, James. "Why Islamic State Militants Care So Much About Sykes-Picot." Radio Free Europe/Radio Liberty. www.rferl.org, May 16, 2016; accessed September 5, 2018.

Muir, Jim. "Sykes-Picot: The Map That Spawned a Century of Resentment." www.bbc.com, May 16, 2016, accessed March 2, 2019.

Lesson 14

Mansfield, Stephen. *The Miracle of the Kurds*. Brentwood, Tennessee: Worthy Publishing, 2014.

"Who Are the Kurds?" www.bbc.com, October 31, 2017, accessed January 16, 2019.

El Deeb, Sarah. "It's Not Independence, But Syria's Kurds Entrench Self-rule." Associated Press, www.apnews.com, October 8, 2017, accessed October 8, 2017.

Lesson 16

Dixon, Glenn. "The Age-Old Tradition of Armenian Carpet Making Refuses to Be Swept Under the Rug." www.smithsonianmag.com, July 6, 2018, accessed April 6, 2019.

Lesson 18

"Meet the Students of Saudi Arabia's First Driving School for Women." www.cbsnews.com, March 14, 2018, accessed March 15, 2018.

Mollman, Steve. "Women in Saudi Arabia Now Must Be Informed if They've Been Divorced." Quartz News, www.qz.com, January 6, 2019, accessed April 10, 2019.

"Saudi Arabia: 10 Reasons Why Women Flee." Human Rights Watch, www.hrw.org, January 30, 2019, accessed April 10, 2019.

"Saudi Arabia Lifts Ban on Women Drivers." www.voanews.com, June 24, 2018, accessed June 25, 2018.

"Saudi Arabia: Why Weren't Women Allowed to Drive?" www.bbc.co.uk, January 13, 2018, accessed April 10, 2019.

Lesson 20

Trueblood, Elton. *A Place to Stand*. New York: Harper and Row, 1969.

Lesson 21

"The Arab Spring: A Year of Revolution." www.npr.org, December 17, 2011; accessed September 1, 2018.

Amara, Tarek, with Patrick Markey and Toby Chopra. "Tunisia Southern Gas Protests Tense as Negotiations Falter." www.reuters.com, May 18, 2017, accessed September 3, 2018.

Lesson 22

McCullough, David. *The Path Between the Seas*. New York: Simon and Schuster, 1977.

Suez Canal Authority. www.suezcanal.gov.eg. Accessed November 20, 2018.

Rogers, J. David. "Construction of the Suez Canal." web.mst.edu (Missouri University of Science and Technology). Accessed November 20, 2018.

Lesson 23

"Barbary Pirates." Encyclopedia Britannica, 1911 edition. www.penelope.uchicago.edu, accessed April 17, 2019.

"Barbary Wars, 1801-1805 and 1815-1816." Office of the Historian, U.S. Department of State, www.history.state.gov/milestones, accessed April 17, 2019.

Lesson 24

"Allied Military Operations in North Africa." U.S. Holocaust Memorial Museum, www.encyclopedia.ushmm.org, accessed April 19, 2019.

Huxen, Keith. "The US Invasion of North Africa." www.nationalww2museum.org, accessed April 19, 2019.

Matanle, Ivor. World War II. Godalming, Surrey, United Kingdom: Quadrillion Publishing Ltd., 1998.

Sulzberger, C. L. *The American Heritage Picture History of World War II*. Rockville, Maryland: American Heritage Publishing Company, 1966.

Taylor, Alan. "World War II: The North African Campaign." www.theatlantic.com, September 4, 2011; accessed April 19, 2019.

Lesson 26

Gansler, Katrin. "Ivory Coast Sweetens Up with First Locally Made Chocolate." Deutsche Welle, www.dw.com, June 30, 2016, accessed July 10, 2019.

Mavhunga, Columbus. "Zimbabwe Pushing to Sell Its $600 Million Ivory Stock." www.voanews.com, June 26, 2019, accessed July 18, 2019.

"The New Queens of Cocoa." The Fairtrade Foundation. www.stories.fairtrade.org.uk. Accessed July 10, 2019.

Pearce, Fred. "The Real Price of a Chocolate Bar: West Africa's Rainforests." Yale Environment 360. www.e360.yale.edu, February 21, 2019, accessed July 10, 2019.

Sowell, Thomas. *Conquests and Cultures: An International History*. New York: Basic Books, 1998.

Lesson 27

"At Least 95 Killed in Central Mali Village Attack." www.reuters.com. June 10, 2019, accessed June 25, 2019.

"Cliff of Bandiagara (Land of the Dogons)." UNESCO World Heritage Center. whc.unesco.org, accessed June 22, 2019.

"Dogon People of Mali." www.youtube.com, National Geographic TV, natgeotv.com, posted July 1, 2009; accessed June 26, 2019.

"Fact Box: Mali, Remote Land of Deserts and Gold." www.reuters.com, posted March 22, 2012, accessed June 22, 2019.

Hammer, Joshua. "Looting Mali's History." www.smithsonianmag.com. November 2009, accessed June 23, 2019.

Wikle, Thomas. "Living and Spiritual Worlds of Mali's Dogon People." focusongeography.org, January 2016, accessed June 22, 2019.

Lesson 28

Anderson, Becky and Leif Coorlim. "'This Breaks Our Hearts': Ghana Promises Action After CNN Child Slavery Report." www.cnn.com, March 7, 2019, accessed June 18, 2019.

Darko, Sammy. "Eight Surprising Consequences of Ghana's Power Outages." www.bbc.com, May 15, 2015, accessed June 18, 2019.

Fuller, Katy. "As Assessment of the Underwater Timber Salvation Project on the Volta Lake" (2017). MA Thesis submitted to Leiden University, Leiden, Netherlands. www.openaccess.leidenuniv.nl. June 30, 2017, accessed June 18, 2019.

Ofosu-Boateng, Nana Raymond Lawrence. "Underwater Timber Harvesting on the Volta Lake: Implications for the Environment and Transportation" (2012). MS Thesis submitted to World Maritime University, Malmo, Sweden. www.commons.wmu.se. Accessed June 18, 2019

Lesson 29

Peretti, Burton. *Lift Every Voice: The History of African American Music*. New York: Rowman and Littlefield, 2009.

Southern, Eileen. *The Music of Black Americans: A History, Third Edition*. New York: W. W. Norton and Company, 1997.

Lesson 30

Olupona, Jacob K. "African Religions: A Very Short Introduction." Oxford University Press, www.oup.com. May 16, 2014, accessed May 29, 2018.

Turaki, Yusufu. "Africa Traditional Religious System as Basis of Understanding Christian Spiritual Warfare." Lausanne Movement. www.lausanne.org. 22 August 2000, accessed May 28, 2018.

"Traditional African Religious Beliefs and Practices." Pew Forum on Religion and Public Life, www.pewforum.org, n.d., accessed May 29, 2018.

Lesson 31

Kindzeka, Moki Edwin. "CAR Refugees Sing for Peace at Camp in Cameroon." www.voanews.com, July 9, 2018, accessed January 24, 2019.

_____. "More Refugees Flee Carnage in Central African Republic." www.voanews.com, May 24, 2018, accessed January 24, 2019.

_____. "UNHCR Launches $430M Plan for CAR Refugees." www.voanews.com, January 7, 2019, accessed January 24, 2019.

The United Nations High Commissioner for Refugees, www.unhcr.org

Lesson 32

Bergman, Jerry. "Ota Benga: The Pygmy Put on Display in a Zoo." *Journal of Creation*, April 2000, pp. 81-90; reproduced online by Creation Science Ministries, www.creation.com, accessed July 18, 2019.

Di Campo, Therese. "For Congo's Pygmies, Expulsion and Forest Clearance End a Way of Life." www.reuters.com, January 12, 2017, accessed July 18, 2019.

Newkirk, Pamela. "The Man Who Was Caged in a Zoo." www.theguardian.com, June 3, 2015, accessed July 18, 2019.

Raffaele, Paul. "The Pygmies' Plight." www.smithsonianmag.com, December 2008, accessed July 18, 2019.

Lesson 33

Caldwell, Mark. "Cameroon: Colonial Past and Present Frictions." www.dw.com. January 31, 2017, accessed July 22, 2019.

"Cameroon: History." www.thecommonwealth.org, accessed July 22, 2019

"Lake Nyos, Cameroon." www.earthobservatory.nasa.gov. December 18, 2014, accessed July 23, 2019.

"Lake Nyos—Silent But Deadly." Department of Geosciences, Oregon State University, www.oregonstate.edu, accessed July 23, 2019.

Sources

"Swiss Government to Mediate Cameroon Peace Talks." www.reuters.com, June 27, 2019, accessed July 22, 2019.

Lesson 34

"DR Congo Ebola Outbreak: More Than 2,000 Cases Reported." www.bbc.com. June 5, 2019, accessed June 5, 2019.

"Ebola Response Failing Communities in DRC as Epidemic Continues." www.doctorswithoutborders.org. March 7, 2019, accessed May 31, 2019.

"Ebola Survivor Program: From Patient to Caregiver." World Health Organization. www.who.int. No date, accessed May 31, 2019.

Hayden, Erika Check. "Ebola Survivors Still Immune to Virus After 40 Years." www.nature.com. December 14, 2017, accessed May 31, 2019.

Prentice, Alessandra. "Ebola Survivors Comfort Sick and Frightened in Congo Outbreak." www.reuters.com, April 17, 2019, accessed May 31, 2019.

"Tackling Ebola in DRC: 'We Knew We Had to Act Fast.'" www.doctorswithoutborders.org. September 13, 2018, accessed May 31, 2019.

Lesson 35

Brantly, Kent and Amber Brantly with David Thomas. *Called for Life: How Loving Our Neighbor Led Us into the Heart of the Ebola Epidemic*. Colorado Springs: WaterBrook Press, 2015.

"Love, Not Fear, Should Drive Us." *The Christian Chronicle*. www.christianchronicle.org, July 31, 2019, accessed August 10, 2019.

Ross, Bobby, Jr. "Brantly Returning to Africa." *The Christian Chronicle*, August 2019, pp. 1, 8-9.

Lesson 36

Shivani Vora. "Threading Needle of Hope in Rwanda." www.nytimes.com, April 4, 2018, accessed April 20, 2018

Lauren Gambino. "'It's About Our Dignity': Vintage Clothing Ban in Rwanda Sparks US Trade Dispute." www.theguardian.com, December 29, 2017, accessed April 20, 2018.

Chellie Ison. "Clothing Designer Brings Hope to Rwanda." www.christianchronicle.org, April 9, 2018, accessed April 20, 2018

Lesson 37

Kamkwamba, William and Bryan Mealer. *The Boy Who Harnessed the Wind*. New York: Morrow, 2009

www.movingwindmills.org

Lesson 38

caringforkenya.org

Lesson 39

"All Haile the King." www.spikes.iaaf.org, May 11, 2015, accessed July 25, 2019.

Allison, Simon. "Worshipping at the 'High Temple' of Ethiopia's Long Distance Runners." www.theguardian.com. March 23, 2015, accessed April 22, 2019.

Bisceglio, Paul. "The Greatest, Fakest World Record." www.theatlantic.com, October 13, 2019, accessed October 14, 2019.

Denison, Jim. The Greatest: The Haile Gebrselassie Story. Halcottsville, New York: Breakaway Books, 2004.

"Haile Gebrselassie Biography." www.biographyonline.net. Accessed July 25, 2019.

Hattenstone, Simon. "The Ethiopian Town That's Home to the World's Greatest Runners." www.theguardian.com. April 6, 2012, accessed April 30, 2019.

Onywera, Vincent O. "Scientists Are Closer to Pinning Down Why the World's Best Marathon Runner Is So Good." Quartz, www.qz.com. May 26, 2019, accessed May 26, 2019.

"Q & A with Haile Gebrselassie." www.cnn.com. November 7, 2007, accessed July 25, 2019.

Lesson 41

Cahill, Petra. "A Diamond's Journey: Grim Reality Tarnishes Glitter." www.nbcnews.com. June 26, 2009, accessed July 25, 2019.

"Diamond History and Lore." Gemnological Institute of America. www.gia.edu. Accessed July 25, 2019.

Gibb, Michael. "Whether It's Mexico's Gold or Zimbabwe's Diamonds, Mining is Riven with Violence and Business Is Complicit." www.theguardian.com, February 29, 2016, accessed July 25, 2019.

Kohn, David. "Diamonds: A History." www.cbsnews.com. May 8, 2002, accessed July 25, 2019.

"Labor and Community." www.brilliantearth.com. Accessed July 25, 2019.

Raden, Aja. *Stoned: Jewelry, Obsession, and How Desire Shapes the World.* New York: HarperCollins, 2015.

Scott, Katy. "Diamonds in the Deep: How Gems Are Mined from the Bottom of the Ocean." www.cnn.com. September 4, 2018, accessed July 25, 2019

Lesson 42

Derby, Ron. "South African Bride Price Moves from Cattle to Cash." www.reuters.com. January 20, 2007, accessed July 30, 2019.

Greaves, Adrian. *The Zulus at War: The History, Rise, and Fall of the Tribe That Washed Its Spears.* New York: Skyhorse Publishing, 2013.

Josephy, Alvin M., editor. *Africa: A History.* New York: American Heritage/Horizon/New Word City, 2016.

South African History Online, www.sahistory.org.za

Lesson 43

Erik Tryggestad. "A Church of Christ on Wheels." *Christian Chronicle*, Nov. 15, 2017, www.christianchronicle.org

gospelchariot.blogspot.com

Lesson 44

Scott, Katy. "South Africa Is the World's Most Unequal Country. 25 Years of Freedom Have Failed to Bridge the Divide." www.cnn.com, May 7, 2019; accessed May 8, 2019.

Lesson 45

Smith, John C. P. "What Is Truth?" answersingenesis.org. April 17, 2015, accessed June 11, 2018.

Lesson 46

"Basque Whalers." Canadian Museum of History, www.historymuseum.ca, accessed August 4, 2018

Barnes, Bingo. "A Short Basque History," www.boiseweekly.com, July 27, 2005, accessed August 4, 2018.

Belanger, René. "Basques." www.canadianencyclopedia.ca, February 6, 2006, accessed August 4, 2018.

Bochman, Chris. "The Island That Switches Countries Every Six Months." www.bbc.com. January 28, 2018, accessed April 17, 2020.

Douglass, William A. "Basque Immigration in the United States." Basque Studies Consortium Journal, Volume 1, Issue 1, Article 5. www.scholarworks.boisestate.edu, October 2013, accessed August 4, 2018.

Lesson 47

"Balkan Nation Is North Macedonia Now." Associated Press via www.voanews.com, February 12, 2019; accessed February 20, 2019.

Chrepa, Eleni and Slav Okov. "The Bitter Battle over the Name 'Macedonia' Explained. www.bloomberg.com, January 30, 2019, updated February 15, 2019; accessed February 20, 2019.

"Greek Government Crisis Over Macedonia Name Change." www.bbc.com, January 13, 2019; accessed January 13, 2019.

Kalkissis, Joanna. "For Two Countries, The Dispute Over Macedonia's Name Is Rooted In National Identity." www.npr.org, February 4, 2018; accessed February 20, 2019.

Labropoulou, Elinda. "Macedonia Will Change Its Name. Here's Why It Matters." www.cnn.com, January 25, 2019; accessed January 25, 2019.

"The Man Who Has Focused on One Word for 23 years." www.bbc.com, August 2, 2017; accessed February 21, 2019.

Lesson 48

Baldwin-Edwards, Martin. "Migration between Greece and Turkey: from the 'Exchange of Populations' to non-recognition of borders." South East Europe Review, Third Quarter 2006.

Clark, Bruce. *Twice a Stranger: How Mass Expulsion Forged Modern Greece and Turkey.* London: Granta Books, 2006.

Sussman, Paul. "Greece and Turkey: History of Hate." www.cnn.com, September 16, 2001; accessed March 1, 2019.

"Why Turkey and Greece cannot reconcile." www.economist.com, December 14, 2017; accessed March 1, 2019.

Lesson 49

worldatlas.com

Embassy of the Principality of Liechtenstein, http://www.liechtensteinusa.org, accessed August 30, 2018.

Adriano Rosoni. "The Condition of Europe's Curious Microstates." www.worldview.stratfor.com, April 26, 2015, accessed August 27, 2018.

Lesson 50

Richard Oster. *The Acts of the Apostles Part II*, 13:1-28:31. Austin: Sweet Publishing Company, 1979.

Lesson 51

Lear, Linda. "About Beatrix Potter." www.beatrixpottersociety.org, accessed May 16, 2019.

Lesson 52

Dash, Mike. *Tulipomania: The Story of the World's Most Coveted Flower and the Extraordinary Passions It Aroused.* New York: Broadway Books, 2010.

Schuetze, Christopher J. "Dutch Flower Auction, Long Industry's Heart, Is Facing Competition." New York Times, www.nytimes.com, December 16, 2014, accessed February 21, 2018.

Lesson 53

Ambrose, Stephen. *Citizen Soldiers: The U.S. Army from the Normandy Beaches to the Bulge to the Surrender of Germany, June 7, 1944-May 7, 1945.* New York: Simon & Schuster, 1997.

Burnes, Brian. "Seventy Years Later, Battle of the Bulge Looms Large in Overland Park Veteran's Memory." *Kansas City Star*, December 15, 2014. www.kansascity.com, accessed January 30, 2019.

Lesson 54

Charles L. Mee Jr., *Saving a Continent: The Untold Story of the Marshall Plan*, New York: New Word City LLC, 2015.

Lesson 56

Pollard, Niklas and Jussi Rosendahl. "In East-West Diplomatic Drama, Helsinki Punches Above Its Weight." www.reuters.com, July 11, 2018, accessed July 10, 2019.

"The Sami of Northern Europe—One People, Four Countries." United Nations Regional Information Centre for Western Europe. www.unric.org, accessed July 10, 2019.

Wall, Tom. "The Battle to Save Lapland: 'First They Took the Religion. Now They Want to Build a Railroad.'" www.theguardian.com. February 23, 2019, accessed July 10, 2019.

Watts, Peter. "The Dark History of Santa's City: How Rovaniemi Rose from the Ashes." www.theguardian.com, December 19, 2018, accessed July 15, 2019.

Lesson 57

Levy, Clifford J. "In Estonia, Jiggling and Bog-Trekking." www.nytimes.com. August 19, 2010, accessed September 4, 2019.

"Peat." www.turbaliit.ee. Accessed September 4, 2019.

Walt, Vivienne. "Is This Tiny European Nation a Preview of Our Tech Future?" www.fortune.com. April 17, 2017, accessed September 4, 2019.

Lesson 58

"A History of the Settlement of the Faroe Islands." www.icelandictimes.com. September 14, 2016, accessed September 5, 2019.

Amos, Owen. "One Rower, Two Cats, 900 Miles." www.bbc.com. July 27, 2015, accessed September 5, 2019.

Coldwell, Will. "Faroe Islands Fit Cameras to Sheep to Create Google Street View." www.theguardian.com. July 12, 2016, accessed September 5, 2019.

Ecott, Tim. "Sustainable Tourism: Why the Faroe Islands Closed for Maintenance." www.theguardian.com. May 8, 2019, accessed September 5, 2019.

www.faroeislands.fo

Matzen, Erik. "Small Fry: Faroe Islands Seek Fish Export Pledge with Russia Trade Deal." www.reuters.com. June 12, 2018, accessed September 5, 2019.

Lesson 59

"Struve Geodetic Arc." whc.unesco.org, accessed September 6, 2019

"Struve Geodetic Arc: The 2,820 Km Line That Produced the First Accurate Measurement of the Earth's Size." www.amusingplanet.com. Accessed September 6, 2019

"Triangulation" in "The Nature of Geographic Information". Pennsylvania State University, www.e-education.psu.edu. Accessed September 6, 2019.

Lesson 60

Howard, Thomas. *Chance or the Dance? A Critique of Modern Secularism*. Originally published 1969. Second edition: San Francisco: Ignatius Press, 2018

Lesson 61

Giles, Joseph and Frances Giles. Life in a Medieval City. New York: Thomas Y. Crowell, 1969

Howe, Irving. World of Our Fathers. New York: Open Road Media, 1976.

Sources

Lipka, Michael. "The Continuing Decline of Europe's Jewish Population." www.pewresearch.org. February 9, 2015, accessed August 13, 2019.

www.myjewishlearning.com
www.jewishvirtuallibrary.org
www.jewish-history-online.net
www.galiciajewishmusem.org

Lesson 62

"The Dual Monarchy: Two States in a Single Empire." ww1.habsburger.net. Access August 2, 2019.

Michener, James. *The Bridge at Andau*. New York: Random House/Bantam, 1957.

"Raoul Wallenberg and the Rescue of Jews in Budapest." United States Holocaust Memorial Museum. www.encyclopedia.ushmm.org, accessed August 1, 2019.

"Return of the Holy Crown of St. Stephen." hu.usembassy.gov, accessed August 2, 2019.

"Soviets Put Brutal End to Hungarian Revolution." www.history.com. November 24, 2009, accessed August 1, 2019.

Lesson 63

"Listening Guide: Smetana: The Moldau." W. W. Norton & Company. www.wwnorton.com. Accessed August 22, 2019.

"The Moldau." www.many-strings.com. Accessed August 22, 2019.

Sitler, Jiri. "From Bohemia to Czechia." Radio Prague. www.radio.cz. July 12, 2016, accessed August 22, 2019.

Toher, Mackenzie. "A Musical Painting of a Beloved Bohemian Landscape: 'The Moldau' Lifts a Nation." www.pages.stolaf.edu, May 20, 2014, accessed August 22, 2019.

Lesson 64

Heritage Foundation, heritage.org

"After Two Years in Captivity, Minister Released." *The Christian Chronicle*, February 2018, pages 1, 14.

Lesson 67

Goetschel, Samira. "'The Graveyard of the Earth': Inside City 40, Russia's Deadly Nuclear Secret." www.the guardian.com. July 20, 2016, accessed August 29, 2019.

Quartly, Alan. "Siberia's Dying Mansi People." www.bbc.co.uk. September 5, 2002, accessed August 29, 2019.

Lesson 68

www.baikal-marathon.org

Bland, Alistair. "Lake Baikal and More of the Weirdest Lakes of the World." www.smithsonianmag.com, August 7, 2012, accessed July 16, 2019.

Garrels, Anne. "Russia's Troubled Waters Flow With the Mighty Volga." www.npr.org, November 1, 2010, accessed July 17, 2019.

MacFarquhar, Neil. "An Ice Marathon Across a Frozen Russian Lake: 'I Ran Twice as Fast.'" www.newyorktimes.com. March 24, 2019, accessed July 16, 2019.

Parfitt, Tom. "Action Man Vladimir Putin Turns Submariner at Lake Baikal." www.the guardian.com, August 2, 2009, accessed July 16, 2019.

Starinova, Yulia and Farangis Najibullah. "Lake Baikal Faces New Crisis as Russia Lowers Eco-Standards." Radio Free Europe/Radio Liberty, www.rferl.org. March 27, 2019, accessed July 17, 2019.

Lesson 69

Ambrose, Stephen. *Nothing Like It in the World: The Men Who Built the Transcontinental Railroad 1863-1869*. New York: Simon & Schuster, 2000.

"The Trans-Siberian Route." www.wondersofworldengineering.com, April 1937, accessed March 6, 2019.

Lesson 71

Anderson, William, Captain and Don Keith. *The Ice Diaries: The Untold Story of the Cold War's Most Daring Mission.* Nashville: Thomas Nelson, 2008.

Lesson 72

Anderton, Kevin. "The North Magnetic Pole Has Moved. Here's What You Need to Know." www.forbes.com, January 16, 2019, accessed March 2, 2019.

Bryce, Emma. "Who Owns the Arctic?" www.livescience.com, October 13, 2014, accessed October 14, 2019.

"Russia Plants Flag on Arctic Floor." Reuters via www.cnn.com, August 4, 2007, accessed May 9, 2019.

'Russia Plants Flag Under N. Pole." www.bbc.co.uk, August 2, 2007, accessed May 9, 2019.

"Russia Says Floating Nuclear Plant Embarks on First Sea Voyage." www.cbsnews.com, April 30, 2018, accessed April 30, 2018.

Lesson 73

Brockman, Alex. "Life North of 80: Meet the People Living at the Top of the World." Canadian Broadcasting Corporation. www.cbc.ca, September 3, 2018, accessed May 14, 2019.

Jones, Lindsay. "As Nunavut Turns 20, Inuit Rethink Their Own Government." www.macleans.ca, December 12, 2018, accessed May 13, 2019.

Kikkert, Peter. "Nunavut." www.thecanadianencyclopedia.ca, August 9, 2007, updated May 2, 2019, accessed May 13, 2019.

Lesson 74

Alexander, Caroline. *The Endurance: Shackleton's Legendary Antarctic Expedition.* New York: Knopf Doubleday Publishing Group, 2008.

"Antarctica." National Geographic Resource Library. www.nationalgeographic.org/encyclopedia/antarctica/. January 4, 2012, accessed February 6, 2020.

Dixon, Emily. "Antarctic Expedition to Find Shackleton's Lost Endurance Loses Its Own Submarine to the Ice." www.cnn.com. February 15, 2019, accessed February 5, 2020.

Lansing, Alfred. Endurance: Shackleton's Incredible Voyage. New York: Carroll and Graf, 1959. Reprint edition: New York: Basic Books, 2014.

"Shackleton's Voyage of Endurance." NOVA Online. www.pbs.org/wgbh/nova/shackleton/1914/timeline.html. February 2002, accessed February 5, 2020.

Vizcarra, Natasha. "Unexpected Ice." www.earthdata.nasa.gov. Updated January 14, 2020, accessed February 6, 2020.

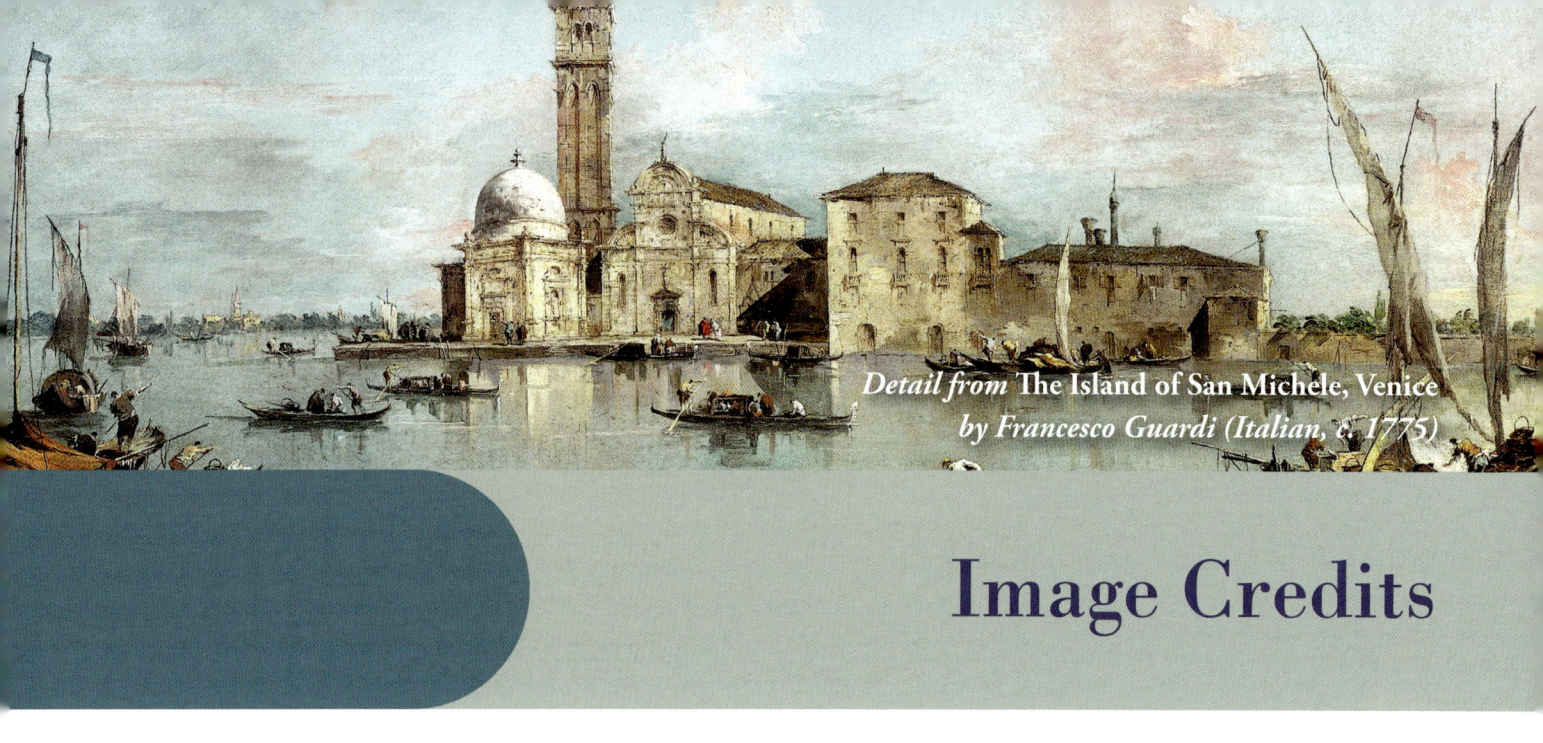

Detail from The Island of San Michele, Venice *by Francesco Guardi (Italian, c. 1775)*

Image Credits

Images marked with one of these codes are used with the permission of a Creative Commons Attribution or Attribution-Share Alike License. See the websites listed for details.

CC BY 2.0	creativecommons.org/licenses/by/2.0
CC BY-SA 2.0	creativecommons.org/licenses/by-sa/2.0
CC BY-SA 2.0 DE	creativecommons.org/licenses/by-sa/2.0/de/
CC BY 2.5	creativecommons.org/licenses/by/2.5
CC BY 3.0	creativecommons.org/licenses/by/3.0
CC BY-SA 3.0	creativecommons.org/licenses/by-sa/3.0
CC BY 4.0	creativecommons.org/licenses/by/4.0
CC BY-SA 4.0	creativecommons.org/licenses/by-sa/4.0

iii	Davor Flam / Shutterstock.com
iv	Dietmar Temps / Shutterstock.com
v	Leonid Andronov / Shutterstock.com
vi	Kriangkrai Thitimakorn / Shutterstock.com
vii	Alec Favale / Unsplash
ix	Anupam hatui / Shutterstock.com
xiii	Dadiolli: Tilman Schalmey / Wikimedia Commons / CC BY-SA 3.0
xv	Digital Content Writers India / Unsplash
xx	Carrastock / Shutterstock.com
1	Shahee Ilyas / Wikimedia Commons / CC BY-SA 3.0
2	Notgrass_Logo_RGB
3	Denis Belitsky / Shutterstock.com
4	NASA
5	NASA
6	NASA Lunar and Planetary Institute
7	Earth Science and Remote Sensing Unit, NASA Johnson Space Center ISS013-E-54329
8	Aerostato / Shutterstock.com
10	Egypt: Earth Science and Remote Sensing Unit, NASA Johnson Space Center ISS036-E-11050; Eratosthenes: Wikimedia Commons
11	Library of Congress
12	National Portrait Gallery, Smithsonian Institution
13	Tharp / Heezen: marie tharp maps / Flickr / CC BY 2.0; Globe: Captain Albert E. Theberge, NOAA Corps (ret.) NOAA Photo Library
14	Japan: Takashi Images / Shutterstock.com; Kenya: Kristof Kovacs / Shutterstock.com
16	Sinop: Kobby Dagan / Shutterstock.com; Map: NASA Worldview
17	Erturac / Wikimedia Commons / CC BY-SA 3.0
18	Encyclopaedia Biblica / Wikimedia Commons
20	Goodpairofshoes / Wikimedia Commons / CC BY-SA 3.0
21	Map © OpenStreetMap contributors / CC BY-SA 2.0 / openstreetmap.org; Flood: SSgt Paul Griffin / U.S. Army
22	1943: K. Segerstrom, U.S. Geological Survey; Modern: LBM1948 / Wikimedia Commons / CC BY-SA 4.0 (cropped)
23	Christopher Michel / Flickr / CC BY 2.0
24	MathKnight / Wikimedia Commons / CC BY-SA 3.0
25	Frame China / Shutterstock.com
27	Fotosr52 / Shutterstock.com
28	hxdbzxy / Shutterstock.com
29	LightField Studios / Shutterstock.com
31	m16brooks / Shutterstock.com
33	Wikimedia Commons
35	IgorGolovniov / Shutterstock.com
36	U.S. Central Intelligence Agency
37	Wikimedia Commons
38	Map: Wikimedia Commons; Globe: Alexander Franke / Wikimedia Commons / CC BY-SA 2.0 DE
39	Library of Congress
41	Herrieynaha / Shutterstock.com
42	GPS: Staff Sgt. Matthew Coleman-Foster / U.S. Air Force; Navigational: © OpenStreetMap contributors / CC BY-SA 2.0 / openstreetmap.org; Topographical: Jordevi / Wikimedia Commons; Thematic: USDA

C-1

43	Maps ETC
44	Wikimedia Commons
45	Strebe / Wikimedia Commons / CC BY-SA 3.0
46	Alexander Lukatskiy / Shutterstock.com
48	GMaple Design / Shutterstock.com
49	Jailbird / Wikimedia Commons / CC BY-SA 3.0
50	VectorMine / Shutterstock.com
51	Don Mammoser / Shutterstock.com
53	ambient_pix / Shutterstock.com
54	Internet Archive
55	Joni Hanebutt / Shutterstock.com
57	National Archives (U.S.)
58	Noraphat Vorakijroongroj / Shutterstock.com
59	AlexAnton / Shutterstock.com
61	Rostislav Ageev / Shutterstock.com
63	Jakob Fischer / Shutterstock.com
64	NormanEinstein / Wikimedia Commons / CC BY-SA 3.0
65	thomas koch / Shutterstock.com
66	Dave Primov / Shutterstock.com
67	lkpro / Shutterstock.com
68	Igor Grochev / Shutterstock.com
70	Arthur Simoes / Shutterstock.com
71	Wikimedia Commons
72	Wikimedia Commons
73	British Library
75	Roman Yanushevsky / Shutterstock.com
76	Shelly Bychowski Shots / Shutterstock.com
77	United Nations / Wikimedia Commons
78	Tank: United States Army Heritage and Education Center; Shelter: Lahava Netivot, Yehoshua Neumann, from the Pikiyaki website / CC BY 2.5
79	National Archives (U.S.)
80	Roman Yanushevsky / Shutterstock.com
81	shimriz / Flickr / CC BY 2.0
83	Flag: Felix Friebe / Shutterstock.com; Map: Peter Hermes Furian / Shutterstock.com
84	Jan Smith / Flickr / CC BY 2.0
85	Elena Odareeva / Shutterstock.com
86	thomas koch / Shutterstock.com
88	Andrzej Lisowski Travel / Shutterstock.com
89	Wikimedia Commons
90	Wikimedia Commons
91	R. de Bruijn_Photography / Shutterstock.com
92	Nancy Anderson / Shutterstock.com
93	Yale Beinecke Library
95	Sun_Shine / Shutterstock.com
97	Artem Avetisyan / Shutterstock.com
98	Gor Davtyan / Unsplash
99	Manuscript: Wikimedia Commons; Cathedral: Gromwell / Shutterstock.com
100	meunierd / Shutterstock.com
102	Gallipoli: Everett Collection / Shutterstock.com; Cartoon: Library of Congress
103	Boule / Shutterstock.com
104	NASA Worldview
105	Everett Collection / Shutterstock.com
106	National Army Museum (UK)
106	Clay Gilliland / Flickr / CC BY-SA 20
108	Road: Andrew V Marcus / Shutterstock.com; Manal al-Sharif: Manal al-Sharif / Wikimedia Commons / CC BY-SA 3.0
109	Shaybah Saudi Arabia / Shutterstock.com
110	imrankadir / Shutterstock.com
112	Yunqing Shi / Shutterstock.com
113	MetMuseum
114	Sony Herdiana / Shutterstock.com
115	Notgrass Family
116	Notgrass Family
117	Oleg Bakhirev / Shutterstock.com
118	Grytviken: Thomas Barrat / Shutterstock.com; Harbin: Lawrie Cate / Flickr / CC BY 2.0
119	Rohan Reddy / Unsplash
120	India: Duttagupta M K / Shutterstock.com; Ethiopia: alfotokunst / Shutterstock.com
121	Big Joe / Shutterstock.com
122	kissdaybreak / Shutterstock.com
123	BkhStudio / Shutterstock.com
125	Chris Belsten / Flickr / CC BY 2.0
126	Mahmoud Bundesministerium für europäische und internationale Angelegenheiten / Flickr / CC BY 2.0
127	NASA Worldview
128	Sergii Nagornyi / Shutterstock.com
129	Dates: Dmitry Chulov / Shutterstock.com; Monastir: cpaulfell / Shutterstock.com
130	Kekyalyaynen / Shutterstock.com
131	Birdiegal / Shutterstock.com
132	Visualizing Culture
133	NASA Worldview
134	Frank Mason Good / Library of Congress
135	1865: Marzolino / Shutterstock.com; 1900: Photoglob Co / Library of Congress
136	byvalet / Shutterstock.com
137	Wikimedia Commons
138	Yavuz Sariyildiz / Shutterstock.com
139	RudiErnst / Shutterstock.com
140	Uploadalt / Wikimedia Commons / CC BY-SA 3.0
141	Wikimedia Commons
142	U.S. Naval History and Heritage Command
144	Wreckage: Moussar / Shutterstock.com; Rommel: Everett Collection / Shutterstock.com
145	Plane: Everett Collection / Shutterstock.com; Tanks: Thomas Wyness / Shutterstock.com
146	Montgomery: Wikimedia Commons; Map: NASA Worldview
147	Algiers: National Archives (U.S.); Flyer: Wikimedia Commons
149	Qur'an: Adli Wahid / Unsplash; Kaaba: Sufi / Shutterstock.com
150	Tashkent: monticello / Shutterstock.com; Edmonton: 2009fotofriends / Shutterstock.com
151	Umer Arif / Shutterstock.com
152	DanKe / Shutterstock.com
153	Amsterdam: www.hollandfoto.net / Shutterstock.com; Jeddah: AFZAL KHAN MAHEEN / Shutterstock.com
154	Erico setiawan / Shutterstock.com
155	Guadalupe Polito / Shutterstock.com

Image Credits

157 Peek Creative Collective / Shutterstock.com
159 Paola_F / Shutterstock.com
160 Roger Brown Photography / Shutterstock.com
161 BOULENGER Xavier / Shutterstock.com
163 Field: Claudiovidri / Shutterstock.com; Mosque: trevor kittelty / Shutterstock.com
164 trevor kittelty / Shutterstock.com
165 Granaries: Scott S. Brown / Shutterstock.com; Toguna: Torsten Pursche / Shutterstock.com
166 Hieroglyphs: Claudiovidri / Shutterstock.com; Dance: Torsten Pursche / Shutterstock.com
168 Linda Hughes Photography / Shutterstock.com
169 Nkrumah: Abbie Rowe. White House Photographs. John F. Kennedy Presidential Library and Museum, Boston; Dam: Sopotnicki / Shutterstock.com
170 Sopotnicki / Shutterstock.com
171 Veennema / Wikimedia Commons / CC BY-SA 3.0
172 Ajibola Fasola / Shutterstock.com
173 Jack.Q / Shutterstock.com
174 New York Public Library
175 Fela Sanu / Shutterstock.com
176 Old Plantation: Wikimedia Commons; Brazil: *Life in Brazil*
178 Robert Szymanski / Shutterstock.com
179 tera.ken / Shutterstock.com
180 Ceremony: gracindojr / Shutterstock.com; Altar: Gilles MAIRET / Wikimedia Commons / CC BY-SA 3.0
181 Erikapajama / Wikimedia Commons / CC BY-SA 3.0
183 Sergey Gaydaburov / Shutterstock.com
185 sandis sveicers / Shutterstock.com
186 mbrand85 / Shutterstock.com
187 mbrand85 / Shutterstock.com
189 Village: natacabo / Shutterstock.com; Hunting: Radio Okapi / Flickr / CC BY 2.0
190 Garry Walsh Trócaire / Flickr / CC BY 2.0
191 Wikimedia Commons
192 Library of Congress
194 Homo Cosmicos / Shutterstock.com
195 Bundesarchiv, Bild 163-161 / CC BY-SA 3.0
196 Steve Mvondo / Wikimedia Commons / CC BY-SA 3.0
197 Fabian Plock / Shutterstock.com
199 3ffi / Shutterstock.com
200 Fabian Plock / Shutterstock.com
201 Nicole Macheroux-Denault / Shutterstock.com
203 Mark Fischer / Flickr / CC BY 2.0 / CC BY-SA 2.0
204 llucky78 / Shutterstock.com
205 Pete Souza / White House
207 Radek Borovka / Shutterstock.com
209 Tetyana Dotsenko / Shutterstock.com
210 Rytc / Wikimedia Commons / CC BY 3.0
211 StreetVJ / Shutterstock.com
212 Used clothing: Sarine Arslanian / Shutterstock.com; Seamstresses: Oscar Espinosa / Shutterstock.com
214 Sunset: Martin Mwaura / Shutterstock.com; Village: africa924 / Shutterstock.com
215 Karl Beeney / Shutterstock.com
216 Books: Mike Lee / Flickr / CC BY 2.0; Windmill: Tom Rielly

217 Erik (HASH) Hersman / Flickr / CC BY 2.0
219 Nick Fox / Shutterstock.com
220 Caring for Kenya
221 Caring for Kenya
222 Caring for Kenya
223 Caring for Kenya
224 Runners: aman ahmed ahmed / Shutterstock.com; Rome: Comitato organizzatore dei Giochi della XVII Olimpiade
225 360b / Shutterstock.com
226 Pete Lewis Department for International Development / Wikimedia Commons / CC BY 2.0
227 Shahjehan / Shutterstock.com
228 Cristian Teichner / Shutterstock.com
230 alarico / Shutterstock.com
231 GagliardiPhotography / Shutterstock.com
232 PierreSelim / Wikimedia Commons
233 National Library NZ on The Commons
235 Kelly Ermis / Shutterstock.com
237 Lucian Coman / Shutterstock.com
238 Uncut diamonds: Imfoto / Shutterstock.com; Jewelry: Daderot / Wikimedia Commons
239 Vladislav Gajic / Shutterstock.com
240 Lemonreel / Shutterstock.com
242 Mark Dumbleton / Shutterstock.com
243 The National Archives UK
244 Cattle: Leonard Zhukovsky / Shutterstock.com; Home: meunierd / Shutterstock.com
245 Nationaal Archief (Netherlands)
247 Burundi: Juriz / Shutterstock.com; Malawi: Gospel Chariot Missions
247 Gospel Chariot Missions
248 Gospel Chariot Missions
249 Gospel Chariot Missions
251 Stellenbosch: ModernNomad / Shutterstock.com; Silhouette: Shoot Digital / Shutterstock.com
252 mbrand85 / Shutterstock.com
253 LSE Library
254 Library of Congress
255 Carol M. Highsmith Archive, Library of Congress, Prints and Photographs Division
256 Sunshine Seeds / Shutterstock.com
257 Library of Congress
258 Brais Seara / Shutterstock.com
261 Zvonimir Atletic / Shutterstock.com
263 Balate Dorin / Shutterstock.com
265 ARK NEYMAN / Shutterstock.com
266 Map: NASA Worldview; Shepherd: Jesus Keller / Shutterstock.com
267 EQRoy / Shutterstock.com
268 Sign: Wirestock Images / Shutterstock.com; Festival: Laiotz / Shutterstock.com
270 Ohrid: Leonid Andronov / Shutterstock.com; Map: NASA Worldview
271 National Archives (U.S.)
272 Milos Djokic / Shutterstock.com
273 Freddie Everett / U.S. Department of State
274 Library of Congress

275	Wikimedia Commons	334	Smit / Shutterstock.com
276	Wikimedia Commons	335	Andrew Mayovskyy / Shutterstock.com
277	Ottoman Archives	336	Vincent van Zeijst / Wikimedia Commons / CC BY-SA 3.0
278	Basketball: American National Red Cross / Library of Congress; Bakery: Frank and Frances Carpenter Collection / Library of Congress	338	stilrentfoto / Shutterstock.com
		339	Russia: Islander / Wikimedia Commons / CC BY 3.0; Norway: Clemensfranz / Wikimedia Commons / CC BY-SA 3.0
279	Noradoa / Shutterstock.com		
280	Passion Travel Fruit / Shutterstock.com	340	Sean Killen
281	Laborant / Shutterstock.com	342	NASA JPL-Caltech Space Science Institute
283	tichr / Shutterstock.com	343	Wikimedia Commons
284	Rostislav Glinsky / Shutterstock.com	344	Swellphotography / Shutterstock.com
285	Eszter Szadeczky-Kardoss / Shutterstock.com	347	romeovip_md / Shutterstock.com
286	jorisvo / Shutterstock.com	349	isparklinglife / Shutterstock.com
287	RossHelen / Shutterstock.com	350	Wikimedia Commons
289	Areopagus: ArtMediaFactory / Shutterstock.com; Acropolis: Trajan 117 CE / Wikimedia Commons	351	Wikimedia Commons
		352	Fred Romero / Flickr / CC BY 2.0
290	Vasilii L / Shutterstock.com	353	Jewish Encyclopedia / Wikimedia Commons
292	Renata Sedmakova / Shutterstock.com	355	Mor65_Mauro Piccardi / Shutterstock.com
293	Logan Bush / Shutterstock.com	357	FOTO : FORTEPAN Nagy Gyula / Wikimedia Commons / CC BY-SA 3.0
295	shutterupeire / Shutterstock.com		
297	Helen Hotson / Shutterstock.com	358	givaga / Shutterstock.com
298	Hill Top: A D Harvey / Shutterstock.com; Sheep: Johann Knox / Shutterstock.com	360	FOTO : FORTEPAN Pesti Srác / Wikimedia Commons / CC BY-SA 3.0
299	Seungwon Lee / Shutterstock.com	361	Nationaal Archief
300	Charlesy / Shutterstock.com	362	czb / Shutterstock.com
301	Neirfy / Shutterstock.com	364	DaLiu / Shutterstock.com
302	Daderot / Wikimedia Commons	365	Stefan Rotter / Shutterstock.com
303	Eduard Pop / Shutterstock.com	366	1968: Central Intelligence Agency; 1989: MD / Wikimedia Commons / CC BY-SA 3.0
304	Wikimedia Commons		
305	Walters Art Museum	368	Berilova Irida / Shutterstock.com
307	101st Engineers: U.S. Army; Germans: Bundesarchiv, Bild 183-J28477 / Göttert / CC-BY-SA 3.0	369	Л.П. Джепко / Wikimedia Commons / CC BY-SA 3.0
		370	Pani Garmyder / Shutterstock.com
308	Notgrass Family	371	The Presidential Office of Ukraine / CC BY 4.0
309	Library of Congress	372	Creative Travel Projects / Shutterstock.com
310	Bel Adone / Wikimedia Commons	374	Melinda Nagy / Shutterstock.com
312	Nationaal Archief	375	Vlad Sokolovsky / Shutterstock.com
313	Bundesarchiv, Bild 183-B0527-0001-753 / CC-BY-SA 3.0	376	Sergey-73 / Shutterstock.com
314	U.S. Department of State	377	Bitomovski / Shutterstock.com
315	Tufts Archives (Pinehurst, N.C.)	379	Kichigin / Shutterstock.com
316	George C. Marshall Foundation	381	Vaclav Sebek / Shutterstock.com
319	FamVeld / Shutterstock.com	382	Wikimedia Commons
320	Noord-Hollands Archief	383	Library of Congress
323	TTstudio / Shutterstock.com	384	Cornell University – PJ Mode Collection of Persuasive Cartography
325	Reindeer: Vachonya / Shutterstock.com; Map: Reto Stöckli, NASA Earth Observatory NASA's Earth Observatory		
		385	Starikov Pavel / Shutterstock.com
326	Traditional: Marek Rybar / Shutterstock.com; Modern: V. Belov / Shutterstock.com; Soldiers: Military Museum on the Finna service hosted by the Finnish Ministry of Education and Culture / CC BY 4.0	387	Mountains: Yuri Kabantsev / Shutterstock.com; Necklace: Daderot / Wikimedia Commons
		388	Sergey Nemanov / Wikimedia Commons / CC BY-SA 3.0
		389	Wikimedia Commons
327	Smelov / Shutterstock.com	391	Strelyuk / Shutterstock.com
328	Roman Babakin / Shutterstock.com	392	Tilpunov Mikhail / Shutterstock.com
329	Andrea Hanks / White House	393	VarnakovR / Shutterstock.com
330	Tallinn: Dmitry Tkachenko Photo / Shutterstock.com; Trail: Ija Reiman / Shutterstock.com	395	ALEKSANDR RIUTIN / Shutterstock.com
		396	Library of Congress
331	FotoHelin / Shutterstock.com	397	Library of Congress
332	Skype: Evgenia Bolyukh / Shutterstock.com; Park: F-Focus by Mati Kose / Shutterstock.com	398	Library of Congress
		399	Library of Congress

Image Credits

400 NASA
401 PardoY / Shutterstock.com
402 Discover Marco / Shutterstock.com
403 Kerk Phillips / Wikimedia Commons
405 posteriori / Shutterstock.com
407 National Archives (U.S.)
408 John Notgrass
409 National Archives (U.S.)
410 Naval History and Heritage Command
411 National Archives (U.S.)
413 Helicopter: Maximillian cabinet / Shutterstock.com; Submersible: RMS Titanic Team Expedition 2003, ROI, IFE, NOAA-OE
414 Iceberg: Mikhail Varentsov / Shutterstock.com; Ships: AMFPhotography / Shutterstock.com
415 shuttermuse / Shutterstock.com
417 Max Forgues / Shutterstock.com
418 Sophia Granchinho / Shutterstock.com
419 RUBEN M RAMOS / Shutterstock.com
421 Sean M Smith / Shutterstock.com
422 Mitchell Library, State Library of New South Wales
423 Mitchell Library, State Library of New South Wales
424 Mitchell Library, State Library of New South Wales
425 Map: Sean Killen
426 Dene' Miles / Shutterstock.com
427 robert mcgillivray / Shutterstock.com
428 Jiri Kulisek / Shutterstock.com
429 AndreAnita / Shutterstock.com
S-1 EvrenKalinbacak / Shutterstock.com
C-1 Metropolitan Museum of Art

Find Your Next Curriculum

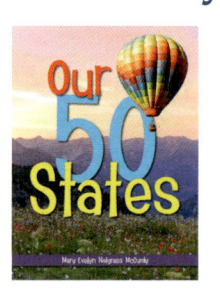

Elementary
Middle School
High School

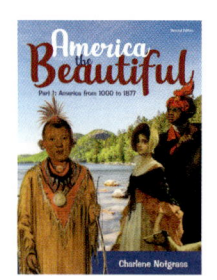

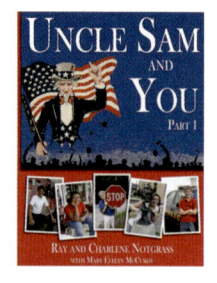

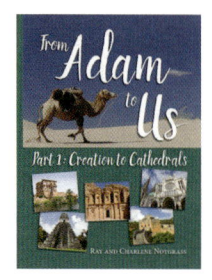

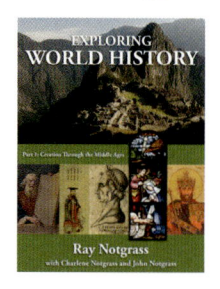

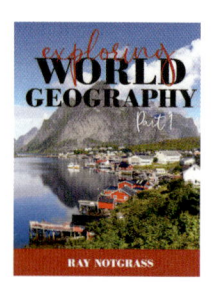

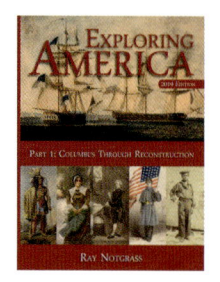

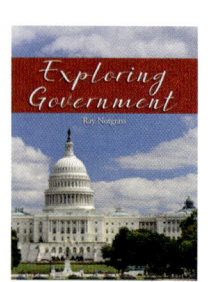

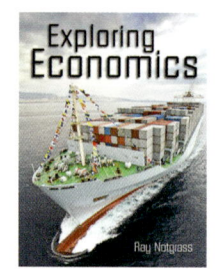

NOTGRASS.COM/SHOP

Homeschool History

Suggested videos, virtual tours, games, and more to enhance your studies.

NOTGRASS.COM/HH

Support

Bonus downloads, an encouraging blog, and a community for moms.

NOTGRASS.COM/SUPPORT

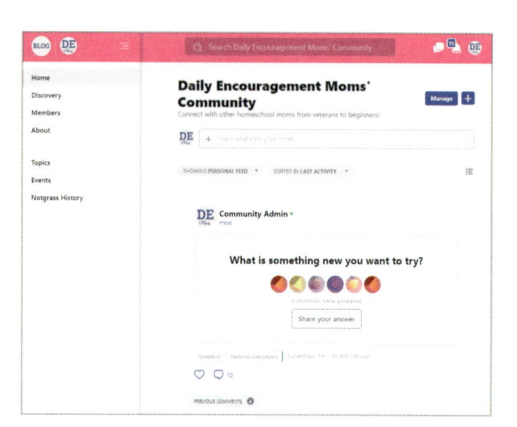